Proceedings of Symposia

IX International Congress of Plant Protection

Washington, D.C. U.S.A.

August 5-11, 1979

Volume II
Integrated Plant Protection
for Agricultural Crops
and Forest Trees

Editor
Dr. Thor Kommedahl
Department of Plant Pathology
University of Minnesota

Published for the CONGRESS by Burgess Publishing Company, Minneapolis, Minnesota. Additional copies
available from Entomological Society of America, 4603 Calvert Road, College Park, MD 20740, U.S.A.

Printed in United States of America

Table of Contents

Chapter 37   Integrated Plant Protection in Forestry

28.1 Integrated Control of Diseases of Corn and Sorghum in the United States

A. J. Ullstrup

Professor Emeritus, Purdue University
Presently Plant Pathologist, FFR Cooperative
West Lafayette, Indiana

The diseases of corn (*Zea mays*) and sorghum (*Sorghum bicolor*), as those of other crops, fluctuate from year to year and from one locality to another within a given year. This variation depends largely on the vicissitudes of the weather, to some extent on cultural practices, and on relative host resistance. Each of these crops is subject to a number of diseases, a few are, or have been of economic importance; a greater number seldom cause appreciable damage. The amount of loss sustained is related to time of onset of the disease in relation to host development, and to the level of severity attained before harvest.

There are three fundamental methods in integrated control of the diseases of corn and sorghum: application of fungicides, cultural practices and host-plant resistance.

Fungicides are employed mostly as seed treatments to control seed rot and seedling blights, caused mainly by soil-borne pathogenic fungi, that often become aggressive when soils are cold and wet following planting. Practically all corn and sorghum seed planted in the United States is treated with a seed protectant(s) to reduce losses from this disease complex. The practice is an inexpensive insurance. Sorghum seed, in addition to being treated to protect it against seed-rot pathogens, may also be treated to control certain smut disease fungi.

Fungicide sprays are used to control leaf and ear-rot pathogens in seed fields and in winter nurseries of breeding stocks in southern Florida. This is also done on Florida-grown sweet corn intended for the green-market in the North during the winter. The use of such sprays in this area is essential for the production of economic yields and high quality seed. *Helminthosporium turcicum* Pass. and *H. maydis* Nisik. & Miyake are the primary pathogens involved. However, it amounts to a very small fraction of the corn and sorghum grown in the United States. Control of diseases in these crops by fungicide sprays is feasible only where there is a high monetary return on the harvested product, i.e., for seed or sweet corn.

Insecticide sprays are rarely used to reduce populations of the vectors of certain diseases of these crops—the corn flea beetle, vector of bacterial wilt, and aphid and leafhopper vectors of some virus diseases. The practice is very limited and usually applied only to a high-value product.

During the southern leaf blight epiphytotic of 1970-1971, some spraying of corn was done with fungicides, but this was not completely successful. Timing and frequency of application are critical, and these factors were not often considered.

A second thrust in integrated control of corn and sorghum diseases is the employment of a number of cultural practices. Delay in spring plantings until soil temperatures are high enough to favor rapid germination and seedling growth is an integral practice along with seed treatment to reduce losses in stand owing to seed rots and seedling diseases. The use of full-season hybrids tends to lower severity of stalk rotting and consequent breakage. Because stalk rots are, in most instances, associated with senescence of stalk tissue, hybrids that mature too early in a given area become targets for some of the weak pathogens that invade stalks during the remaining warm, wet fall weather. Properly balanced fertility will also tend to reduce losses from stalk rots. Low levels of potassium and/or relatively high amounts of nitrogen appear to favor early onset of senescence of stalks resulting in infection, premature death and breakage.

Rotation and crop sanitation are two other cultural practices used to minimize losses from disease. Crop rotation can help, in that few diseases are common between corn and other crops or between sorghum and other other crops in a rotation. A few are common with these two crops and with others; these can be avoided. Head smut of corn and of sorghum, each of which is incited by a specific form of *Sphacelotheca reiliana,* may be controlled in part by rotation. In limited areas of southeastern Texas, where sorghum downy mildew is a problem, corn should not follow grain sorghum or broom corn (*Sorghum bicolor* var. *technicum*) in the rotation. It is on these latter two crops where the overwintering oospores of the pathogen are formed; these provide a source of soil-borne inoculum. Some diseases are little affected by rotation. In many of the wide expanses of river bottom lands, corn has been grown continuously for many years without noticeable increase in disease severity. Crop sanitation, in which inoculum on over-seasoned debris is destroyed by deep, clean plowing, can be effective in delaying onset of some fungus diseases of leaves and ears. One bacterial pathogen of corn over-winters on such debris. Within the past two decades "minimum tillage" in preparation of seed beds, as opposed to clean plowing, has become increasingly popular. With plant debris left on the soil surface, wind and/or water erosion of soil is reduced. Cost of seed-bed preparation is also lowered. The decision as to which practice to follow—clean plowing or minimum tillage with potential plant pathogens on the soil surface—remains entirely up to the farmer. Economy of seed-bed preparation and soil conservation may outweigh the risk of disease. If he has integrated his plant-health protection program with host-plant resistance, he may decide that the latter is sufficient to avoid a serious disease problem.

Eradication of an over-seasoning host, in which a pathogen may reside, is another resource in the arsenal of integrated disease control in these two crops. Destroying johnsongrass (*Sorghum halepense*), in which some of the newer herbicides are effective, is resorted to not only to reduce the severity and prevalence of certain strains of maize dwarf mosaic (MDM) and maize chlorotic dwarf (MCD), but also to eliminate the competitive effects of the weed.

Improving soil drainage to reduce root rots is a helpful practice. It is the only known means, at present, to control crazy top of corn and sorghum.

In its broadest interpretation, cultural practices, designed to reduce disease losses, may include harvesting at full maturity to avoid continued decomposition of stalks and consequent lodging, and to circumvent deterioration of grain quality that ensues when ears are exposed to warm, wet weather that may follow after the crop is mature. Provision for adequate drying and protection from moisture in storage will further ensure checking the development of so-called storage molds; some of which may elaborate toxins affecting man and animals.

Among the different methods that may be integrated into disease control in corn and sorghum, that employing host-plant resistance can be the most effective and efficient. It is generally resorted to in controlling those diseases not easily suppressed by fungicides or cultural means. Genetically determined resistance to disease-inciting organisms of these crops may be under monogenic or polygenic control. Resistance to some pathogens may involve both types. Genetic resistance has been successfully used in the control of bacterial wilt of sweet corn, of sorghum downy mildew in both sorghum and corn, of the virus disease complex in corn, involving MDM, MCD [3, 6], of corn stunt [4], of the milo disease of sorghum, and of common rust of corn. During the recent past, northern corn leaf blight has *not* been severe in the Corn Belt of the United States. This may be due to the incorporation of both the monogenic and polygenic types of resistance in a large number of hybrids. Resistance to stalk rot has been markedly improved in the past two decades through the employment of some of the newer breeding methods to incorporate genes controlling resistance in the inbred lines and hybrids [5, 7].

In a very general way, host-plant resistance may be looked upon as a kind of biological control; it is the suppression of a disease-inciting agent by another organism—the host. The use of genetic resistance as a component of integrated disease control in corn and sorghum has come under criticism in some quarters because of its tendency, as used, to narrow the spectrum of germplasm, i.e., to incorporate too much genetic uniformity among the hybrids grown. This was underscored following the southern corn leaf blight outbreak in 1970-1971. At that time, about 85% of the corn grown in the United States was of Texas male-sterile cytoplasm, which is hypersusceptible to Race T, to apparently new physiologic race of *H. maydis*. The susceptibility was determined by this type of cytoplasm and only slightly modified by nuclear genes. Texas male-sterile cytoplasm corn was used to facilitate seed production. It eliminated the expensive and onerous task of detasseling "seed rows" in seed fields. The reversion in 1972 to *normal cytoplasm* corn, which is highly resistant to Race T of *H. maydis*, completely controlled the disease.

Examination of the genetic constitution of current corn hybrids does show some uniformity—some germplasm is common to many of them [8]. This is due, in part, to the popularity of relatively few high-yielding hybrids that contain a number of inbred lines in common. This, it would seem, makes them potentially vulnerable to pathogens now present, or which may arise in the future as mutant races possessing enhanced or specialized virulence. Greater genetic diversity is certainly desirable as a means of coping with this, as well as other hazards. A number of practices are now being employed by breeders, and will find increasing use in the future, to mitigate against pitfalls occurring from excessive genetic uniformity in corn and sorghum hybrids [1]. These include: (1) Changes in relatively short periods (5-7 years) in the hybrids grown. This is owing to improvements in new hybrids, as they are developed, in several of the attributes sought for in these crops. (2) Differences in genetic make-up of hybrids among different areas of the country. This is dictated mainly on the bases of maturity and adaptability of hybrids to diverse geographical locations. (3) Increasing diversity among hybrids within a given region. A fourth factor is the exploitation, on part of breeders and pathologists, of exotic germplasm. These materials are often not directly usable in the Corn Belt of the United States, but such genes contributing to disease resistance and other desirable characters can be transferred to adapted inbred lines and hybrids.

In sorghum there has, as yet, been no widespread epiphytotic as occurred on corn in 1970-1971. The cytoplasmic base in this crop is relatively narrow, being confined to a single source [2]. New and different sources of cytoplasm that confer male sterility are being explored and some of these may reduce the risk inherent in a single uniform cytoplasm.

The newer cytoplasms conferring male sterility, now being used or examined in corn and sorghum, should be carefully monitored in strategic places throughout the world, in order to observe at the earliest time, possible susceptibilities among them as compared with their *normal cytoplasm* counterparts. This requires close cooperation, not only between the pathologist and the breeder in public and private organizations, but also a well-coordinated international effort in this regard.

Summary

The three main methods employed in integrated control of corn and sorghum diseases have been brought out and a few examples of these have been briefly delineated.

Chemical control of diseases of these crops, to which chemical control is applicable, is, from the farmers' standpoint, "already in the bag" in the fungicide that has been applied to the seed he will plant. The application of sprays to minimize leaf and ear diseases amounts to an infinitesimal fraction of the corn grown in the United States. Except for protection of green-market sweet corn from diseases, fungicide sprays are used by seed producers, not farmers.

Cultural practices, on the other hand, are almost entirely farmer-operations. He must decide whether to employ these or not; his decisions are dictated mainly on past experience with one or more diseases versus the economy of land preparation and soil conservation. Eradication of wild hosts of a pathogen, proper planting time, proper harvest time, and a number of other cultural measures are of the grower's own choice to adopt or not.

Genetically governed host-plant resistance has been, and will be, the main means of controlling diseases of corn and sorghum. This is within the province of the breeders of these crops.

An attempt has been made here to show how these three methods, means or operations designed for disease control in these two crops are, or can be, integrated.

Literature Cited

1. Duvick, D. N., 1975. "Using host resistance to manage pathogen populations." *Ia. State Jour. Res.* 49:505-512.
2. Duvick, D. N., 1977. "Major United States crops in 1976." *Ann. N.Y. Acad. Sci.* 287:86-96.
3. Findley, W. R., *et al.,* 1976. "Breeding corn for resistance to virus in Ohio." *Proc. Internat'l. Maize Virus Disease Colloquium and Workshop.* 123-128.
4. Grogan, C. O. and E. Rosenkranz, 1968. "Genetics of host reaction corn stunt virus."* *Crop Sci.* 8:251-254.
5. Jinahyon, S. and W. A. Russell, 1969. "Evaluations of recurrent selection for stalk rot resistance in open-pollinated varieties of maize. Effects of recurrent selection for stalk rot resistance on other agronomic characters in an open-pollinated variety of maize." *Ia. State Jour. Sci.* 43:229-251.
6. Loesch, P. J., Jr. and M. S. Zuber, 1967. "An inheritance study of resistance to maize dwarf mosaic in corn (*Zea mays* L.)." *Agron. Jour.* 59:423-426.
7. Sprague, G. F., 1967. "Plant Breeding." *Annu. Rev. Genetics* 1:269-294.
8. Zuber, M. S., 1975. "Corn germplasm base in the US—is it narrow, widening or static?" *Proc. Thirtieth Ann. Corn and Sorghum Conf.*, pp. 277-286.

*Corn stunt has since been shown to be incited not by a virus, but by a *Spiroplasma.*

28.2   Integrated Control of Maize and Sorghum Diseases in Tropical and Subtropical Areas

B. L. Renfro and R. A. Frederiksen

The Rockefeller Foundation
GPO Box 2453, Bangkok, Thailand
and
Department of Plant Sciences, Texas A&M University
College Station, Texas 77843 USA

Introduction

Maize (*Zea mays* L.) and sorghum (*Sorghum bicolor* (Linn) Moench.) are the third and fourth most important cereal crops of the world. They are cropped for food, feed, fodder and industrial purposes and are important to the tropical and subtropical regions. Classified as coarse grains, they account for about 35% of the world's cereal grain production. The low value of the produce affects the inputs of labor, husbandry and pest control.

The warm humid environment of the tropical zone affords more favorable conditions for the overseasoning and multiplication of pathogens and their vectors than those found in temperate areas. However, in many locations a dry season partially compensates for the quiescent winter months of temperate areas.

Diseases are a major factor in determining adaptation of maize and sorghum. Most of the sorghum therefore is grown in the more arid climates in order to escape disease loss. Tropical maize types, on the other hand, which are indigenous to Latin America and possess much greater resistance to the prevalent diseases of the area, are grown throughout the tropical lowlands of the world. The common cultivars of both crops are generally tall in stature with good seedling vigor enabling them to compete much better with weeds.

In the past, disease control methods of the tropical and temperate regions have shown great variation. Tropical agriculture has conventionally been carried out on a subsistence level and as a result cultural control tactics have been determined by low-cost labor, tradition and experience. Crop improvement programs established during the last 20-25 years have made progress in developing improved agronomic types with disease resistance. Consequently production has increased; during the 10-year period from 1961-1965 to 1971-1975 the annual production of maize and sorghum in the developing countries increased 38 and 35%, respectively, compared to increases of 27, 50, 11 and 8% for rice, wheat, millet and barley, respectively. (CIMMYT Review 1978).

Nevertheless, achievements in the agricultural sciences have surpassed the developments in seed production technology, delivery systems, extension farm credit and marketing programs. To capitalize on the gains made in the agricultural production sector, scientists must broaden their perspective and work more closely with government planners to alleviate the constraints affecting the transfer of technology to actual food production.

The major diseases of maize in tropical and subtropical areas vary with regions of the world. Of five downy mildew diseases causing major economic losses, sorghum downy mildew is the most widespread while Java downy mildew is essentially confined to Indonesia. Maize streak is considered the most damaging disease in sub-Sahara Africa; it has also been found in India. In Latin America corn stunt and sorghum downy mildew are the major diseases. Late wilt is a major problem of maize production in Egypt and is also important in India. Colletotrichum stalk rot is important in some Southeast Asia countries.

Areas having seasons with high temperatures coupled with high atmospheric and soil moisture are prone to bacterial stalk rots which can often be severe. Other stalk rots including black bundle, charcoal and fusarium are quite prevalent although their relative economic importance has not been determined. Banded leaf and sheath blight occasionally including the stalk rot and ear rot phases, is also widespread. Several of the major foliar diseases vary with climate, for example, Northern leaf blight and common rust prevail in cooler areas while Southern leaf blight and Southern rust are common in warmer climates.

In tropical and subtropical regions sorghum diseases are more widespread and damaging than those of maize. Humid growing conditions generally cause significant yield reductions due to anthracnose, one of the most damaging sorghum diseases. Weathering and the accompanying grain molds and head blights are present when maturation occurs during moist weather conditions; they result in a qualitative and quantitative reduction of harvested grain. Sorghum downy mildew, a very prevalent and damaging disease, has caused alarm in some countries. The major stalk diseases, caused by species of *Fusarium* and *Macrophomina,* are most severe when the crop undergoes a drought stress as it matures. Foliage diseases are also common although the severity of any particular disease will vary among sorghum growing regions. The more important foliage diseases include anthracnose, zonate leaf spot, Helminthosporium leaf blight, rust, gray leaf spot, sooty stripe and bacterial stripe.

The deterioration of stored grain and seed of both maize and sorghum is very extensive. Fungi and insects are the biotic agents involved.

Several non-parasitic diseases prevail throughout warmer regions. Soil nutrient deficiencies as well as toxicities caused by aluminum and sodium are omnipresent.

Of the many ways by which diseases of maize and sorghum can be controlled or their losses minimized, host plant resistance and avoidance have, in the past, been most effective. Resistance to the major diseases can and indeed, should be incorporated into commercial and local cultivars; this can be done efficiently through the use of modern plant breeding and pathological procedures. Disease avoidance embraces a number of cultural techniques including alteration of planting dates and rotational sequences.

Chemical control for the most part is restricted to fungicidal seed treatment, since the value of the crops does not warrent foliar or broadcast applications, except in situations involving high capital input such as in sweet corn or seed-production. Exclusion on a national basis is the responsibility of government quarantine/plant introduction authorities. Although stringent laws have been enacted by many governments during the past 30 years, the benefits were questionable. Prior to that time most of the seed-borne diseases were already present and the movement of maize and sorghum seed was done with little or no restriction. Subsequent restrictions produced a reduction of germplasm exchanges.

We know of no disease of maize or sorghum being eradicated from countries within the tropical and subtropical regions in the few attempts made. However, primary inoculum can be substantially reduced by: the elimination of collateral wild hosts, e.g. Johnson grass, roguing diseased plants, crop rotation, sanitation, and the growing of resistant cultivars in the area.

Seed Rot and Seedling Blight

The problems of seed rot and seedling blight are less severe in tropical than in temperate areas. They are caused by fungi, particularly *Pythium* spp. The disease generally can be avoided by planting good quality seed in warm soil where soil moisture is adequate for rapid germination and seedling establishment. Fungicidal seed treatment provides adequate control and could be profitably integrated in the future with seed treatment to control downy mildew of both crops, some smuts and shootfly of sorghum.

Root and Stalk Rots

Little is known about the severity of root diseases of maize and sorghum in these areas. While we believe they reduce yields, no control measures are specifically followed. Pythium root rot is widely distributed, considered the most important, and is most severe in poorly drained and physically compacted soils where oxygen is in low supply.

Stalk rots are widespread and one or more are usually important where the crops are grown. Pythium and bacterial rot are severe on corn in hot, humid regions, attacking plants from pre-silk to the early milk stages. The application of chlorine, e.g. to irrigation water, is an effective control of bacterial stalk rot even though it is seldom practiced. No effective control has been found for Pythium stalk rot, although it has been observed that some cultivars are more resistant than others.

The other stalk rots do not ordinarily attack actively growing plants. However, during seasons when they are severe, plants die prematurely resulting in blasting, light grain weight and lodged plants. Fusarium stalk rot and charcoal rot are widely prevalent on both maize and sorghum, particularly in the dry, warm areas. Well-balanced soil fertility, based on soil and tissue tests, and proper plant density tend to reduce the effects of these diseases as well as late wilt of maize. Host resistance, of a relatively low order, (for this complex of stalk-affecting diseases) is known but very few deliberate efforts have been made to utilize it. Disease development is favored by high (38-42 C) soil temperature and low soil moisture after the flowering stage. It is not a problem under irrigated conditions.

Nematodes and insects are injurious, and in addition, weaken and predispose plants to root and stalk rots, provide avenues of entrance for the pathogens concerned and often desseminate the pathogens. The development of a functional management-control program for insects and nematodes would reduce losses from root and stalk rots.

Ear Rots and Grain Molds

Maize ears and sorghum heads are attacked by many fungi, which manifest themselves during grain development and after maturity. Infection takes place in seed shortly after flowering. Disease avoidance by maturing the crop after the rainy season is commonly practiced in tropical areas; photosensitive varieties are usually grown which mature under short days after rainy periods. In sorghum, open-headed cultivars develop less grain mold. Tight husk cover and stalk rot resistance helps exclude several pathogens of maize. High lysine corn is much more susceptible to kernel rots, especially those caused by *Fusarium* species, than normal kernel types. Cultivars of both crops differ quite significantly in reaction and the selection for resistance is a regular feature of most improvement programs.

Foliar Diseases

The foliage of both crops is attacked by many fungal and bacterial pathogens. The early infection by these organisms is often related to cropping history. Disease avoidance systems frequently practiced by farmers include crop rotation and destruction of debris to reduce initial infection; the cultivation of tall, sparse-leafed genotypes and low plant densities which facilitate air movement; eradication of wild *Sorghum* species; no intercropping or adjacent cultivation of the susceptible sorghum-sudan hybrids; and the selection of a planting date on the basis of experience to coincide with lower severity.

High to moderately high levels of host resistance are available for these diseases. A moderate level of resistance is most often sufficient to prevent economic loss although the development of generalized or nonspecific resistance is wisely preferred in most improvement programs. Chemical control is possible for some of these diseases, but the low value of the produce renders the practice non-economical except in special circumstances.

Downy Mildews

Brown stripe downy mildew of maize is economically important only in areas of Central and Northern India receiving 75 cm or more rainfall. It has been largely controlled by varietal resistance of a nonspecific type. It can also be controlled by fungicides, and its incidence reduced by rotation to avoid infection from oospores. Crazy top infection can be eliminated by proper soil drainage since infection occurs only when soil is flooded at any time from germination until seedlings are 10 to 15 cm tall.

The *Peronosclerospora*-incited downy mildews are soil-borne, air-borne and seed-borne. Internal mycelia in seed represents little danger of introduction to new areas or as primary inoculum because it is inactivated within about 6 weeks when seed moisture content is reduced to below 20%. Oospores carried on seed surfaces, however, represent threats of dissemination with sorghum seed, as do internal mycelia of sugarcane downy mildew in sugarcane sets. Treatment of propagules with Apron (Ridomil), or another appropriate systemic fungicide, is now deemed an effective control.

Today, host resistance appears to be the single most effective way to control the *Peronosclerospora*-incited downy mildews. Many improved cultivars possess high levels of generalized resistance. Resistance to one species of *Peronosclerospora* generally conveys resistance to other species. However, host resistance needs to be integrated with the chemical and cultural means of control. Several new systemic fungicides have recently been developed that are effective as seed dressings or foliar sprays. The diseases can also be partially excluded by deep tillage, early planting and isolation from neighboring fields. Current information reveals that growing a trap crop, including non-hosts, significantly reduces the number of soil-borne oospores and is as effective as crop rotation. Prolonged cultivation of a resistant cultivar greatly reduces the quantity of residual oospore inoculum so that a susceptible crop; e.g., a sorghum-sudan forage crop, can be grown with less risk. The eradication of wild hosts, e.g., *Sorghum* and *Saccharum* spp., is helpful in reducing the prevalence of these diseases. Loss can be reduced, or compensated for, in part by increasing plant density since most systemically infected plants are poor competitors for light, soil moisture and soil nutrients. High quality seed, rapid germination and rapid growth during the first 3 weeks help plants escape systemic infection.

Virus and Mycoplasma-Like Diseases

Strains of sugarcane mosaic virus are present throughout the world on maize and sorghum. In general, losses are light, but severity is dependent on time of infection, cultivar reaction and the prevailing environment. Control of mosaic is dependent on the eradication of collateral hosts, avoidance of the several aphid vectors and host resistance. Since the vectors are migratory and the wild species of *Sorghum* and *Saccharum* are often highly prevalent in most regions, host resistance is considered the most feasible control measure.

28.3   Integrated Control of Nematodes of Corn and Sorghum in the United States

J. M. Ferris

Purdue University, USA

Prior to the introduction of soil insecticides which also had nematicidal properties, it was almost impossible to demonstrate yield reductions in field corn caused by nematodes. The reasons were many and included the fact tht fumigant type nematicides did not give good control of nematodes in the heavy clay soils in whch corn is frequently grown. Also, these materials were expensive and difficult to use. Often elaborate tests were set up only to experience an ideal growing season where all corn yielded well, in spite of attack by pests.

With the widespread use of soil pesticides which were effective against both insects and nematodes, entomologists began to note corn yield increases where no injurious insects were present. Nematologists who joined these investigations soon learned how much damage nematodes were able to cause in field corn. This came about because large acreages were treated every growing season, with all possible variations in growing conditions. Thus in some years, when environmental conditions were not ideal and stress was placed upon the plants, it was possible to correlate high nematode populations with corn yield reductions. It has now been demonstrated that *Pratylenchus* spp. (including *P. hexincisus, P. zeae*), *Hoplolaimus galeatus, Longidorus breviannulatus, Belonolaimus* (*nortoni?*), and *Meloidodera charis* are capable of causing injury to corn which is severe enough to result in yield reductions.

With the availability of relatively inexpensive nematicides which can be applied at planting time, a new problem has arisen. Although mean counts of plant parasitic nematodes are reduced by these treatments, differences in yields may in fact be caused by other factors. Often damage to corn roots by low populations of soil insects may be sufficient to account for any yield differences observed. Thus careful evaluation of the effect of pest management strategies on all crop pests present is necessary for correct interpretation of yield results.

It is my opinion that future work must be directed toward the clarification of those factors which enhance the damage caused by parasitic nematodes to the point where yield reductions occur. With our present knowledge it is not possible to predict whether a particular nematode species is present in sufficiently high numbers to cause significant plant injury. Depending on weather conditions during the growing season, competition from weeds, or undefined stress factors, a given population level may cause severe damage one year and not the next. The combination of factors causing plant damage must be determined for each species of nematode on the various soil types on which corn is grown.

We will also need to know the control efficiency of the pesticides we have available, as well as their cost. in this way we can maximize returns to the grower. The effects of new tillage practices on soil nematode populations and the resultant damage to crops needs to be investigated. Some studies have indicated that more nematodes are available to attack seedling plants when reduced tillage methods are used. Other studies have indicated exactly the opposite effect.

Additional research is needed to determine the extent to which nematodes interact with other organisms to produce plant damage. We know that the beneficial effects of crop rotation in reducing numbers of a nematode species damaging to corn may be reduced if a weed species on which the nematode is able to reproduce abundantly is present when the field is not planted to corn. Other relationships, however, are less obvious. For instance, we do not know whether nematode feeding makes corn plants more susceptible to stalk rot fungi. Nor do we know whether the damage caused by corn rootworms and nematodes is additive. The answers to these and other questions are necessary before truly integrated control of nematodes and other pests can become a reality in corn and soybean crops in the USA.

28.4 Integrated Approach to Control of Nematodes Infesting Corn and Sorghum in the Tropics and Subtropics

A. R. Seshadri

Indian Agricultural Research Institute, New Delhi 110012, India

The nematode problems of corn and sorghum have received only sporadic attention so far though many potentially pathogenic species are known to be associated with these crops. This is particularly so in the tropics and subtropics.

Maize

Over 27 species of nematodes belonging to 13 genera have been reported to be associated with corn in the tropics and subtropics, especially from India. Of these *Heterodera avenae, H. zeae, Meloidogyne incognita, Pratylenchus zeae, P. brachyurus, P. delattrei, Rotylenchulus reniformis, Hoplolaimus indicus* and *Tylenchorhynchus vulgaris* are well known pathogens and cause considerable damage to this crop [1, 2, 11, 25, 39, 41, 50, 56, 76, 78, 82, 87, 90, 92, 94]. The other species belong to the genera *Aphelenchus, Ditylenchus, Gymnotylenchus, Helicotylenchus, Hoplolaimus, Longidorus, Paralongidorus, Rotylenchulus, Telotylenchus* and *Tylenchorhynchus,* however nothing is known about their effects on corn [3, 11, 19, 20, 21, 24, 32, 57, 59, 61, 70, 71, 72, 84, 85, 91].

The serious corn problems of the temperate zones caused by *Belonolaimus longicaudatus, Ditylenchus dipsaci, Trichodorus christiei* etc. have not so far been reported from the tropics.

Work done on corn nematode control particularly in the tropics and subtropics is very limited, hence the experience gained on similar problems in the temperate regions could also be considered in formulating integrated approaches to nematode management in this crop.

Cultural Approach

Tillage

Among the several cultural practices, ploughing, fallowing and crop rotation have been found useful in managing nematode populations in corn cultivation. Fall ploughing was found to bring down the root-knot nematode *Meloidogyne incognita* population to an extent of 73% as compared to untilled plots in Tennessee, USA [78]. Similarly in Nigeria, Caveness [11] found that *M. incognita* and *Helicotylenchus pseudorobustus* were more numerous in non-tilled plots. However, *Pratylenchus* spp. in soil and corn roots were more in tilled than in untilled soils. Recently, the effects of seven tillage treatments, namely, fall plow, spring plow, chisel plow, offset disk, till flat, no-till flat and no-till ridge on nematode densities in corn growing fields in Iowa were studied by Thomas [86]. These seven treatments brought significant differences in nematode population densities of *H. pseudorobustus, Pratylenchus* spp. (*P. scribneri* and *P. hexincisus* were most common), *Xiphinema americanum,* other dorylaimids and total numbers of nematodes. With the exception of members of the Tylenchinae, highest densities occurred in untilled plots and the lowest number in spring and fall plowed plots. Possibly the tillage treatments influence nematode population densities in one or more of the following ways: (i) delay in egg hatching because soil warmed more slowly after plowing, (ii) tillage may lead to increased spread of fungal pathogens and (iii) reduced tillage may lead to accumulation of surface debris which in turn acts as a mulch affecting moisture retention, warming rate and depth of root growth.

Cropping Pattern

The influence of tillage on nematode population is often superimposed by the cropping sequence. Corn is an inefficient host for *M. incognita* [78] and even to *H. avenae* [36, 38]. However, corn is reported to be damaged by *H. avenae* in India [94, 95]. Under such circumstances wheat, barley and corn being hosts of *H. avenae,* the cropping pattern should not include all the three crops in sequence or more frequently in four or 5-year rotation programmes. According to Hirling [29], corn could suffer an estimated loss of 40% in Germany even though no cysts were formed on the roots, with only males developing to maturity. As against this, Obst and Diercks [53] found that growing winter wheat or spring barley in Germany three times in four years did not increase the cyst numbers to dangerous levels and the continuous growing of corn had little effect on intensifying the cereal cyst nematode problem. Monoculturing of corn may lead to increased population of *D. dipsaci,* a potential pathogen of several field crops including corn. On the contrary, corn was found to be sensitive to even very small populations of *H. avenae* in France [64]. Nonhosts like beet and lucerne may be included in the cropping sequence in cereal cyst nematode infested fields. In India other nonhosts for this nematode include fenugreek, mustard, cumin, radish, and carrot commonly grown in rotation with wheat and barley. Southards[78] reported the effects of fall plowing, fallowing and selected hosts on the population of *Pratylenchus zeae* and observed that in a rotation with corn and tobacco, the nematode did not build up in tobacco, but increased in population on corn. It was suggested that tobacco should precede corn to check the nematode population.

Polyspecific populations are a rule in nature and often several species may be recovered from the rhizosphere of a single corn plant. Thus Johnson *et al.* [35] studied the influence of different cropping sequences including cotton, peanut and soybean along with corn on polyspecific nematode populations in south-eastern USA. *M. incognita, Trichodorus christiei, Criconemoides ornatus* and *Pratylenchus* sp. increased rapidly on corn, while populations of *X. americanum* and *Heliootylenchus dihystera* were not much influenced by it. Most of the above species were suppressed by monocropping of peanut. They recommended corn-peanut-cotton-soybean and cotton-soybean-corn-peanut as effective cropping systems which suppressed nematode densities.

Adequate Nutrient Supply

Application of fertilizers adequately improves the host tolerance against nematodes. Primavesi and Primavesi [58] concluded from their experiments that in the tropics and subtropics, nematode infestation should not limit the yields of corn provided an adequate supply of fertilizers is given. Besides, the previous crop will have an important bearing and a tropical legume *Cajanus indicus* is good as crop to precede corn. Crop growth, rates of applied N and higher numbers of ring nematodes were found related [65].

Chemical approach

Soil fumigants. Several workers in USA [17, 18, 34, 97] have found soil fumigants like DD, Vorlex and EDB very effective in controlling nematode populations leading to increases in plant growth and yield of corn. Soil fumigation with D-D at 300 L/ha increased corn yields in Nigeria in field plots infested with *Pratylenchus* sp. and *Helicotylenchus pseudorobustus* [10]. DD-MENCS and DD were found effective fumigants in controlling *P. zeae, dihystera* and *Criconemoides* sp. infesting corn in Trinidad [73]. In the United Arab Republic, Oteifa and Taha [56] obtained increased yields of corn following effective control of *P. zeae* by using metham sodium.

Systemic chemicals

Field application. Rhoades [62] reported that sting nematodes attacking field corn could be controlled economically with low rates of several nonfumigant organophosphate and carbamate nematicides. *B. longicaudatus* and *T. christiei* were controlled more effectively by nonvolatile chemicals than by soil fumigants [34]. Among several systemic nematicides, carbofuran was reported to be very effective in controlling corn nematode populations in Canada and USA [7, 14, 17, 52, 89] and Germany [42]. Application of systemic nematicides at the planting time was as effective as preplant treatment and the normal dose ranged from 1.12 to 2.24 kg a.i./ha depending on the formulation. The control of tylenchid nematodes at the beginning of the growing season using carbofuran increased yields even up to 200%. An added advantage of these systemic organochemicals is that they are effective insecticides also. Fensulfothion, ethoprop, dyfonate, phenamiphos and phorate also gave good results and increased the yields up to 46% where the soil was heavily infested with sting nematodes [17]. Terbufos, aldicarb, CGA-12223 have also been found effective in Iowa at the rates ranging from 1.12 to 3.36 kg a.i./ha and increased corn yields up to a maximum of 20% over untreated control [52]. In a 3-year study conducted by Rhoades [63] in Central Florida, fensulfothion, phenamiphos, ethoprop, carbofuran, aldicarb, oxamyl, sulfocarb, CGA-12223 and AC 64475 applied at 2.2 kg a.i./ha in the row just ahead of planting reduced populations of *B. longicaudatus* and significantly increased yields of corn.

In a field trial against *Pratylenchus delattrei* infesting corn in South India, aldicarb was found to be the most effective nematicide at 2.5 kg a.i./ha applied in the seed hole at the time of sowing. The nematode population was checked throughout the crop growth and maximum grain yield obtained in this treatment. Tocis residue of the chemical in the harvested grain was below tolerance level [49].

Seed Treatment. Many of the systemic chemicals have been tried recently for seed treatments. An effective seed treatment has several advantages over traditional method of application of nematicides to soil. The quantity required being less makes the treatment very economical and possibilities of environmental and other health hazards are also relatively remote. Besides, it permits growing of nonresistant but desirable plant varieties under conditions where adequate soil treatment might be too expensive. Thus Truelove *et al.* [89] found that treatment of corn seeds with oxamyl at 5 g/100 mL solution decreased *H. dihystera* population considerably. However the period of effectiveness of this treatment remains to be worked out.

Studies now in progress at New Delhi using aldicarb, aldicarb sulfone, carbofuran and fensulfothion at 0.25-1.5% a.i. (w/w) for seed treatment of corn against *Heterodera zeae* have shown that aldicarb and aldicarb sulfone controlled the nematode effectively resulting in maximum yield of cobs [79].

Quarantine to Prevent Entry of Exotic Nematodes

Some of the important nematode species like *B. longicaudatus* and *D. dipsaci* have not so far been reported from corn growing areas in the tropics including India. Even though the cereal cyst nematode *H. avenae* is reported from Morocco and S. Africa, its impact on corn cultivation is not known. Similarly *D. dipsaci* is also reported from North and South Africa, not much being known about its damage to corn. Species of *Pratylenchus* have been reported from almost all corn growing areas. However, the dominant pathogenic species may vary from place to place, and thus their spread should be watched. The cyst nematode *Punctodera chalcoensis* is reported only from UK [81] and *Heterodera zeae* apparently only from India. Many of these nematodes can be carried over long distances as contaminants of seed material inside soil clods, root bits and other debris [66, 69]. In the context of large scale exchange of germplasm and other seed material within as well as between countries, it is very essential to take minimum precautionary measures to prevent spread and establishment of dangerous species of nematodes in areas where they are not known.

Sorghum

Several species of nematodes belonging to at least 17 genera are reported to be associated with sorghum in different countries in the tropics. These include species of *Aglenchus, Aphelenchoides, Aphelenchus, Gymnotylenchus, Helicotylenchus, Hoplolaimus, Longidorus, Meloidogyne, Neopaurodontus, Paratylenchus, Pratylenchus, Rotylenchulus, Teolotylenchus, Tylenchorhynchus, Tylenchus* and *Xiphinema* [6, 12, 13, 15, 24, 31, 37, 41, 43, 46, 60, 61, 67, 68, 70, 71, 75, 77, 88]. The only cyst nematode reported on sorghum is *H. gambiensis* from Gambia [46]. The problems caused by *Trichodorus allius, Longidorus elongatus* and *Meloidogyne naasi* reported from temperate regions like USA and Canada are not known to occur in the tropics. However *Meloidogyne incognita* occurs on sorghum in Arizona, USA [8] as well as Kenya [5].

Control

Experimental data on the control of nematodes associated with sorghum are very scanty. Application of carbofuran and aldicarb at the time of planting as 7 in. furrow treatment significantly increased grain sorghum yields in Oklahoma [80].

Grain sorghum was found to be a good host of *M. incognita* in Arizona with population increase in four months at a rate comparable with cotton as host. However barley was as effective as fallowing in reducing the population, the numbers declining to zero in about eight months [8].

The population of *Helicotylenchus* spp. was markedly reduced in fine sandy soil planted with *Sorghum bicolor* and *Avena sativa* as test plants following the application of 8 to 32 tons/ha of composted municipal refuse. This reduction in population was attributed to the toxic effects of organic substances emanating from the material [30].

The search for resistance in sorghum germplasm in Kenya was partially successful and cultivars E 5765, E 6250, E 6255, E 6256, E 6263 and E 6487 were found to be immune (no larval penetration) to *M. incognita, M. javanica* and *M. hapla* [4].

A review of literature reveals that even though several methods of control like fallowing, ploughing, crop rotation, use of chemicals, etc. have been tried individually against nematodes of corn and sorghum, an integrated approach to nematode control is yet to be developed. Such an approach has given good results against *Globodera rostochiensis* on potato [22, 26, 27, 28, 45, 54], *Pratylenchus penetrans* on fruit trees [44], *Meloidogyne* spp. on vegetable crops, soybean etc. [47, 93] and a few other nematode species.

While nematicides are powerful tools to achieve immediate kill of nematode populations or suppression of their multiplication, their efficacy is short-lived and their use beyond the economic means of the average farmer, especially in the developing countries. The prospects of any of these chemicals getting wide acceptance would therefore depend upon a favourable cost benefit ratio. Apart from their prohibitive cost, the difficulties of application of chemicals over large areas, the need for repeat treatment every year, health hazards involved etc. are serious disadvantages. Even so, chemical control if judiciously used could still form a part of any overall control programme against nematodes infesting corn and sorghum. There is an urgent need to intensify research on seed treatment techniques so as to minimize the use of chemicals and reduce the costs. Work done in India has amply demonstrated that summer fallowing and deep ploughing two or three times at intervals of 10-15 days reduces nematode populations to the minimum. Based on available information and experience, an integrated approach involving the use of two or more methods like crop rotation, fallowing and deep ploughing during the peak summer months, use of nematicide treated seed, and nutritional care of the host (corn and sorghum in this case) offers good possibilities of managing nematode populations in these two crops. Use of resistant varieties if and when available will further strengthen this approach.

The integrated approach of course would remain a theoretical proposition until widely tested through carefully laid out field experiments. This is especially true of the tropics and subtropics where nematode problems of corn and sorghum have received hardly any attention.

References

1. Ahmed, M. A. (1970). Studies on the root lesion nematode, *Pratylenchus delattrei* Luc, 1958. Ph.D. thesus, University of Madras, India.
2. Anonymous (1970). Report of the Secretary for Agriculture, 1968-1969, Salisbury, Rhodesia, 82 pp.
3. Anonymous (1973). Tobacco Research Board, Abridged Annual Report for the year ended 30th June, 1973, Salisbury Rhodesia, 26 pp.
4. Anonymous (1975). Federal Department of Agricultural Research, Annual Report 1973-1974. Moor Plantation Ibadan, Nigeria, 178 pp.
5. Anonymous (1975). East African Agriculture & Forestry Research Organization, Annual Report, 194 pp.
6. Aytan, S. and Dickerson, O. J. (1969). *Plant Dis. Reptr.,* 53:737.
7. Bergerson, G. B. (1978). *Plant Dis. Reptr.* 62:295-297.
8. Careter, W. W., Nieto, S. Jr. (1975). *Plant Dis. Reptr.* 59:402-403.
9. Caubel, G. and Rivoal, R. (1971). *Phytoma,* 24:15-18.
10. Caveness, F. E. 1973). *Nematropica,* 3:1.
11. Caveness, F. E. (1974). *J. Nematol.,* 6:138 (Abstr.)
12. Chandrasekharan, J. and Seshadri, A. R. (1969). *All India Nematology Symposium, New Delhi,* India p. 10-11.
13. Chandwani, G. H. and Reddy, T. S. N. (1967). *Indian Phytopath.,* 20:383-384.
14. Cohick, A. D. (1975). *Pflanzenschutz-Nachrichten Bayer,* 28:80-91.
15. Das, V. M. (1960). *Z. parasiten K.* 19:553-605.
16. Deshmukh, M. G. (1967). Studies on *Hoplolaimus indicus* Sher, 1963, Ph.D. Thesis, IARI, New Delhi, India.
17. Dickson, D. W. and Johnson, J. T. (1974). *Proc. Soil Crop Sci. Soc. Fla,* 33:74-77.
18. Edmunds, J. E., Bothroyed, C. W. and Mai, W. F. (1967). *Plant Dis. Reptr.* 57:15-19.
19. Edward, J. C., Mishra, S. L., Naim, Z. and Singh, R. A. (1963). *Allahabad Farmer,* 37:1-7.
20. Edwards, J. C. and Mishra, S. L. (1963). *Nematologica,* 9:218-221.
21. Edward, J. C. (1966). *Allahabad Farmer,* 40:269-271.
22. Efremenko, V. P. and Klimakova, E. T. (1971). *Byull. Vses. Inst. Gelm.,* 6:13-21.
23. Egunjobi, O. A. (1974). *Nematologica,* 20:181-186.
24. Fortuner, R. and Amougou, J. (1973). Cahiers de l'office de la Recherche Scientifiqui et Technique Ontre-Mer, Biologie, No. 21.
25. Graham, T. W. (1951). *S. Car. Agric. Expt. Sta. Bull.* 390, 35 pp.
26. Guskova, L. A., Sreshnikova, N. M. and Gladkaya, R. M. (1969). *Zaschch. Rast. Vredit. Bolez.,* 8:7.
27. Guskova, L. A. and Gladkaja, R. M. (1974). *J. Nematol.* 6:185-186.
28. Hijink, M. J. (1972). *Bull. OEPP.* 7:41-48.
29. Hirling, W. (1974). *Gesunde Pflanzen,* 26:58-52.
30. Hunt, P. G., Hortenstine, C. C. and Smart, G. C. Jr. (1973). *Journal of Environmental Quality* 2:264-266.
31. Husain, Z. (1967). *Proc. Natn. Acad. Sci., India,* 37:184-185.
32. Janarthanan, R., Seshadri, A. R. and Subramaniam, TR (1969). *All India Nematology Symposium,* New Delhi, India, p. 34.
33. Johnson, A. W. and Burton, G. W. (1977). *Plant. Dis. Reptr.* 61:1013.
34. Johnson, A. W. and Chalfant, B. B. (1973). *J. Nematol.,* 7:158-168.
36. Johnson, P. W. and Fushtey, S. C. (1966). *Nematologica,* 12:630-636.
37. Khan, E., Singh, D. B. (1974). *Indian J. Nematol.,* 4:191-211.
38. Kort, J. (1972). In *Economic Nematology* (Ed. J. M. Webster), Academic Press, London, New York, pp. 97-126.
39. Koshy, P. K. Swarup, G. and Sethi, C. L. (1970). *Nematologica,* 16:511-516.
40. Koshy, P. K. and Swarup, G. (1971). *Indian J. Nematol.,* 1:106-111.
41. Krishnamurthy, G. V. G. and Elias, N. A. (1967). *Indian Phytopath.* 20:374-377.
42. Küthe, K. (1975). *Pflanzenschutz-Nachrichten Bayer,* 28:67-69.
43. Lamberti, F. (1969). *Plant Dis. Reptr.,* 53:421-424.
44. Mai, W. F. (1972). *Proc. Ohio State Hort. Soc.* 125:95-97.
45. McKenna, L. A. and Winslow, R. D. (1973). Records of Agril. Research, Dept. of Agri., N. Ireland 23:63-64.
46. Merny, G. and Netscher, C. (1976). *Cahiers O.R.S.T.O.M., Serie Biologie, Nematologie,* 11:209-218.
47. Minton, N. A. and Parker, M. B. (1975). *J. Nematol.,* 7:60-64.
48. Naganathan, T. G. and Sivakumar, C. V. (1975). *Indian J. Nematol.,* 5:162-169.
49. Naganathan, T. G. and Sivakumar, C. V. (1976). *Indian J. Nematol.,* 6:32-38.
50. Nath, R. P. (1970). Studies on the genus *Pratylenchus.* Ph.D. Thesis, Bihar Univ., India.
51. Norse, D. (1973). *Tropical Agriculture,* 49:335.
52. Norton, D. C., Tollefson, J., Hinz, P. and Thomas, S. H. (1978). *J. Nematol.,* 10:160-166
53. Obst, A. and Diercks, R. (1972). *Bull. OEPP,* No. 377-386.
54. Oostenbrink, M. (1974). *Bull. OEPP,* 4:475-480.
55. Oteifa, B. A. (1962). *Plant Dis. Reptr.,* 46:572-575.
56. Oteifa, B. A. and Taha, A. (1964). *Tech. Bull. Bahtim. Exp. Sta. Egypt agric. Org.* No. 14:pp. 26.
57. Prasad, S. K., Khan, E. and Chawla, M. L. (1965). *Indian J. Ent.,* 27:182-184.
58. Primavesi, A. M. and Primavesi, A. (1973). *Agro. chimica,* 17:511-521.
59. Raina, R. (1966). *Indian J. Ent.,* 28:438-441.
60. Raski, D. J. (1975). *J. Nematol.,* 7:15-34.
61. Raski, D. J., Prasad, S. K. and Swarup, G. (1964). *Nematologica,* 10:83-86.

62. Rhoades, H. L. (1969). *Proc. Soil Crop. Sci. Soc. Fla.* 28:262-265.

63. Rhoades, H. L. (1978). *Plant Dis. Reptr.* 62:91-94.

64. Rivoal, R. (1973). *Phytoma* 25:17-18.

65. Robertson, W. K., Hammond, L. C., Lundy, H. W. and Dickson, D. W. (1974). *Proc. Soil Crop Sci. Fla.* 33:80-82.

66. Sanwal, K. C. and Mathur, V. K. (1975). *Indian J. Nematol,* 5:251-254.

67. Seshadri, A. R. (1970). Agricultural Yearbook ICAR, New Delhi, India.

68. Seshadri, A. R. and Sivakumar, C. V. (1963). *Madras agric. J.,* 50:134-137.

69. Sethi, C. L., Nath, R. P., Mathur, V. K. and Ahuja, S. (1972). *Indian J. Nematol.,* 2:89-92.

70. Sethi, C. L. and Swarup, G. (1968). *Nematologica,* 14:77-88.

71. Siddiqi, M. R. (1961). *Nematologica,* 6:59-63.

72. Siddiqi, M. R. (1963). *Z. parasitenkunde,* 23:397-404.

73. Singh, N. D. (1976). *J. Nematol.* 8:302-303.

74. Singh, N. D. (1976). *Plant Dis. Reptr.* 60:783.

75. Sitaramaiah, K., Singh, R. S., Singh, K. P. and Sikora, R. A. (1971). *Exptl. Stn. Bull.* No. 3, *U.P. Agric. Univ. Pantnagar,* India, pp. 70.

76. Siyanand (1975). Studies on the relationships of the nematode parasites *Hoplolaimus indicus, Tylenchorhynchus vulgaris* and *Pratylenchus thornei* to maize (*Zea mays*). Ph.D. Tgesis, IARI, New Delhi, India.

77. Simolik, J. D. (1977). *Plant Dis. Reptr.* 61:855-858.

78. Southards, C. J. (1971). *Plant Dis. Reptr.* 55:41-44.

79. Srivastava, A. N. (1979). Div. of Nematology, IARI, New Delhi, India (Personal communication)

80. Sturgeon, R. V. Jr. & Jackson, K. (1975). Oklahoma Agril. Expt. St. Res. Report pp. 728.

81. Stone, A. R., Sosa Moss, C. Mulvey, R. H. (1976). *Nematologica,* 22:381-389.

82. Swarup, G. Prasad, S. K. and Raski, D. J. (1964). *Pl. Dis. Reptr.* 18:234.

83. Swarup, G. and Seshadri, A. R. (1974). In *Current trends in Plant athology* Department of Botany, Lucknow Univ., Lucknow, India, pp. 303-311.

84. Swarup, G. and Sethi, C. L. (1968). *Bull. Ent.,* 9:76-80.

85. Tandon, R. S. and Singh S. P. (1973). *Zool. Anz.,* 191:139-150.

86. Thomas, S. H. (1978). *J. Nematol.,* 10:24-27.

87. Thorne, G. (1961). Principles of Nematology, McGraw Hill, 553 pp.

88. Tikyani, M. G., Khera, S and Bhatnagar, G. C. (1969). *Labdev. J. Sc. Technol.* 7B:176-177.

89. Truelove, B. Rodriguez-Karrana, R. and King, P. S. (1967). *J. Nematol.,* 2:129-138.

91. Verma, R. S. (1973). *Zool. Anz.,* 190:170-174.

92. Verma, S. K. and Prasad, S. K. (1969). *Indian J. Ent.,* 31:36-47.

93. Waldeman, H. (1971). *Gesunde Pflanzen,* 23:227-228, 230-232.

94. Yadav, B. S. and Verma, A. C. (1970). *Proc. 57th Indian Sci. Congress,* pp. 551-552.

95. Yadav, B. S. and Verma, A. C. (1971). *Indian J. Nematol.,* 1:97-98.

96. Yassin, A. M. (1973). *Cotton Growing Review,* 50:161-168.

97. Young, P. A. (1964). *Plant Dis. Reptr.* 48:122-123.

28.5   Integrated Control of Corn and Sorghum Arthropods in the U.S. (A Brief Summary)

B. R. Wiseman

Southern Grain Insects Research Laboratory
Agricultural Research, Science and Education Administration, USDA
Tifton, GA 31794

W. P. Morrison

Extension Entomologist
Texas A & M University, Lubbock, TX 79401

Field crops of corn and grain sorghum are grown and harvested annually on more than 102 million acres in the United States. In 1976 the farm value of these crops amounted to over $15.9 billion. Corn is the principal cash grain crop in the U.S. and is ranked as teh 3rd most important cereal food crop of the world. Sorghum ranks 3rd as a U.S. csh grain crop and is 5th in importance as a world cereal crop. One of the major costs to the farmer is the loss incurred because of pests and/or resultant expense of protecting the crop from damage. It is estimated that annual losses in corn and sorghum from arthropod pest damage amount to approximately 10% of the total crops.

In 1952 only 1% of the total corn acreage was treated with insecticides, but pesticide usage has increased steadily so that in 1976, 38% of all corn acreage was treated. As a result, the total poundage of insecticides used for control of corn insects has more than doubled since 1964, from 15.7 million lb to an estimated 32 million lb in 1976. In sorghum, insecticides were used sparingly until 1971 when an estimated 39% of the total acreage received treatment, and 5.7 million lb of insecticides were used, most of it apparently was applied for control of the new biotype of the greenbug (*Schizaphis graminum* (Rondani)) that was attacking sorghum. By 1976, only 27% of the total sorghum acreage was being treated with an insecticide (an estimated 4.6 million lb), in part perhaps because new greenbug-resistant hybrids were being grown and grain sorghum pest management programs had been implemented.

Corn

The corn earworm (*Heliothis zea* (Boddie)) which is distributed over the entire corn growing areas of the U.S. is responsible for an estimated 2.5% of the crop losses. No economic thresholds have been established to date, which represents a serious gap in our overall research efforts. Present integrated controls include early planting of resistant varieties and reliance on increases of natural predator-parasite populations to reduce the earworm populations.

The European corn borer (*Ostrinia nubilalis* (Hubner)) is distributed throughout the corn growing areas from the east coast through the corn belt, into the Great Plains, and as far south as the Texas Panhandle and northern Florida. Overall losses due to European corn borers have been estimated at 2%. The several economic thresholds for the corn borer range from 35 to 75% of the plants damaged with live larvae present. Integrated controls involve the use of resistant varieties, cultural methods, and chemical control. Overwintering corn borers can be reduced by early harvest and stubble destruction. In some years predators, parasites, and pathogens reduce corn borer populations.

Corn rootworms (*Diabrotica virgifera* LeConte, *Diabrotica longicornis* (Say), and *Diabrotica undecimpunctata howardi* Barber), are generally a problem only in the corn belt. Overall losses due to the rootworm complex have been estimated at 2%. Little economic threshold work has been accomplished to date, but two states have reported economic thresholds of 3 larvae/plant; and the adult thresholds range from 1 to 5 adults/plant at silk when the effort is made to suppress overwintering populations. Cultural control, plant resistance, and insecticidal control are the present components of integrated control. However, some hybrids, because of a large, deep root system, are tolerant to corn

rootworm, and fall plowing has reduced the percentage of stalks affected by the rootworms more than spring plowing. Also, the application of manure assisted in the build-up of populations of predaceous mites. Crop rotation is the primary cultural control and it is very important because it is a practical measure and can also be used in conjunction with the production of other crops such as soybeans. Insecticide applications are usually a preventive treatment.

The fall armyworm (*Spodoptera frugiperda* (J. E. Smith)) is another important pest of corn. It is a special problem in the south, but periodically it spreads to the southern part of the north central region. For example, crop loss in 1975 in Georgia alone was estimated to be over $20 million. Overall losses attributed to the FAW have been estimated at 2%. Few economic threshold studies have been conducted, but most states treat when 50% of the plant show damage in the whorl stage. No threshold exists for small plants or mature plants. Efforts at integrated control should take into account early whorl feeding, ear feeding, and protection of the entire plant when grown as a late crop. Early plantings in the south can help avoid much of the fall armyworm damage. However, when heavy populations are present, insecticides are used extensively. Few commercial hybrids at present possess resistance to fall armyworm. Natural predators and parasites assist somewhat in limiting damage.

The southwestern corn borer (*Diatraea grandiosella* (Dyar)) is found in 16 states, from Arizona to Tennessee. Overall losses due to southwestern corn borers have been estimated at 1%. Economic thresholds range from no chemical control to treatment when 35% of the plants are infested with eggs and or small larvae. Certainly, this diversity of opinion indicates the need for more research in this area. Integrated controls have largely centered on combinations of cultural and chemical controls. Stubble destruction exposes overwintering larvae to cold temperatures and thus increases mortality. However, the soil erosion that can occur in some sandy soil means that stubble destruction may not always be practical. The use of early planting dates has been advocated. Also, the use of early harvest dates to escape "girdling" by the borers and eventual lodging is recommended. In Kansas, as high as 82% of southwestern corn borer eggs have been reported parasitized by *Trichogramma* spp. The recent development and release of resistant germplasm will enhance varietal selection by growers as commercial companies utilize this material.

The black cutworm (*Agrotis ipsilon* (Hufnagel)) is becoming more of a problem every year throughout the corn belt. Overall losses due to this pest are estimated at 1%. Few clear-cut thresholds have been developed. The only methods presently used for control of black cutworm are early plowing to control spring weeds and the use of insecticides.

Spider mites, primarily the two-spotted spider mite (*Tetranychus urticae* Koch) and Banks grass mite (*Oligonychus pratensis* (Banks)) are becoming increasingly important to the production of ca. 2 million acres of corn in the drier and warmer areas of the western Great Plains. Overall losses caused by the mites on corn have been estimated at 0.5%. The economic threshold for spider mites is not well defined, but several state guides indicate that chemical control should begin when the mites have colonized up to the middle 1/3 of the plant. The only current method of control is the use of miticides and the hope that several predators can prevent the mite populations from increasing. Delayed planting is recommended, and this does result in lower mite populations, less damage, and significant yield increases.

Overall reduction in insect losses in corn to the several corn pests will depend to a large extent on the individual grower and especially on his ability to manage the insect problems. Success depends on defining economic thresholds and his ability to select the best planting dates, cultural practices, and resistant hybrids and also on his willingness to enlist the assistance of private or public pest management specialists. However, these pests must be managed on an area-wide basis by using a systems approach. Therefore, research must be directed at increasing the level of insect resistance in corn and, more especially, in developing inbreds that possess multiple pest resistance. Other components of pest management should be evaluated on an area-wide basis, and those that can be used should be recommended and promoted. If the technology has not been developed, the research must be increased to see that they are developed.

In 1978, 17 states participated in some phase of an insect pest management program involving corn. Several of these Cooperative Extension Service programs are pointing out deficiencies in research in such areas as economic threshold or injury levels, population estimation techniques, selective pesticides and rates, and gaps in basic biological information. These programs have served to demonstrate and, in some cases, generate corn pest management technology. Also they have caught the attention of farmers and are creating an awareness of the benefits of using sound pest management practices.

Sorghum

The sorghum midge (*Contarinia sorghicola* (Coquillet)) is the most damaging of all sorghum insects, and this pest is generally found wherever sorghum is grown. Overall losses caused by the midge have been estimated at 4%. Economic thresholds are generally given as 1 or 1-2 adults per head. When populations reach this level, insecticidal control is recommended. However, in the advent of resistant hybrids, new thresholds will be required, especially if the resistance mechanism involves high levels of nonpreference or tolerance. Cultural controls are generally the most frequently recommended strategy for escaping midge damage. However, if late plantings of sorghum are necessary, as often happens in the southeast, then reliance on insecticidal controls may be necessary.

Biotype C of the sorghum greenbug (*Schizaphis graminum* (Rondani) occurs in most areas of the U.S. where sorghum is produced. Overall losses caused by the greenbug are estimated at 2.5%. Several states have reported economic thresholds. For seedling plants, treatment is recommended when damage is visible and colonies are present on the lower leaf surface. When plants are in the preboot stage, they should be treated before any entire leaves are killed. When they are in the head stage, treatment should be applied when numbers are sufficient to cause death of two lower normal-sized leaves. The integrated controls for reducing greenbug damage are as complete as those for any insect. Plant resistance, cultural control, insecticidal controls, and adequate predators and parasites have all been used together in managing greenbug populations.

In the case of grain sorghum, the fall armyworm and corn earworm are considered together. The geographical distribution coincides generally with the areas of sorghum production, especially in years of high populations. Overall losses caused by FAW-CEW in sorghum are estimated at 1.5%. Economic thresholds reported are a mean of 2 larvae/head, but no thresholds have been reported for seedling or whorl-stage sorghum. Current integrated controls are not clearly recognized. However, early planting of hybrids with open heads is used; and natural predator-parasites and especially *Orius insidiosus* (Say) are very effective against corn earworms.

The distribution of sorghum webworm (*Celama sorghiella* (Riley)) is of particular importance throughout the humid south; however, it may range as far north as Illinois and Nebraska. Overall webworm losses in sorghum are estimated at 0.5%. The economic threshold has been reported as a mean of 5 larvae/head. Current controls center around the use of early plantings and the use of insecticides to keep the webworm population in check.

The two-spotted spider mite and Banks grass mite are important pests of the sorghum crop. The mites are distributed from eastern New Mexico, through the High Plains of Texas, and north into eastern Colorado and western Oklahoma, Kansas, and Nebraska. Overall losses due to mites are estimated at 0.5%. No economic threshold has been reported. However, if the population builds up rapidly away from the leaf midrib, treatment has been recommended. Currently, damaging populations can only be controlled by miticides.

Probably insect losses in sorghum could be reduced more than the losses in corn. This view is based on the advances made since the introduction of greenbug-resistant hybrids in 1976. Since the new midge-resistant germplasm is being utilized by commercial companies, the grower will be provided with the flexibility needed to adjust his plantings as dictated by moisture conditions at planting time or to avoid damaging populations of greenbugs, especially in states such as Georgia. The remaining biggest hurdle is, therefore, multiple pest resistance. This should be coupled with multi-genic resistance for each pest. If research is increased to any degree, it will be possible to add this as the "power-packed" punch needed for pest suppression, protection of our environment, and reduction of costs to the grower.

In 1978, seven states and the Virgin Islands participated in pest management programs for sorghum. However, many showed up the deficiencies in current research programs. For real progress, the farmer must have an option of more than one control method; then pests, crops, and systems can be truly managed.

Conclusion

How do you get there from here? Obviously, some basic research remains to be done, and there is still much work to be done. Many field tests must be conducted, and tactics must be developed for using different pest control methods. The areas that should receive the greatest emphasis are those that appear to avoid increasing the cost to the farmer and reducing the quality of our environment.

Problems alone cannot justify a pest management program; however, solutions to those problems are still essential. Until research can provide the solutions, pest management programs have nothing to offer.

Bibliography

Andrilenas, P. A. 1975. Farmers' use of pesticides in 1971: extent of crop use. USDA, Agric. Econ. Rep. No. 265. 25 pp.

Anonymous. 1976. ARS National Research Program. NRP No. 20240. Insect control: grains, forages, sugar crops, and oilseeds. USDA. 202 pp.

Chiang, H. D. 1978. Pest management in corn. Annu. Rev. Entomol. 23: 101-23.

Dicke, F. F. 1977. The most important corn insects. *In* Corn and Corn Improvement. G. F. Sprague (ed.) Agronomy 18: 501-90.

Holcomb, R. W. 1970. Insect control: alternatives to the use of conventional pesticides. Science 168: 456-58.

Kendrick, J. B., Jr. 1978. Agriculture's most important battle. Calif. Agric. 32: 6-7.

Luckman, W. H. 1978. Insect control in corn-practices and prospects. Pages 137-55 *In* Pest Control Strategies. E. H. Smith and D. Pimentel (eds.) 334 pp. Academic Press, N. Y.

Teetes, G. L. 1976. Integrated control of arthropod pests of sorghum. Pages 24-41 *In* Proc. U.S.-USSR Symposium: The integrated control of the arthropod, disease, and weed pests of cotton, grain, sorghum, and deciduous fruit. Tex. Agric. Exp. Stn. Misc. Publ. 1276. 216 pp.

Young, W. R., and G. L. Teetes. 1977. Sorghum entomology. Annu. Rev. Entomol. 22: 193-218.

28.6 Integrated Control of Corn and Sorghum Insect Pests in the Tropics and Subtropics

William R. Young

Entomologist and Foundation Representative, The Rockefeller Foundation
G.P.O. Box 2453, Bangkok, Thailand

Both corn (*Zea mays*) and sorghum (*Sorghum bicolor*) rank among the six most important cereal crops of the world in area and production. According to FAO 1975 statistics, approximately one-half the world's corn area, and one-third the production, was in tropical and subtropical regions of Africa, Central and South America and Asia, while three-fourths the sorghum area and 40% of the production came from the same regions. Average yields/ha of both crops from these tropical regions were a little more than one-half the world average, with the exception of sorghum in South America where the yield was twice the world average. Maximum country average temperate zone yields of corn and sorghum were 6.5 and 4.2 T/ha, respectively, compared with only 1.5 and .75 ton yields in the African and Asian tropics. [1] The generally lower production levels of these two important cereals in tropical and near tropical zones tend to limit the insect control measures that can be applied to them over an extensive area.

A perusal of the literature on insect pests of corn and sorghum reveals some 100 or so species that have been reported as damaging. All stages of growth are attacked from seedling to fruiting, as well as the stored seeds. Many occur on both crops, and earlier literature tends to group corn and sorghum insects, although, a number of them are more specific to one of the crops. Recently published international reviews, or listings, of corn insects include those by Ortega and De Leon [2] and Chiang (1978) [3], the latter dealing with cooperative research efforts on corn pests, and pest management programs at high and minimal input levels of production. A review by Young and Teetes [4] covers major sorghum pests. An FAO Guideline to Integrated Control of Sorghum Pests by Teetes et al [5] is in press. I would refer you to these publications as an entry to the literature for more detailed information.

The time alloted does not permit a species by species treatment of all the major corn and sorghum pests in tropical regions. However, I would like to call your attention to a few of the more important considerations related to effective management of pests on these crops in the tropics. A brief discussion of the unique features of some of the more important groups of pests on corn and sorghum also may serve to highlight present status or future research needs for more effective control.

First, as mentioned earlier, current levels of production in the tropics are generally low. The crops are often grown at a subsistance level under marginal conditions for agriculture, on hill sides, or areas subject to drought. They usually are grown under rain fed conditions with little fertilizer application, and limited weed control. Average field size per farmer is small and technology provided by extension services is frequently meagre or nonexistent. Under these situations costly inputs for pesticides do not pay even if they are available. Special attention needs to be given to the management of cultural practices, preservation of natural enemies, and the use of resistant varieties that are of lower cost to the farmer. In many instances the farmers themselves have evolved a system of rotation or cultivation that tends to limit losses to both pests and diseases. It behooves us to take a careful look at these situations, to see what is actually happening in the pest-host plant-environmental interaction both before and after recommending changes.

Second, it is commonly stated that insect pest problems tend to be more serious in the tropics, since the warm climate allows for year around feeding and reproduction of pest species. This is not generally true, especially during the production or growth and fruiting stages of these crops. In most of the tropics there is a prolonged dry (and hot) season that is just as decimating to insect populations as is the cold period in temperate climates. Where there is irrigation in the tropics, crops can be grown the year around and may provide a substrate for both the harmful species and their natural enemies, limiting pest outbreaks. However, if natural enemies are lacking continuous cropping can lead to pest built up.

It is certainly true that losses to insect pests tend to be more serious in seed or grain storage in warmer humid climates. Field infestations of the rice (*Sitophilus oryzae*) and maize (*Sitophilus zeamais*) weevils and of the Angumois grain moth (*Sitotroga cerealella*) are more wide spread and intensive, leading to higher levels of infestation in storage. The technology of limiting storage losses to insects is directly transferable from temperate to tropical zones. However, maintenance of required storage conditions such as low levels of temperature, humidity, and good aeration of stored grain is more difficult and requires more energy and higher cost. As the human population increases and we attempt to produce more food in the tropics we must make provision for better storage of buffer stocks. This is now becoming apparent in India where some 20 million tons of food grains is being stored under less than ideal conditions.

An additional pressing concern for future insect pest control on corn and sorghum in the tropics is the necessity and opportunity of producing more than one or two crops a year on the same land, or in the same region, as more irrigation becomes available. Research has been expanded greatly on multiple and relay cropping and various cropping systems. A second or third crop following the first of corn by corn, or sorghum by sorghum, or of either of these crops following a crop affected by the same pest species can lead to pest outbreaks. We need to be on guard and closely monitor the pest situation as cropping frequency and levels of production per unit area per year are increased. Entomology should be a major input in these cropping systems research programs.

Fortunately the numbers of more important pests, that tend to be problems on corn and sorghum in any one region of the tropics year after year, are relatively few. The nature of these crops also is such that they usually produce foliage and roots in excess of that needed to produce satisfactory yields. They are able to tolerate a certain amount of insect damage without suffering serious economic loss. The exception of course is sweet corn that is more susceptible to ear worms and borers, and, has a lower threshold of acceptable injury.

An abbreviated listing of more important pests has

Table 1. The Most Important Insects Pests Affecting Corn and Sorghum in the Field in Tropical and Subtropical Areas of the World

Insect	Geographical distribution	Crop affected	Nature of Injury
ON SEEDLINGS			
Sorghum shoot fly			
Atherigona soccata	AF, AS	sorghum	larvae kill central shoot
ON FOLIAGE BEFORE FLOWERING			
Plant hoppers			
Peregrinus maidis	C	corn + sorghum	nymphs and adults suck sap, are disease vectors
Dalbulus spp.	NW	corn + sorghum	nymphs and adults suck sap, are disease vectors
Cicadulina mbila	AF, AS	corn	nymphs and adults suck sap, are disease vectors
Armyworms			
Spodoptera frugiperda	NW	corn + sorghum	larvae feed in whorl and on leaves
Mythimna separata	AS	corn + sorghum	larvae feed in whorl and on leaves
Spodoptera exempta	AF, NW	corn + sorghum	larvae feed in whorl and on leaves
Spodoptera exigua	AF, NW	corn + sorghum	larvae feed in whorl and on leaves
Pseudaletia convecta	AF, O	corn + sorghum	larvae feed in whorl and on leaves
Stem borers			
Diatraea grandiosella	NW	corn + sorghum	larvae feed on leaves, in stems and ears
Diatraea saccharalis	NW	corn + sorghum	larvae feed on leaves, in stems and ears
Zeadiatraea lineolata	NW	corn + sorghum	larvae feed on leaves, in stems and ears
Chilo partellus	AF, AS	sorghum + corn	larvae feed on leaves, in stems and ears
Ostrinia furnacalis	AS	corn	larvae feed on leaves, in stems and ears
Busseola fusca	AF	corn + sorghum	larvae feed on leaves, in stems and ears
Sesamia calamistis	AF	corn + sorghum	larvae feed on leaves, in stems and ears
Sesamia inferens	AS	corn	larvae feed on leaves, in stems and ears
EAR OR PANICLE PESTS			
Earworms			
Heliothis zea	NW	corn + sorghum	larvae feed on ears or immature panicles
Heliothis armigera	AF, AS	corn + sorghum	larvae feed on ears or immature panicles
Sorghum midge			
Contarinia sorghicola	C	sorghum	larvae feed on ovaries destroying seed

Distribution Key: C = Cosmopolitan, AF = Africa, NW = New World, AS = Asia, O = Oceana been prepared in Table 1 showing the geographical distribution, crops affected and the nature of injury of each. These are the species I consider to be most important round the world in the tropics at the present time. A brief description of these insects or groups of insects follows below including the present status of control measures for some of them with an attempt to point out unique features that provide opportunities for integrated approaches to their control.

The Sorghum Shoot Fly

The muscid fly, *Atherigona soccata,* is presently considered to be the predominate shoot fly species attacking seedlings of sorghum in Africa and Asia. It does not occur in the new world. The adult, a small gray fly with 3 pairs of black spots on the dorsum of its pale pink abdomen, is diurnal, and places red-shaped white eggs on the under surface of the leaves of seedlings through about the 6th leaf stage. The maggot emerges from the egg in about 2 days and thereafter tunnels toward the base of the plant through or behind the leaf sheath to ultimately cut the growing point. A dead heart results above the cut, and the larva completes its development within the plant, pupating either in the plant or in the soil. Small seedlings may be killed, and larger seedlings continue to produce tillers that are successively attacked. Losses result from stand redution and reduced size and delayed maturation of tillers.

The shoot fly, like many other pests and diseases, tends to be more serious on experiment stations where continuous cropping is carried on for research purposes. Relay cropping or ratooning of sorghum would tend to increase shoot fly attack, as natural enemies appear to play a minor role. Approaches to control have included a combination of cultural practices, systemic insecticides, and resistant varieties. Earliest planted crops in farmers' fields tend to escape injury. Later planted crops are attacked by fly populations produced on early crops. Adjustment and control of planting date in Israel has reduced fly attack below economic levels. Seed treatment with carbofuran is effective and is being used to some extent commercially in India. Through systematic screening of the World Sorghum collection varieties resistant to the fly have been identified. These varieties have so far demonstrated resistance to fly populations in West and East Africa, India and Thailand. They are local, low yielding varieties that have been developed over decades in Africa and India where the fly has been a continuous problem. Since it is a seedling pest, with predictable infestation levels in the field, screening and selection

for resistance has not required elaborate laboratory facilities and techniques. Through a research network involving The International Crops Research Institute for the Semi Arid Tropics (ICRISAT) in Hyderabad, India and The International Center for Insect Physiology and Ecology (ICIPE) in Nairobi, Kenya, and national research programs, an attempt is being made to incorporate fly resistance in higher yielding varieties, and to develop other control practices.

Plant and Leaf Hoppers

These sucking insects are of principal importance on corn and sorghum because they are vectors of virus and mycoplasma caused diseases, rather than for the physical injury nymphs and adults cause by sucking plant sap. They are all small hoppers, less than 5 mm in length. Eggs are deposited by adults in the leaf tissue and nymphs emerge to feed on the leaves. There is continuous reproduction when suitable hosts are available. Injury is most severe from the late seedling to mid-whorl stage, since disease innoculation at that time has greatest effect on the plants. The corn plant hopper, a Delphacid, *Peregrinus maidis* is a widely distributed cosmopolitan species in the tropics feeding on a number of the Gramineae. It is a vector of mosaic on corn and freckled yellow virus on sorghum. The Cicadellids, *Dalbulus spp.* that occur primarily on corn, though also infest other grasses, are vectors of corn stunt caused by mycoplasma in the new world. Another Cicadellid, *Cicadulina mbila* is the vector of corn streak in Africa and the Middle East. The insect is also present in India.

In general the application of insecticides for hopper control has not brought about sufficient reduction in disease incidence. Therefore, approaches to control both these insects and the diseases they transmit have been through the use of host plant resistance to both the disease organisms and their vectors. The International Corn and Wheat Improvement Center (CIMMYT) in Mexico, and The Institute for Tropical Agriculture (IITA) in Nigeria have active research programs to develop higher levels of resistance in agronomically acceptable corn varieties. This research requires laboratory cultures of the pests and special facilites for a collaborative effort by entomologists, plant pathologists and plant breeders.

Armyworms

Five species of Noctuid armyworms have been listed, that cover a distribution from the tropics of the new world across Africa and Asia to Australia. Injury to both corn and sorghum by armyworms is caused by defoliation and weakening of plants, from seedling through flowering stage of growth. Eggs are deposited by adult moths at night. They are sometimes placed between two touching leaves and are covered by scales by the female moth. The small larvae that emerge spread over the plants, usually feeding first in the whorl, where there is some cannibalism, and then feeding on the leaf margins as the larvae grow larger. On occasions the plants are completely defoliated and larvae may move like an army across a field. Pupation occurs in the soil. Other species of Graminae as well as corn and sorghum are attacked.

Armyworm infestations are sporadic in nature and do not occur year after year at most locations. Where outbreaks do take place, loss of the crop can be severe. There are a number of Dipterous and Hymenopterous parasites that are very effective in checking armyworm populations. These probably account for the sporadic nature of outbreaks. Insecticides have been used to control the larvae where outbreaks occur. Differences in varietal resistance to armyworm have been observed in both corn and sorghum, but, as yet a sustained effort has not been made to concentrate and utilize this resistance in varietal improvement programs. Research with host plant resistance will require laboratory rearing facilites for these insects, since field infestations are so sporadic.

Stem Borers

The listing of more important stem borers, of corn and sorghum, includes five Pyralid and three Noctuid species. While the habits of these borers are similar there are certain unique differences that are of importance in developing control measures for them. All of the borers are nocturnal in habit, but the ovipositional and larval feeding sites vary. For example, the Pyralids all deposit their eggs on the under side of the leaves of plants, starting in the whorl stage. On hatching the small larvae usually move to the whorl or leaf bracts to begin feeding. As they grow larger they tunnel in the mid veins of the leaves, the stems, and ears. Pupation usually occurs in the plant stems. Of the Noctuid species, the moths of *Busseola fusca* deposit their eggs behind the leaf sheath and the newly hatched larvae emerge, crawl over the plant to the whorl where they began to feed, much as do the Pyralids. The two *Sesamia* species also place their eggs behind the leaf sheath but on hatching, tunnel directly into the stem. These larvae are less exposed to parasites, predators, and to

insecticides, if applied. Losses by all borers on both corn and sorghum are caused by stem tunneling and tunneling in the ears and ear peduncles. Borer infestations in the tropics tend to be sporadic like armyworms. Borers are also subject to attack by a rich fauna of parasites and predators.

Following the success of intensive research on resistance to the European corn borer in the United States, that has produced varieties resistant to both 1st and 2nd broods, considerable research has been initiated on host plant resistance to other borer species around the world. This includes work on *Diatraea saccharalis* and *Zeadiatraea lineolata* on corn in Mexico, *Chilo partellus* on sorghum and corn in India and *Ostrinia furnacalis* on corn in Thailand and the Philippines. Although both corn and sorghum varieties with resistance to a number of these borers have been identified, few agronomically suitable varieties with acceptable levels of resistance have been distributed to farmers in the tropics. This work has been handicapped by the need to develop adequate borer rearing and artificial infestation techniques for each species concerned. The diversity of habits of different borer species makes it likely that a variety with resistance to one will not necessarily have resistance to other species, as has been the case with resistance to European corn borer and southwestern corn borer. Nevertheless, the most promising approach to borer control in the tropics appears to be a combination of naturally occuring parasites and predators and some level of host plant resistance. Selective use of insecticides may be feasible, such as the whorl application of granular formulations for *Chilo partellus* control on sorghum and corn in India.

Earworms

The corn earworm, *Heliothis zea,* in the new world, and the American bollworm *Heliothis armigera,* in Africa and Asia, both injure corn and sorghum by feeding on the maturing seeds on the tips of the ears or on the panicle. Eggs of these Noctuid moths are deposited at night on the silks of corn, or, on sorghum panicles shortly after flowering. The larvae then feed on and destroy the maturing grain. Full grown larvae pupate in the soil. The amount of damage caused on both crops, in most regions of the tropics, does not appear to be severe. There are numerous predators that attack the earworm eggs and small larvae. In most instances chemical control on field corn and sorghum is not economical. Corn varieties with long tight husk cover are more resistant and suffer less damage, as do loose or open panicle type sorghums that expose larvae to predators.

Research on host plant resistance to earworms in the tropics is not well advanced, since other problems have had a higher priority.

The Sorghum Midge

Probably the most destructive insect pest on sorghum, wherever the crop is grown, is the Cecidomyiid Midge, *Contarinia sorghicola*, that is thought to have originated with the crop in Africa, and spread with it throughout the world on wild and cultivated sorghum. Extensive losses in grain production have been reported from Texas, Argentina, Nigeria, India and Australia. The small orange colored females, about 2 mm in length, oviposit in the flowering spikelets. The flies live just a day or two and may not be noticed in the field. The injury is caused by larval feeding on the ovary, which prevents normal seed development. The midge may complete a generation in a little over 2 weeks, permitting 9-12 generations in a single season, and a build up of high populations where flowering times are extended by a range of planting dates.

The control of midge by cultural practices requires good knowledge of the ecological relationships of a particular region. A few individuals of each generation enter a larval diapause that is broken by exposure to warm, humid conditions. Since build up early in the season takes place on wild sorghums, early planting of the main crop over a wide area to shorten the flowering period has been found to reduce midge injury. Intermixing of plantings of an early hybrid with late traditional sorghum varieties in India created a severe problem on later flowering local varieties. Insecticidal control has been difficult since timing of applications to peak flowering periods is essential. The use of resistant varieties has been slow in developing since breeding programs have been handicapped by a lack of techniques for rearing the insect for uniformly infesting sorghum varieties. Selection in segregating populations, that flower at different times, also presents difficulties. Some progress is being made, however, in selecting for host resistance. As yet planting date management is the best means of control. Agronomists working with sorghums need to beware of intensive cropping systems that will require a series of successive sorghum crops to maximize water or land use. These are ideal conditions for midge.

In closing, I would repeat, it is fortunate that both corn and sorghum in the tropics have a high tolerance level to insect damage and few major pests. Nevertheless, there is need for better evaluation of the economic thresholds for insect pests both common and specific to corn and sorghum. Indicated control measures for these pests include cultural practices, the development of resistant varieties, and, the preservation by proper management of the rich complex in the tropics of natural enemies, expecially parasites, and predators on some of the pests. There is a continuing need to more effectively monitor the actual pest situation in farmers' fields as mixed and multiple cropping systems are promoted to increased production levels. Strategic use of carefully selected insecticides, as seed treatments or soil and foliage applications, may become economically feasible. Cooperative international research networks should hasten the development of appropriate controls for both corn and sorghum pests. The International Agricultural Research Centers, CIMMYT and ICRISAT should take the lead in coordinating this activity with national based researchers.

References

1. FAO. 1976. Monthly Bulletin of Agricultural Economics and Statistics. 25:2.
2. Ortega, A., and C. DeLeon. 1974. Insects of Maize. Proc. Symp. World-Wide Maize Improv. 70's, Role CIMMYT, El Batan, Mexico. 36 p.
3. Chiang, H. C. 1978. Pest Management in Corn. Ann. Rev. Entomol. 23:101-23.
4. Young, W. R. and G. L. Teetes. 1977. Sorghum Entomology. Ann. Rev. Entomol. 22:193-218.
5. Teetes, G. L., et al. (In Press). Introduction to the Integrated Control of Sorghum Pests. FAO. Rome.

28.7   Integrated Plant Protection for Corn and Sorghum in the United States—Weeds

V. M. Jennings

Associate Professor of Plant Pathology, Seed and Weed Science and
Extension Coordinator, Integrated Pest Management
Iowa State University, Ames, Iowa

Weeds as a Pest

Agriculturalists recognize that left uncontrolled, weeds will be a major limiting factor to efficient crop production in corn and sorghum growing regions of the United States. Most weed problems in corn and sorghum are annual species. The most troublesome are about 20 grasses and broadleaf weeds. Johnsongrass, a perennial, is the most troublesome weed in sorghum growing regions of the United States. It is also a serious corn pest in southern corn growing regions. In corn, the foxtails (*Setaria* sp.) are the most serious grassy weed problems. [1].

Losses from Weeds

Determining losses due to weeds has been a problem for weed scientists. Yield losses from weeds are largely a result of competition for sunlight, water and soil nutrients. Varying mixes of weed populations, the degree of competition between the weeds themselves and the size and shape of the crop canopy creates a complex interacting crop-weed modeling problem. Losses due to weeds are greater in shorter less competitive genotypes, yet in the absence of weeds, these genotypes may produce high yields equal to or greater than taller-growing, more weed competitive cultivars. Lower-growing genotypes may also be desired by the grower to reduce lodging problems from stalk rot and also to improve harvesting efficiency.

Research studies in Oklahoma show that in sorghum, one pigweed (*Amaranthus* sp.) per eight feet of row may reduce yields by 800 pounds per acre. [2] In corn, Illinois experiments show an average of 81% yield reduction in corn due to weeds in unweeded plots. Minnesota studies show 51% less yield than weeded plots with reductions ranging from 16 to 93% [3].

Weeds and Integrated Pest Management

Integrated pest management (IPM) is a concept offered as an alternative to improving efficiency of crop production by limiting damage caused by pests like weeds. Its application to weeds doesn't fit as well as for insects. The concept of economic thresholds, for instance, does not apply well to weeds. Populations of weeds are often located sporadically and in mixed communities. Weed control strategies, therefore, must be based on a broad based mixed species population rather than on a specific pest as with insects.

Specifically, the concept of IPM can be defined for corn and sorghum applications as—the environmentally sound use of combined pest control strategies to most effectively and efficiently limit damage to crops from pests.

For successful application of IPM to weeds, the following pest control strategies need to be considered.

Determining the Most Effective Pest Control Strategies for Weeds

The application of good cultural and crop management practices has always been the backbone of a weed control program. The use of herbicides alone has not been a solution. The combined strategies of good cultural and crop management supplemented by judicious use of herbicides has been highly successful in minimizing crop losses due to weeds.

The question now asked is—are there other IPM control strategies which will achieve acceptable weed control more efficiently?

The available strategies are:

1.   Cultural and Crop Management Strategies
a.   *Tillage* directly or indirectly has been the major weed control strategy used by gowers. Conversely, the effect of tillage enhances germination of weeds and each tillage practice results in a new flush of weeds to be controlled. An increased emphasis on conservation tillage has resulted in an increased dependence on herbicides for weed control and less reliance on the mechanical effects of tillage. By not tilling the soil as much, fuwere but different species of weeds become a problem and result in a different type of

weed control challenge. The trend to conservation tillage is increasing due to improved erosion control benefits, the need for improving water quality and efficiency of reduced tillage practices, all of importance to the grower.

Crop producers still rely heavily on a combination of herbicides and some type of cultivation to control weeds.[4] Most producers now cultivate corn only once when using a herbicide. If they do not use a herbicide, they tend to cultivate more than once.

b. *Crop rotation* historically has been widely used for pest control. For weed control, row crop rotations provide little benefit as a weed control strategy. Most all weeds are common problems in all row crops. Rotation to non-row crop crops for several years well provide limited control of some annual weeds and moderate to good control of some perennial weeds.

c. *Earlier planting* of full season corn genotypes has resulted in higher yields. Earlier planted corn has less vegetative growth and is not as competitive against weeds as later planted corn. Along with earlier planting, more reliance has been placed on herbicides to supplement cultural practices for effective weed control. Delaying planting dates would reduce but not eliminate weed problems. Later planting of less than full season varieties would result in a lower yield potential.

d. *Water management* may be utilized in weed control. By draining wet areas, certain weeds like yellow nutsedge *(Cyperus esculentus),* smartweed *(Polygonum* sp.) and others are less of a problem. The improved aeration of the soil enhances crop growth and makes it more competitive.

2. *Chemical Control Strategies*

Herbicides are widely used on corn and sorghum to potentially offset disastrous crop losses due to weeds. Weed controlling chemicals also provide other economic and efficiency benefits to the grower. In Iowa, of the 13.5 million acres planted to corn in 1977, 12.7 million acres or 93.8% were treated with herbides for weed control.(4) Similarly, in other states, most corn and sorghum growers rely heavily on herbicides to supplement cultural and crop management practices to cntrol weeds.

a. *Soil applied herbicides* are most widely used by most growers. In Iowa, 41.6% and 45.6% of treated corn acres in 1977 received preplant incorporated and preemergence herbicides respectively.(4) Most applications are broadcast treatments applied to the soil before or following planting but before crop emergence. Most herbicides are applied as combinations to provide broad spectrum control.

b. *Foliar applied herbicides* or postemergence treatments are often used as supplementary control, often as a second treatment following a soil-applied treatment.

Since most weed seedlings in corn and sorghum emerge with the crop, post emergence treatments are presently used to reduce weed escapes from soil-applied treatments. Used as an IPM strategy, postemergence treatments can be beneficial in a field monitoring program. If susceptible weed populations are excessive, postemergence herbicides can be utilized as a control measure. Limited reliability of post emergence herbicides due to environmental variables and spectrum and size of weeds controlled by these treatments is currently a limiting factor in increasing their use as an IPM method.

3. *Biological Control Strategies*

As a week control strategy, the use of biological variables like predators, parasites and microbial controls appear to have limited application to annual weed population management in corn and sorghum.

4. *Field Monitoring as a Strategy*

The application of field monitoring or scouting has not been widely used as a weed management strategy. The use of monitoring techniques has numerous possibilities in IPM.

Early season monitoring can be useful in evaluating soil applied herbicides and cultural and crop management control of weed populations. If weed escapes are observed in sufficient numbers, appropriate additional controls can then be used. An adaptation of this would be to use only soilapplied grass herbicides as an initial control strategy followed by monitoring for broadleaf weeds and applying appropriate postemergence herbicides or cultural practices as needed.

Late season monitoring can be useful in preparing weed maps and providing a rating of the mix of weeds present. This will be useful in preparing a weed control program for the following year.

5. *Predictive Modeling as a Strategy*

The use of predictive modeling as a weed populations management tool is in its infancy. There are many possibilities of using this technique by weed scientists to predict weed problems.

It is obvious there is a differential response to temperature and degree day requirement of various weed species. Yet, little research has been done to develop climatological relationships with weeds for use in IPM strategies. In the future "real time" climatological data will be available and could be useful in predicting severity and the mix of weeds in monitored crop fields. A greater emphasis in this area is needed.

What Is the Best Mix of IPM Strategies for Weed Control?

It is impossible to identify the best possible mix for all situations involving corn and sorghum production. It is obvious that the future will bring about change. Increasing concern about energy and a need to improve the environment will be important issues. These concerns will result in an increase in conservation tillage, an improvement in water quality and a strong producer desire to lower input production costs.

By utilizing the above IPM strategies, increased efficiency of corn and sorghum will occur. The continued use of cultural practices and crop management along with judicious use of herbicides will be used in weed control programs. New application of field monitoring and predictive modeling will improve the efficiency of weed control practices.

The end result will be utilization of efficient IMP programs that will in the short run save and make the grower more money than his present system. In the long run, IPM strategies will result in a more environmentally sound agriculture.

References

1. National Academy of Science 1975. Pest Control: An Assessment of Present and Alternative Technologies. Vol. II Corn/Soybeans Pest Control. Washington, D.C.
2. Greer, H. A. L., C. E. Denman. Chemical Weed Control in Grain Sorghum, Cooperative Extension Service, Oklahoma State University, Stillwater, Oklahoma.
3. Behrens, R., D. C. Lee. 1966 Weed Control. Advances in Corn Production, Iowa State University Press. pp. 331-352.
4. Jennings, V. M., H. J. Stockdale. Herbicides and Soil Insecticides Used in Iowa Corn and Soybean Production, 1977. Pm-845, Cooperative Extension Service, Iowa State University, Ames, Iowa.

29.1 Wheat in the Developing Countries from the Atlantic to the Pacific

E. E. Saari

CIMMYT-Wheat Program, P.O. Box 2344, Cairo, Egypt

Wheat represents one of the oldest and most important food crops for many of the developing countries in Asia and Africa. In countries where wheat is not cultivated, its importance in the daily diet has been increasing through importation. In a few instances, wheat has risen from a relatively minor crop to one of major importance. Bangladesh is an example of such a change with wheat hectarage increasing from 30,000 to more than 500,000 hectares in a period of less than a decade [1].

In any production program a number of factors need careful evaluation and the plant protection elements are a major consideration. In the developing countries the supply of inputs, particularly for plant protection measures, is usually limited and the infrastructure for the application of pesticides is frequently not well developed. In addition, many farmers lack experience in the use of pesticides.

The application of pesticides to wheat is not a scheduled input in many countries. The importation of pesticides uses scarce foreign reserves and they are not considered unless there is a high probability of need and economic benefit. If an unexpected problem arises under such circumstances, the prospects for chemical control are limited. The lack of knowledge and evaluation of the plant protection elements has on more than one occasion resulted in reduced yields and failure for production campaigns. Such a failure can be expensive because of the loss of inputs and the need to supplement food supplies with additional imports.

Throughout the Indo-Pak subcontinent the majority of the wheat is fall sown with spring habit cultivars. This area is typified by dry, sunny, winter weather and a summer, monsoon climate. About half of the area, or approximately 14 million hectares, is completely or partially irrigated. A high yield situation is possible under such conditions and the sharp increases in wheat production in the subcontinent in the late 1960s and 1970s occurred mostly on this type of area. In the non-irrigated or rainfed areas wheat is grown predominantly on residual moisture. If winter rains occur, they provide the difference between minimal and satisfactory yields. There is also a small area of summer sown wheat in the mountainous area [12, 13, 24, 25].

The diseases represent a major problem to successful wheat production. The most important disease in South Asia are the three rust diseases: *Puccinia recondita, P. striiformis, P. graminis.* The leaf blights caused by *Helminthosporium* spp. and *Alternaria-triticina* are important in many areas. Smuts (*Ustilago* sp., *Urocystis* sp.) and bunts (*Tilletia* spp.) are more localized by represent serious problems where they occur (17, 36).

The insect pests of wheat are considered to be of minor significance and this partially reflects the natural state of insects on wheat when compared with other cereal crops. The minor importance of insects also reflects the multiple cropping patterns that occur in such climatic regions. The heavy summer precipitation and high temperatures decompose the crop residues rapidly. These factors, plus the growing of a different crop in rotation, reduce insect survival and multiplication. Similarly, many disease-causing organisms do not survive and the importance of root rots and soil pathogens is reduced. The most common insects encountered in the irrigated areas are the polyphagous insects (armyworms and cutworms), aphids, and, to some extent, stem borers [4, 5, 20].

The common monsoon crop in these areas is rice and it is traditionally grown in a puddled soil. This procedure of breaking down the soil structure creates a number of agronomic difficulties for wheat cultivation. The puddling technique does, however, help to control weeds and this aspect is carried over from rice to the wheat crop. Other factors that reduce the magnitude of weeds in South Asia are the rotations being used and the extensive amount of hand-weeding practiced because labor is relatively cheap. In the dry, winter months weeds are often considered valuable as a green forage for cattle and have some economic value. Where spraying with herbicides such as 2-4-D is practiced, the grassy weeds become increasingly important, i.e. wild oats (*Avena* spp.) and *Phalaris* spp. [3, 37].

In the areas where irrigation is not available, the rust diseases and occasionally root rots are encountered [17]. Insects are less of a factor probably because of the drying situation and the declining temperature profiles encountered during the early part of the growing season. The insects most often encountered in these situations are white ants (termites), the brown wheat mite (*Petrobia* sp.) and the polyphagous insects [4, 20].

In contrast, the Mediterranean-type weather prevails in many of the countries of the Middle East and North Africa. This area is characterized by a winter rainfall pattern with cool temperatures and warm, dry summers. The rainfall varies from less than 250 to more than 1000 mm per year in the agricultural areas. All of the rain comes between October-November and April-May. Slightly different periods may be involved depending upon elevation. In lower rainfall areas a fallow system is practiced with one crop being harvested every second year [7, 35].

There is a large area where the winter temperatures are below or near freezing for extended periods of time. Depending upon the amount of snow cover varying degrees of winter hardiness will be required and the crop may take 10 to 11 months to mature. In a few instances the crops may overlap. In areas with mild, winter temperatures, spring habit wheats are sown and if winter rainfall is lacking, irrigation is essential e.g. Egypt, Sudan. There is also a small area in the Middle East that has a summer wheat crop. This area is probably less than 5% of the total and it is sown at the higher elevations [35].

Wheat was first cultivated in the Middle East and North Africa and has been the staple food throughout the region for several thousand years. In this area the wild progenitors of wheat and many closely related grasses can be found. There exists a diversity of disease, insect and weed problems. The low average yields for wheat are partially a reflection of the agronomic and pest problems.

The conservation of moisture is important since wheat is often cultivated in areas with less than 600 mm of rain. To conserve moisture, weeds must be controlled during the summer or in the fallow year. In some instances comon-law or community grazing rights limit the practice of a clean fallow. Weeds represent one of the main reasons for low yields because of moisture losses and the direct competition on wheat [2, 10, 20, 26, 38].

Broadleaf weeds are a major problem and as many as 112 weed species have been identified [2]. The crucifers alone infest an estimated 6 million hectares [26]. Other broadleaf weeds which infest more than a million hectares include *Boreava* spp., *Centaurea* spp., *Convolvulus* spp., *Carthamus* spp. The most importnt grassy weeds are *Avena* spp., *Phalaris* spp., *Cynodon dactylon* and *Lolium* spp. The grassy weeds have been increasing in importance as use of chemicals have brought the broadleaf weeds under control [26].

Insect problems are common and diverse, but most of the research in entomology has been descriptive. The main insects appear to be sawfly (*Cephus* spp.), aphids and the stinkbugs (*Aleia* spp., and *Eurygastar* spp.). Although locusts are a major insect in the Middle East, they have not been a serious problem in recent years due to control efforts in areas where they breed [33]. Other important insect pests are the Hessian fly (*Mayetiola* spp.), the cereal leaf beetle (*Oulema* spp.) and white grubs (*Melolontha* spp.) [42].

Diseases still represent a major constraint when weeds are controlled. The rust diseases along with *Septoria* blight and *Helminthosporium* spp. are found throughout the region. At higher elevations, especially where winter wheats are grown, the bunt diseases (*Tilletia* spp.) are serious. Another important disease encountered is the root rots collectively. A number of local disease problems have also been encountered [20, 22, 28, 36].

The acreage of wheat in East Africa with the exception of Ethiopia is relatively small. Much of the wheat area is located at higher elevations and pest problems are a major constraint. Weeds, insects and diseases are considered of equal importance and the great variation in the ecological areas along with the relatively scattered acreage makes it difficult to specify the dominant pest [8]. Certainly, the rusts are the major diseases and stem rust in East Africa is world-known. The black wheat beetle (*Heteronychus* sp.) and the aphids are reported as serious insect problems [8, 42]. Weed problems are diverse but *Boreava* spp., *Nicandra* spp., *Avena* spp. are recognized as some of the more serious [8, 9].

Our knowledge of wheat production and problems in the Far East is much more limited. Wheat is grown in several countries but it is a major crop in China. Winter wheats are grown in the area north of the Yangtze River and south of the Great Wall of China. Other important wheat areas are in north and northwest China where spring wheats are sown. Fall sown spring wheats are also found in the south and southwest following rice, and spring sown wheat in scattered areas in the west [16].

The pest problems are similar to other wheat areas. Weeds tend to be of limited importance because of intensive crop rotations. In much of the area maize is interplanted with wheat to maximize the use of the frost-free days and the summer rains. Wheat is again sown in the fall after maize is harvested. This procedure and extensive hand cultivation with the occasional use of herbicides minimizes the weed problems (16, Saari, *personal observation*).

The most common insects appear to be aphids, leafhoppers and armyworms. Regularized spray programs to control these insects have been developed [16].

The diseases of wheat are varied but the most common are the three rusts. Scab caused by *Gibberella zeae* is a common and serious problem, particularly in Eastern China. The wheat-maize rotation also contributes to this problem. Other diseases encountered are the root rots and take-all caused by *Gaeumannomyces graminis* can be found where wheat is grown each year (Saari, *personal observation*). There are a number of leaf spots but those caused by *Helminthosporium* spp. seem to be most common (Saari, *unpublished data*). The viruses diseases transmitted by leafhoppers and aphids are readily observed. The two most common appear to be barley yellow dwarf virus and rossette dwarf virus. (16, Saari, *personal observation*). The extensive spray programs conducted for insect control help to limit the occurrence of virus diseases [6, 16].

In the developing world the documentation of losses caused by the various wheat pests is not complete or up-to-date. The most extensive and comprehensive review was done by Cramer [8]. This review placed the losses in wheat production for Asia, excluding Russia and China, at 27% of potential production. The losses for Africa were placed at 39%. These percentages were considered conservative at that time. In the intervening years there has been a substantial increase in wheat production in Asia which will influence these values, while the production in Africa has remained somewhat static during this period.

The increase in average yield in South Asia resulted primarily from the use of new cultivars and the additional use of fertilizer inputs. Another contributing factor was the high degree of disease resistance incorporated in the new cultivars. The production increases in West Asia were realized primarily because of the use of better agronomic techniques and, in particular, the control of weeds. In some areas the use of herbicides for controlling broadleafed weeds accounted for more than 60% of the yield increases even when simple technology was used [23].

Losses caused by weeds have received more attention than insects or diseases. In an evaluation of the weed situation Parker and Fryer [27] classified weed losses into three categories. The over-all world loss for all food crops was 11.5%. For countries with intermediate technologies and least developed technologies the losses were placed at 10 and 25%, respectively.

The over-riding importance of weeds in North Africa and the Middle East has resulted in a number of trials and estimates on losses. Cramer's [8] figure of 15% loss due to weeds for Africa appears low, for most studies in the 1970s indicate that losses start at 20% and may be as high as 60% [2, 10, 26, 38]. It has been estimated that only 5% of the wheat area is sprayed for broadleaf weeds and less than 1% of the area is treated for grassy weeds [38]. This would suggest that weed losses are generally understated. In Asia, Cramer's figure of 10% losses caused by weeds seems reasonable [31, 37]. This could change as the grassy weeds increase in importance.

Loss estimates for insects in Africa and Asia have not received adequate attention. Until a more comprehensive study and evaluation is carried out, Cramer's figures of 12 and 5% loss due to insects in Africa and Asia, respectively, remain as valid today as in 1967.

Nematodes are occasionally encountered. *Anguina tritici* can be effectively controlled by good management and sound seed stocks. The cyst nematode *Heterodera avenae* is being recognized as a problem, especially on barley in some areas [16, 36, 40]. Losses caused by other nematodes have not been determined.

The losses caused by pests such as birds and animals vary greatly. Bird and rodent attacks can be major problems. Pre-harvest loss values from these pests are not well established but in some areas the losses can be extensive [8, 39, 42].

Disease awareness in wheat is on the increase but critical evaluation of the losses is still limited. The difficulty of developing average loss figures is complicated by the changes occuring in cultivar composition and their influence on the losses will vary greatly. The evolution of a "new" race or strain that renders a resistant cultivar ineffective must also be taken into account.

We know, for example, that the use of the semidwarf wheats such as Kalyansona and Mexipak in India and Pakistan contributed to more than the doubling of the average yields for the respective countries [24, 30, 41]. Part of this gain was realized because of disease resistance. Loose smut was reduced from about 5% to trace during this period and the area involved probably reached 10 million hectares [17, 24]. The reduction in loose smut can be translated to gains in wheat production. These cultivars were also resistant to bunt and powdery mildew, and the contribution of this resistance has not been calculated.

New races of rust, of course, eventually evolved. Serious losses due to leaf rust were averted by the changing of cultivars in India [41] and Nepal. A change in the cultivated varieties did not occur in Pakistan because an effective seed multiplication and distribution system was lacking [29]. Eventually, the components for an epidemic emerged in the crop year of 1977-1978 [19, 24]. The losses caused by leaf and yellow rust amounted to a short fall in production and resulted in an increase in the amount of wheat imported. The cost of importing 2.2 million tons of wheat was estimated at 400 million U.S. dollars [19], and the amount imported in excess of normal amounts would have cost more than 200 million dollars using the average prices and freight charges for wheat during the 1978-1979 trading year [14].

There is a need to find ways to increase our efforts in quantification of losses. For example, there was a yellow rust epidemic on a newly released variety in Spain in 1978 [24]. There is an extensive system of yield trial data available in Spain and by comparing resistant and susceptible wheat cultivars of similar yield potential some loss equivalents could be developed. A second method of comparison was made by using a 3-year cultivar average yield from rust-free years and comparing it to its yield in the rust year. The results from the two methods in Spain were in good agreement. The resistant cultivar comparison indicated that the loss was 41.4% while the 3-year average indicated the loss at 45.7%. Fortunately, the newly released cultivar was in the early stages of increase and did not occupy a substantial acreage [24].

In most developing countries some sort of yield trial system has been established. With little additional effort or cost, a number of estimates could be developed for the losses caused by the different pests [21]. This would be especially true for on-farm trials. This system, however, does imply that plant protection specialists would have to coordinate their activities with agronomists and breeders. The limited resources in the developing nations does not always promote such cooperative efforts. The encouragement of such cooperation could provide badly needed data to establish research priorities.

Literature Cited

1. Anderson, R. G., E. E. Saari, S. D. Biggs, P. N. Marko and D. Byerlee, 1979. *Bangladesh Wheat 1979—Rapid Change.* Monograph No. 3. Bangladesh Agric. Res. Inst., Dacca. 15 pp.

2. Annual Report 1977-1978. Weed control research section. The International Center for Agricultural Research in Dry Areas (ICARDA), Aleppo, Syria. 52 pp.

3. Bhardwaj, R. B. L. 1978. New agronomic practices. pg. 79-98. In *Wheat Research in India 1966-1976.* Indian Coun. of Agric. Res., New Delhi, 244 pp.

4. Bhatia, S. K. 1978. Insect Pests Pg. 152-167. In *Wheat Research in India 1966-1976.* Indian Coun. of Agric. Res., New Delhi, 244 pp.

5. Chaudhry, G. Q. 1978. Plant protection for field food crops on small holdings in Pakistan pg. 254-258. In *Technology for Increasing Food Production* J. C. Holmes (ed). FAO-TFINT 285, Rome. 695 pp.

6. Chiang, H. C. 1977. Pest management in the People's Republic of China—monitoring and forecasting insect population in rice, wheat, cotton and maize. FAO Plant Prot. Bull. 25:1-7.

7. Clawson, M., H. H. Landsberg and L. T. Alexander, 1971. *The Agricultural Potential of the Middle East.* American Elsevier Co. New York 312 pp.

8. Cramer, H. H. 1967. Plant protection and world crop production. Planzenschutz Nachrichten Bayer 20:1-524.

9. Handbook on Crop Production in Ethiopia (1st Edition) 1979. Institute of Agric. Res., Addis Ababa. 41 pp.

10. Hussain, S. M. and M. H. Kasim. Weeds and their control in Iraq. PANS 22:399-404.

11. Insect Control in the Republic of China. 1977. CSCPRC Report No. 2 Nat. Acad. of Sci., Wash. D.C. 218 pp.

12. International Maize and Wheat Improvement Center, 1977. CIMMYT. Report on Wheat Improvement 1976. El Batan, Mexico. 234 pp.

13. International Maize and Wheat Improvement Center, 1978. CIMMYT. Report on Wheat Improvement 1977. El Batan, Mexico. 245 pp.

14. International Wheat Council Reports. 1978 and 1979. Haymarket House, London, England.

15. Javaid, I. and M Ashraf. 1977. Wheat diseases in Zambia, 1973-1976. Plant Dis. Rep. 61:953-954.

16. Johnson. V. A. and H. L. Beemer, Jr. 1977. Wheat in the People's Republic of China. CSCPRC Report No. 6 Nat. Acad. Sci., Wash. D.C. 190 pp.

17. Joshi, L. M., K. D. Srivastava, D. V. Singh, L. B. Goel and S. Nagarajan. 1978. Annotated Compendium of Wheat Diseases in India. Indian Coun. Agri. Res., New Delhi, 332 pp.

18. Kelman, A. and R. J. Cook. 1977. Plant pathology in the People's Republic of China. *Annu. Rev. Phytopathol.* 17: 409-420.

19. Kidwai, A. 1977. Pakistan reorganises agriculture research after harvest disaster. *Nature* 277:169.

20. Koehler, C. S., R. D. Wilcoxson, W. F. Mai, and R. L. Zindahl, 1972. Plant Protection in Turkey, Iran, Afghanistan, and Pakistan. U.S. Agency for International Development-Univ. Calif. Berkeley. Contract No. AID/csd-3296./ 82 pp.

21. Kranj, J. 1976. Crop loss appraisal. Poljoprivredna Znastevna Smotra 39:11-20.

22. Maladies et Ravagewrs des Plantes Cultivess au Maroc Tome 1. 1976. Ministere de L'Agriculture et de la Reforme Agraire. Direction de La Rechereche Agronomonuque, Rabat. 207 pp.

23. Mann, C. K. 1978. Factors affecting farmers' adoption of new production technology: clusters of practices. pg. 280-289. *In* Barley Vol. II: Proc. Fourth Regional Winter Cereals Workshop, Amman, Jordan, 24-28 April 1977. ICARDA-CIMMYT, Aleppo, Syria. 420 pp.

24. Muhammed, A. 1978. Wheat in Pakistan. *In* Wheat Research and Production in Pakistan. pg. 1-12. M. Tahir (ed). Pakistan Agric. Res. Coun., Islamabad. 227 pp.

25. Murthy, R. S. and S. Pandy. 1978. Wheat Zones of India. pg. 11-19. *In* Wheat Research in India 1966-1976. Indian Coun. Agric. Res., New Delhi. 24400.

26. Nelson, W. 1976. The weed situation in the regions of North Africa and the Middle East. pg. 178-183. *In* Proc. Third Regional Wheat Workshop, Tunis, Tunisia. CIMMYT, Mexico. 388 pp.

27. Parker, C. and J. D. Fryer. 1975. Weed control problems

causing major reductions in the world food supplies. FAO Plant Prot. Bull. 23:83-95.

28. Prescott, J. M. and E. E. Saari, 1975. Major disease problems of durum wheat and their distribution within the region. pg. 104-114. *In* Proc. Third Regional Wheat Workshop. Tunis, Tunisia. CIMMYT, Mexico. 388 pp.

29. Qureshi, J. 1978. Seed industry project. pg. 125-129. *In* Wheat Research and Production in Pakistan. M. Tahir (ed.). Pakistan Agric. Res. Coun., Islamabad. 227 pp.

30. Rao, M. V. 1978. Wheat in India: an overview. pg. 1-10. *In* Wheat Research in India 1966-1976. Indian Coun. of Agric. Res., New Delhi. 244 pp.

31. Report of the Coordinated Experiments of Wheat Agronomy 1977-1978. All India Coordinated Wheat Imp. Project. Indian Coun. Agric. Res., New Delhi. 167 pp.

32. Roelfs, A. P. 1977. Foliar fungal diseases of wheat in the People's Republic of China. Plant Dis. Rep. 61:836-841.

33. Roy, J. L. H. and J. S. Gill. 1977. FAO Desert locust programme. FAO Plant Prot. Bull. 25:167-170.

34. Saari, E. E. 1978. Survey of Wheat Diseases in Southern Spain. Mimeographed Report to Sevico de Defensa Contra Plagas e Insepeccion Fitopatologica, Ministeria de Agriculture, Madrid. 11 pp.

35. Saari, E. E. and J. P. Srivastava. 1977. Improved and stabilization of production of winter cereals: Potentials in a single-crop system. Mimeographed paper. The Middle East and Africa Agric. Seminar, Tunis, Tunisia Feb. 1-13, 1977. The Ford Foundation, New York, 34 pp.

36. Saari, E. E. and R. D. Wilcoxson. 1974. Plant disease situation of high-yielding dwarf wheats in Asia and Africa. *Annu. Rev. Phytophatol.* 12:49-68.

37. Saeed, S. A. 1978. Weeds and weed control in wheat. pg. 103-107. In *Wheat Research and Protection in Pakistan.* M. Tahir (ed.). Pakistan Agri. Res. Coun., Islamabad, 227 pp.

38. Saghir, A. R. 1978. Weed control in wheat and barley in the Middle East. pg. 300-308. *In* Barley Vol. II: Proc. Fourth Regional Winter Cereals Workshop, Amman, Jordan, 24-28 April 1977. ICARDA-CIMMYT, Aleppo, Syria, 420 pp.

39. Sanchez, F. F. 1975. Rodents affecting food supplies in developing countries: problems and needs. FAO Plant Prot. Bull. 23:96-102.

40. Sheshadri, A. R. and C. L. Sethi. 1978. Nematode problems, pg. 169-187. *In* Wheat Research in India 1966-1977. Indian Coun. of Agric. Res., New Delhi. 244 pp.

41. Swaminathan, M. S. 1978. Wheat revolution the next phase. pg. 7-11. Indian Farming. Wheat Special Number. February, 1978. 60 pp.

42. Walker, P. T. 1975. Pest control problems (pre-harvest) causing major losses in world food supplies. FAO Plant Prot. Bull. 23:70-77.

29.2   Breeding for Stable Resistance: A Realistic Objective?

E. L. Sharp

Department of Plant Pathology, Montana State University
Bozeman, MT 59717

It is usually not difficult per se to obtain resistance to specific plant diseases, but it is often difficult to obtain resistance which is of long duration. In the past, emphasis was largely on obtaining clean plants and the resistance genes involved were major in effect. The resistance obtained was usually short term with some notable exceptions. Methods used in utilizing major effect genes have more recently involved gene deployment in time and space and virulence frequencies in the pathogen population. At one time several types of plant disease protection were believed to have potential for durability. These included such types as mature plant resistance, intermediate resistance and resistance obtained via transfer of genes from alien species. Unfortunately, all of these types of resistance may also be overcome readily by the pathogens. They are mainly important from the standpoint of possibly furnishing additional new sources of resistance.

It has been convenient to consider resistance as being major or minor in effect. (Henceforth, these will be referred to simply as major genes or minor genes.) For example, Zadoks (Tydschr. Pl. Zeikt. 67:69) on the basis of field experiments hypothesized two genetic mechanisms operating in resistance of wheat to stripe rust. One consisted of major genes conditioning race specific resistance and the other of minor genes conditioning race non-specific resistance. The range within the latter type showed considerable variation among cultivars. It was determined by Sharp and Lewellen et al. (1st Mont. Symp. Integr. Biol. 1965, 35; Can. J. Bot. 45:2155) that minor genes for resistance to strip rust occur in a number of wheat cultivars, are hypostatic to major genes, are temperature sensitive and can be accumulated to give effective plant protection. Some criticisms which have been directed at their utilization have been that they are difficult to identify, have low heritability, may occur on homologous chromosomes only and will be no more effective than other sources of resistance. It is my intent in the subsequent portion of this manuscript to discuss each of these aspects and to indicate some host-parasite systems other than the stripe rust disease where they may be present and equally as effective.

Our work with minor genes in wheat conditioning resistance to stripe rust began some 15 years ago when the plant introduction P.I. 178383 was determined to have several such genes in addition to a major gene. P.I. 178383 was crossed to a susceptible wheat cultivar, Itana, and plants representing specific infection types, in the absence of the major gene, were selected over several generations. Plants containing stabilized increments of one, two, and three "detectable" minor genes were determined after several generations of inoculation and selection. The fact that this required as many as seven selfing generations to obtain stabilized reaction classes indicated that a background component was being established before genes could be detected and further stabilized for reaction. These developed wheat lines were challenged with all available prevalent cultures of the stripe rust fungus, *Puccinia striiformis* West, and showed similar reaction within lines to all cultures. A commercial wheat cultivar, Crest, was developed containing the major gene plus the minor genes using P.I. 178383 as the non-recurrent parent. The cultivar has been protected from stripe rust by the minor gene components for more than 10 years. Other cultivars containing only the major gene from P.I. 178383 were heavily rusted 3 years after release (Plant Dis. Rep. 53:91). Crosses with a number of commercial wheats indicated that many contained minor genes that combined to give an additive type of resistance. Extreme environments were used for detecting minor genes in wheat cultivars (Phytopathology 66:794). For example, cultivar Lemhi, considered to be a "universal suscept" gave a susceptible reaction at 18/24°C (dark/light) while many other cultivars showed some incompatibility at this temperature regime, although they displayed complete susceptibility at 15/24°C. Such cultivars were then crossed with the original minor gene lines from P. I. 178383 and selections for the best resistance were made in the succeeding segregating generations (Phytopathology 66:794). These lines when stabilized were evaluated to isolates of stripe rust from both the Northwestern U.S. and Europe (Proceedings 4th Eur. & Medit. Cereal Rusts Conf., Interlaken, 1976, 159). Figure 1 illustrates the relative resistance of three minor gene lines selected from a cross of wheat cultivar Lancer and Selection A. Selection A was derived from Itana × P. I. 178383 and contained one detectable minor gene while Lancer was completely susceptible at normal temperature regimes. The *P. striiformis* isolates used

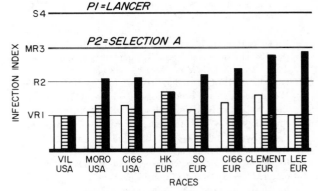

Figure 1. Reaction of 3 F8 wheat lines to 8 races of *Puccinia striiformis*

represented virulence for each differential of the world set. The selections showed a uniform reaction to all isolates but one of the three selections had a lower level of resistance. A total of 49 wheat selections involving 9 parental lines showed no appreciable differential in reaction class when inoculated with the different isolates. Figure 2 illustrates the reaction of 17 wheat lines to one isolate of *P. striiformis* and represents the range obtained among the 49 wheat selections. The greatest resistance was obtained at the higher temperature regime, 15/25°C (dark/light). Mature plant reaction was also more closely associated with 15/25 than with 2/18°C. Furthermore, Stubbs (Cereal Rusts Bull. 5:27) evaluated the original minor gene lines in international rust nurseries over a period of several years to several virulence types and noted no evidence for a specific type of resistance. As a follow-up and a better utilization of minor genes, Krupinsky and Sharp (Phytopathology 69:400) crossed several susceptible wheat cultivars believed to contain minor genes and selected for resistance in succeeding generations. Ten winter wheat cultivars and nine spring wheats were used for various crosses. Thirty-eight winter wheat crosses and 43 spring wheat

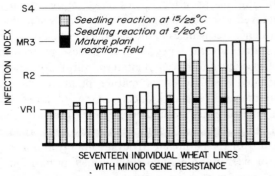

Figure 2. Influence of temperature and growth stage on interaction of wheat lines with the Clement race of *Puccinia striiformis*

crosses using parents of susceptible or intermediate reaction class showed transgressive segregation for greater resistance. Of these, 31 of the winter wheat crosses and 9 of the spring wheat crosses showed no resistance types until the F_4 generation, pointing out the importance of testing several segregating generations to determine minor gene resistance. Typical progress in selection for resistance after crossing normally susceptible wheats is shown in Figures 3 and 4. For both the winter wheat and spring wheat crosses illustrated, the F_1 and F_2 showed only susceptible reaction with evidence of resistance first appearing in the F_3 generation. Once resistance was noted within a segregating population, it was possible to stabilize it by further selfing and selection. The crosses showed significant additive gene action and heritabilities of over 99%. The additive minor gene resistance could thus be manipulated easily in a breeding program. Effective resistance can be obtained by crossing normally susceptible cultivars. Thus, commercially acceptable cultivars can be used as parents to avoid undesirable traits often associated with exotic plant types or various plant introductions which are usually utilized as resistance sources. Accumulation of minor additive genes can be facilitated by crossing the F_1 plants from two sets of parents containing minor additive genes and the increased increments of resistance obtained can be followed readily due to lack of epistasis among the minor genes.

In Krupinsky's research (Ph.D. Thesis, Montana State University, 1977) minor gene resistance to stripe rust was also obtained by crossing the susceptible durum wheat cultivars, Leeds and Wells, and selecting for several generations within the segregating progeny. An appreciable change towards resistance was first noted in the F_4. Amitae (*personal communication*) has also obtained evidence of

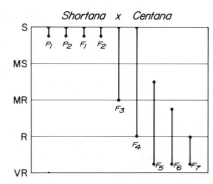

Figure 4. Transgressive segregation for stripe rust resistance in spring wheat

minor gene resistance in wild emmer populations in Israel. In glume blotch of wheat caused by *Septoria nodorum*, it has been difficult to identify any major genes conditioning resistance but transgressive segregation for greater resistance has been noted for several parental combinations (Scharen, Abstr. Can. Phytopathological Soc., Aug., 1979, Lethbridge, p. 49). Finally, evidence has been obtained indicating that minor gene resistance effective against net blotch caused by *Pyrenophora teres* can be selected from progeny developed following crossing of susceptible barley cultivars (Bjarko, M. S. Thesis, Montana State University, 1979). In work with net blotch of barley more than 50 isolates of *P. teres* have been investigated from various parts of the world. The barley parents that indicated transgressive segregation in crosses had major genes effective against other isolates. It is possible that these genes behaved as minor genes with the virulent isolates or as residual genes as advocated by Clifford (Rep. Welsh Plant Breed. Sta. 1974:107). The results show that minor genes for disease resistance occur at all ploidy levels.

The so-called minor genes furnish a readily available and usable source for plant protection. Some special procedures may be required for their detection but these are not limiting. They appear to be available for many host parasite systems and should furnish a broad-based and long lasting resistance.

How long will resistance conditioned by minor genes persist? It is unlikely that any resistance will last forever but it is believed that its use has been largely neglected due to misconceptions. The criticisms advocated against using minor genes as sources of resistance have been either false or greatly overstated.

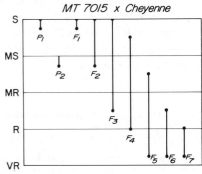

Figure 3. Transgressive segregation for stripe rust resistance in winter wheat

29.3   Control of Wheat Virus Diseases

C. L. Niblett

Department of Plant Pathology, Kansas State University
Manhattan, KS 66506

At least 30 viruses are known to infect wheat and cause significant economic losses. These viruses and the diseases they cause have been reviewed recently by Slykhuis (Annu. Rev. Phytopathol. 14:189-210) and Wiese (Compendium of Wheat Diseases, Am. Phytopathol. Soc., St. Paul, MN 55121). In this limited forum I will discuss only the four most important viruses infecting wheat. These are barley yellow dwarf virus, soilborne wheat mosaic virus, wheat spindle streak mosaic virus and wheat streak mosaic virus. I will discuss the properties of each virus and how they relate to epidemiology and control. Information for this discussion was gathered from the above reviews, C.M.I./A.A.B. Descriptions of Plant Viruses, research papers cited and personal experience with wheat viruses in Kansas. I thank those colleagues who shared slides and unpublished information.

Barley Yellow Dwarf Virus

Barley yellow dwarf virus is probably the most widely distributed and most economically significant of the viruses infecting wheat and other cereals. It was first described in California in 1951 (Oswald and Houston, Plant Dis. Reptr. 35:471-475). The host range of BYDV, though limited to monocotyledons, includes over 100 species and all the major cereals. Symptoms of BYDV infection vary markedly with the host species. They are most obvious in oats where infected leaves turn red or purple, and least obvious in corn and sorghum, which may be symptomless hosts (Stoner, Phytopathology 67:975-981; *personal communication*). Symptoms in wheat often are mild and easily overlooked, unless a very susceptible cultivar is involved. The most common symptoms are a general stunting of some plants in spots, giving the field an irregular appearance, and the presence of yellow-brown to reddish tips on the upper leaves. This discoloration then develops inward from the tip and margins. Symptoms generally are much more severe and winter-killing and yield reductions greater in fall-infected wheat. Yield reductions range from 10-60%, depending on the wheat genotype. BYDV was widespread in Kansas in 1976 and yield loss was estimated at 14.5 million bushels.

BYDV is not mechanically transmissible. At least 14 species of aphids (Rochow, C.M.I./A.A.B. Description No. 32) are vectors. The virus is transmitted in a circulative manner, and there is no evidence for virus replication within the vectors. At least five strains of BYDV have been differentiated on the basis of the vector specificity shown in Table 1 by Rochow.

Rochow (Annu. Rev. Phytopathol. 10:101-124) demonstrated that the viral capsid protein determines the vector specificity. From plants infected with both the MAV and RPV strains *Macrosiphum avenae* transmitted only MAV, whereas *Rhopalosiphum padi* transmitted both MAV and RPV. In other experiments, Rochow treated a mixture of purified MAV and RPV with MAV-specific antiserum, neutralizing the MAV in the mixture. He then membrane-fed the preparation to aphids. No virus was transmitted by *M. avenae*, but *R. padi* transmitted both MAV and RPV. Both of these approaches indicate that in doubly infected plants some of the MAV RNA is encapsidated by RPV protein and thus transmitted by the RPV-specific vector *R. padi*. In the infected assay plants the MAV nucleic acid then "breeds true" to produce MAV protein to encapsidate its homolgous RNA. Thus it can be transmitted by the MAV-specific vector *M. avenae*. This type of heterologous encapsidation is termed transcapsidation. Similar results were obtained from MAV+RMV infected plants. It is interesting that MAV RNA can be transcapsidated by both RPV and RMV proteins, but MAV protein appears only to be able to encapsidate its own RNA. The epidemiological implications of transcapsidation have not been determined.

Table 1. Identification and transmission of strains of barley yellow dwarf virus

Virus Strains	*R. padi*	*M. avenae*	*R. maidis*	*S. graminum*
PAV	+	+	−	+
RPV	+	−	−	+
MAV	−	+	−	−
RMV	−	−	+	−
SGV	−	−	−	+

Brakke and Rochow (Virology 46:117-126; 61:240-248) have purified and characterized BYDV and its RNA. The virus is phloem-restricted, occurs in very low concentrations, is isometric and about 25-30 nm in diameter. It sediments at 118 S and contains a single component of RNA, which sediments at 33 S (2×10^6 daltons). Capsid proteins of the MAV and RPV strains have molecular weights of 23,500 and 24,450, respectively (Scalla and Rochow, Virology 78:576-580). Recent results (Duffus, Phytopathology 67:1197-1201; Rochow and Duffus, Phytopathology 68:51-58) indicate that RPV is serologically related to soybean dwarf and beet western yellows viruses, but MAV and PAV are not.

Relationships among strains of BYDV have been determined by serology and cytology. Rochow and Charmichael (Virology 95:415-420), using highly sensitive enzyme-linked immunosorbent assays (ELISA), demonstrated that BYDV strains may be categorized into two groups. One group includes RPV and RMV, which have some common antigens. They are unrelated to the second group, which includes MAV, PAV, and SGV. The MAV and PAV have some common antigens, but SGV is related only to PAV, and it is a distant relationship. The same groupings were determined independently by comparing the cytopathological alterations induced in infected cells (Gill and Chong, Virology 95:59-69).

Control of BYDV is difficult because of the large number of species of aphid vectors and perennial grass reservoir hosts. One cultural method of control is to plant the crop as late as feasible. This helps avoid fall flights of vectors, and those plants which do become infected are less damaged prior to and during the winter. Chemical control of aphid vectors is sometimes effective with small plots and low aphid populations. But the large acreages and low cash value of cereal crops, the mobility of the vectors and the inability to predict vector incidence made chemical control of BYDV unfeasible.

Useful levels of resistance to BYDV have been identified in barley, oats, and wheat. Oat lines segregating transgressively for tolerance and intolerance have been selected, and lower levels of virus replication occur in the tolerant lines (Jedlinski et al., Phytopathology 67:1408-1411). In wheat, marked differences exist in response to BYDV (Palmer and Sill, Plant Dis. Reptr. 50:234-238), yet no extensive genetic studies have been published. In the Great Plains area the cultivar Hart is the most resistant to BYDV. The mechanism of this resistance is not yet known. Stoner (Plant Dis. Reptr. 60:593-596) identified several grass species which are immune to at least one strain of BYDV. These grasses should be tested with other strains of BYDV, and genes from those which are immune or highly resistant should be transferred to wheat. The development of ELISA so that it is now faster and more accurate than aphid transfers (Rochow, Plant Dis. Reptr. 63:426-430) should greatly facilitate this and other research with BYDV.

Soilborne Wheat Mosaic Virus

Soilborne wheat mosaic virus (SBWMV) was first described by McKinney in 1923. Much of the definitive research on SBWMV has been done by Brakke and his colleagues at the University of Nebraska. They showed that the fungus *Polymyxa graminis*, an obligate parasite, vectors SBWMV (Brakke, et al., Phytopathology 55:79-86). The virus is mechanically transmissible, but with low efficiency. The host range is limited, but includes both monocotyledons and dicotyledons. The American isolates infect wheat, rye, barley, bromegrass and *Chenopodium spp*. The Japanese isolates, in addition, infect corn, tobacco, beet, Swiss chard and *Tetragonia expansa*. Based on wheat symptoms, McKinney designated green, yellow and rosette strains of SBWMV. The Japanese isolates are clearly different from the American isolates, based on both host range and serological data (Tsuchizaki, et al., Phytopatholgoy 63:634-639).

SBWMV occurs in the Great Plains, midwest and eastern regions of the USA and in Argentina, Brazil, Egypt, Italy and Japan. Symptoms range from a mild green mosaic with dark "green islands" to a brilliant yellow, depending on the wheat genotype and environmental conditions. Typical fields contain scattered infected areas (usually in low spots), but whole fields may be infected. Fall-sown wheat is most seriously affected, but early-seeded spring wheat also can be infected. The disease is favored by long cool springs, and symptoms diminish as temperatures exceed 18°C. In severe cases yields can be reduced more than 50%, partly by inducing winter-kill in normally hardy cultivars (Campbell, et al., Plant Dis. Reptr. 59:472-476; Nykaza, et al., Plant Dis. Reptr. 63:594-598).

SBWMV has been purified and partially characterized. All isolates of the virus contain a long particle about 20 × 300 nm. There also is a second, shorter particle whose length varies from 90-160 nm, depending on the isolate. Tsuchizaki, et al., (Phytopathology 65:523-532) demonstrated that both types of particles are necessary for infection, and that the short particle controls the length of the short particles, the serotype and the type of inclusion bodies produced. Brakke (Phytopathology 67:1433-1438) demonstrated that the long particle sediments at 219 S and contains a single-stranded RNA of 2.1 × 10[6] daltons. The sedimentation coefficient of the shorter particle and the size of its RNA vary with the particular isolate, and at least two isolates can exist in some fields. SBWMV contains a single capsid protein species of about 20,000 daltons (L. Lane, *personal communcation*). Powell (Virology 71:453-462) demonstrated a serological relationship between SBWMV and tobacco mosaic virus (TMV), and TMV has been isolated from naturally infected wheat in Kansas and Virginia (Paulsen, *et al.,* Plant Dis. Reptr. 59:747-750; S. Tolin, *personal communication*). Other relationships between these two viruses have not been determined.

SBWMV has been especially serious in Nebraska, Kansas and Oklahoma since 1970. In Kansas about 2 million acres are infested, and yield losses in Kansas alone are estimated at 88 million bushels since 1970. Thus control of this disease receives high priority. Cultural practices such as late planting and avoiding infested fields are helpful in reducing losses. Crop rotation is not helpful, because the vector survives for long periods. Chemical control of the vector with methyl bromide or chloropicrin is effective for a few years, but it is too expensive to be used widely. The most efficient and effective method for controlling SBWMV is the use of resistant cultivars. Examples are Arthur, Cheney, Homestead and Newton. Resistance to SBWMV is known in many cultivars, but few have been adapted to the Great Plains. The resistance is simply inherited and is dominant over susceptibility (Dubey, *et al.,* Crop Sci. 10:93-94). Resistant plants generally are free from symptoms but a low percentage may show symptoms. When resistant cultivars are mechanically inoculated they show symptoms in the same proportion as susceptible cultivars. This suggests that the mechanism of resistance may be resistance to the vector rather than to the virus. This explanation is satisfactory in Nebraska and Kansas where *Polymyxa* cytosori are very infrequently found in resistant cultivars, but it is unsatisfactory in Florida where cytosori are readily found in both resistant and susceptible cultivars (Palmer and Brakke, Plant Dis. Reptr. 59:469-471; Campbell, *et al.,* Plant Dis. Reptr. 58:878-881). This inconsistency is being studied in both Kansas and Florida.

Wheat Spindle Streak Mosaic Virus

Wheat spindle streak mosaic virus (WSSMV) is relatively new among the wheat viruses. It was described in Canada, by Slykhuis (Can. Plant Dis. Survey 41:330-343) in 1961, and by Weise and Hooper (Phytopathology 61:331-332) in the USA in 1971.

Wheat is the only known host of WSSMV, and both winter and spring wheats may be infected. Symptoms of WSSMV are a yellow green mottling on lower leaves and chlorotic spindle-shaped streaks. In prolonged cool weather (~10°C) all leaves may show symptoms. Chlorotic areas coalesce, and necrosis ensues. Infected plants are stunted and produce fewer tillers. Yield losses ranged from 7-59% in Ontario (Slykhuis, Phytopathology 60:319-331) and 3-18% in Michigan (Wiese, *et al.,* Plant Dis. Reptr. 58:552-525), depending on the cultivar. The etiology of WSSMV is very similar to SBWMV, except that it is more uniformly distributed in the infected fields and favored by cooler temperatures (10°C vs. 16°C). It too is difficult to transmit mechanically. No strains of WSSMV have been reported.

WSSMV has been reported in southern Canada, north central and east central USA, northeast India and possibly France (Slykhuis, C.M.I./A.A.B. Description No. 167; Ahlawat, *et al.,* Plant Dis. Reptr. 60:782-783).

The particles of WSSMV are unusual and diagnostic. They are 15-18 nm wide and up to 3 μm long. Only rice hoja blanca and citrus tristeza viruses are similar. Infected cells also contain pinwheel inclusion bodies (Wiese, Compendium). No information is available on the nucleic acid or protein of WSSMV.

Cultural practices such as late planting and avoiding of infested fields help to control WSSMV. Chemical treatments aimed at the vector are also partially effective. These include known chemicals such as urea, uric acid and ammonium nitrate and known fungicides such as captan, mercuric chloride, formalin, etc. Soil amendments such as poultry manure also were effective. However, none of these treatments is feasible over large acreages. The most efficient and effective method of controlling WSSMV is the use of resistant cultivars. Resistance has been reported for Monon, Yamhill, Tecumseh, Halytchanka, Blueboy and Miro (Wiese, *et al.,* Plant Dis. Reptr. 58:522-525; Gates, Can. J. Plant Sci. 55:891-895).

It is interesting to note that wheat cultivars highly resistant to SBWMV such as Concho and Arthur are quite susceptible to WSSMV, whereas Monon is resistant to both viruses.

Wheat Streak Mosaic Virus

Wheat streak mosaic virus (WSMV) is the most destructive of the wheat viruses. It is capable of rendering vast acreages unworthy of harvest. WSMV was described by McKinney in 1937. The vector is the wheat curl mite *Aceria tulipae* (syn. *Eriophyes tulipae*) and in Yugoslavia by *E. tosichella*. The virus is readily mechanically transmissible and infects most cereals and many wild grasses. The host range is restricted to monocotyledons. McKinney designated yellow, green and mild strains of WSMV. I have been able to distinguish four strains by differential pathogenicity on corn lines.

WSMV has been identified in Canada, Jordan, Romania, Russia, Yugoslavia and the USA. It is especially serious in the Great Plains of the USA and the adjacent Canadian provinces. Symptoms of the disease include moderately to severely stunted plants and mottled and yellow-green streaked leaves. With high summer temperatures the infected plants become shiny and look wilted, because WSMV inhibits root development. Grain may not be produced by severely infected plants. Part or entire fields or vast acreages may be infected, depending on mite populations.

Both the virus and the mite oversummer in volunteer wheat, grasses and corn. As newly seeded wheat emerges in the fall, the mites are windblown into the seeded fields, and infection occurs. Volunteer wheat results from normal harvesting operations or severe weather near harvest. Severe outbreaks of wheat streak can usually be traced back to a hail storm in the area near harvest the previous year. WSMV is common in the Great Plains because much of the land is cropped on alternate years to conserve moisture. To conserve energy and prevent soil erosion the fallow land is seldom tilled. This permits volunteer wheat to develop adjacent to seeded wheat, and severe yield losses often result. For example, in 1974 a 600 square mile area of western Kansas was affected. Yield losses in Kansas that year were 30 million bushels. Total yield losses in Kansas since 1970 are estimated at 82 million bushels. The recently established large acreages of irrigated corn in Kansas and Nebraska pose a potential threat to wheat production. For although corn is not seriously damaged by WSMV, it can serve as a reservoir of WSMV and its mite vector, thus maintaining the disease cycle through the wheat-free summer months. Near immunity to WSMV exists in several corn lines (Niblett, unpublished), and this should be incorporated into those corn hybrids being grown in wheat producing areas.

Particles of WSMV are flexuous rods 13 × 700 nm. The virus was purified by Brakke and Ball (Phytopathology 58:963-971). It sediments at 165 S and contains a single-stranded RNA which sediments at 40 S (2.8 × 10⁶ daltons) (Brakke, C.M.I./A.A.B. Description No. 48; Virology 42:699-706). Two protein components frequently are obtained from purified preparations of WSMV. The molecular weights are approximately 29,000 and 43,000, but the smaller component may originate from the larger component (L. Lane, *personal communication*).

There are three cultural practices which are helpful in controlling WSMV. The most important is the destruction of volunteer wheat 2-3 weeks prior to seeding. This removes the reservoir of the virus and its vector. WSMV is seldom serious in the absence of volunteer wheat. The second practice is to plant as late as feasible to permit the fall decline of mite populations. Those plants which do become infected will suffer less damage before winter dormancy. The third practice is to plant WSMV-resistant cultivars. Although no cultivars are highly resistant under severe, early infection, the use of cultivars such as Eagle and Larned can reduce yield losses from 80% to 20%. The mechanisms of this resistance appear to be a reduced receptivity to WSMV and a reduced rate of virus synthesis (Young, Martin and Niblett, unpublished). To supplement this resistance we are attempting to combine several minor genes for resistance into a single cultivar.

Several wheat-*Agropyron* addition, substitution and translocation lines which are highly resistant to WSMV are available. The mechanism of resistance is localization in which the inoculated leaves become infected, react hypersensitively and die, leaving the rest of the plant free of virus (Pfannenstiel and Niblett, Phytopathology 68:1204-1209). The most useful of these lines are CI 15322, a translocation line developed by Sebesta at Oklahoma State University and CI 15092, a disomic substitution line developed by Wells at South Dakota State University. CI 15322 also carries resistance to the mite, but CI 15092 does not (Martin *et al.,* Phytopathology 66:346-349). We have developed a translocation line from CI 15092 (Liang, *et al.,* Crop Sci 19:421) and are using it and CI 15322 in the breeding program. Early progeny of CI 15322 are tall and have poor milling and baking characteristics, indicating that several undesirable *Agropyron* genes are closely linked to the WSMV resistance. Thus this approach has not been entirely successful yet, and is viewed as a long-term approach.

Attempts have been made to control WSMV through the chemical control of the mite vector. Chemicals such as Furadan and Thiodan have been used in test plots in Nebraska and have been effective in controlling the spread of WSMV. However, this has not been tested on large acreages and both chemicals are highly toxic. When Furadan is used cattle grazing on wheat is forbidden for 4 weeks, and this reduces the profit for the farmer.

In summary, virus diseases of wheat continue to be

447

destructive in many areas of the world. All of the viruses discussed can be most efficiently and effectively controlled by resistant cultivars, but additional effort is necessary to develop more highly resistant cultivars. This is especially true for BYDV and WSMV. However, we must avoid genetic vulnerability in developing and deploying such cultivars.

29.4   Chemical Control in Intensive Wheat Cultivation

A. Obst

Bayerische Landesanstalt für Bodenkultur und Pflanzenbau, Abt. Pflanzenschutz
D 8000 München 19, Fed. Rep. of Germany

The highest yields of wheat are obtained in NW Europe. Commercial wheat fields here often yield 6-8 tons and sometimes even more than 9 and 10 tons per ha. High yields are essential for most of the wheat growers, as they derive their income from small farms.

1.   Conditions of Intensive Wheat Cultivation

In NW Europe normally 40 to 70% of the arable land is in cereals. As a rule wheat stands as the first crop within a cereal sequence following a non-cereal crop. The cultivation methods are characterized by high yielding cultivars, early sowing of winter wheat, close spacing and high nitrogen fertilization. These conditions of course favor foot rot (*Pseudocercosporella herpotrichoides, Fusarium spp.*) and leaf and ear diseases (*Erysiphe graminis, Puccinia striiformis, P. recondita, Septoria nodorum, S. tritici, Fusarium spp.* etc.). Total yield losses by fungal diseases on a national average are estimated to be about 10% (Obst, 1975), but in most intensive wheat cultivation they may be 20% and even more. Compared with diseases insect pests are of minor economic importance. The yield losses caused by insects are estimated at 5-10% (Verijken, 1976). Those feeding on flag leaves and ears during the grain-filling period cause the most severe damage. The grain aphid (*Sitobion avenae*) is the most common species.

Increasing the intensity of wheat cultivation therefore increases the benefit from chemical treatments. Weed control is common, but this topic will not be dealt with. Disease control is becoming of increasing significance with generally one seed treatment and one or two fungicidal sprays being carried out. The newly developed fungicides give effective control of many cereal pathogens; their systemic action often permits a curative, postinfectional application. Widely applied are the MBC-generators, which are effective against many endemic cereal diseases. To increase their spectrum of activity still more they are often combined with protectant fungicides such as dithiocarbamates. Against epidemic diseases systemic fungicides with a more specific spectrum of activity are recommended. A prominent example for the powdery mildew and yellow rust control is the newly developed triadimefon. Also, regionally, the growth regulator chlormequat-chloride is used. Insect control, although increasing, is still of minor economic importance. Selective insecticides such as pirimicarb are available. Farmers, however, also apply broad-spectrum insecticides because of their lower costs.

2.   Danger and Difficulties of Excessive Chemical Control Practices

Excessive chemical treatments may raise some problems, which are discussed briefly in the following:

2.1.   Lack of Profitability
Treatments are economical only if they can prevent yield losses which are higher than the expenses for the products and their application, otherwise there is no profit. This investment appraisal approach is the most convincing argument for farmers to reduce excessive treatments.

2.2   Harmful Side-Effects to the Ecosystem
As in intensive cereal cultivation, fungicides are most frequently applied; their role within the ecosystem will be considered briefly. The use of fungicides may cause changes in the disease spectrum or in the disease dominance. For instance spraying

against eyespot with MBC-generators sometimes favors other foot rot pathogens. Fungicides may also influence an insect population by controlling fungi harmful to these pests. Zimmermann (1978) has demonstrated this under laboratory conditions for the cereal aphids and their fungal antagonists, the *Entomophthoraceae*. Finally, the fungicide treatments may have a direct effect on the beneficial fauna in a cereal field. A newer problem is the resistance of fungi to systemic fungicides, mainly the MBC-fungicides and ethirimol-type compounds. In wheat cultivation, however, no actual resistance problems are known at present (Dekker, 1977).

2.3. Pesticide Residues on Grain

Intensive use of fungicides and insecticides mainly after heading may cause residue problems as well. Relevant investigations, however, have revealed that at present there is no danger (Fraselle and Martens, 1973). It has to be emphasized that application of pesticides is only permitted within a range of time defined by law.

3. Methods of Minimizing Chemical Treatments and Their Side-Effects

It has been demonstrated that economic and ecological reasons are the major motives for reducing chemical treatments. This can be achieved by methods of integrated pest control. Although proper husbandry and host resistance are major elements, this paper must concentrate on the chemical aspect of this strategy only.

3.1. Use of Disease Prognosis and Economic Thresholds

3.1.1. Static, Empirical Predictive Systems. A first step in abandoning routine sprays is made when over the years the economic benefit of treatments at specific dates of wheat growth is noted considering the condition of cultivation. From this procedure static predictive systems are derived for pathogens occurring more or less regularly (endemic pathogens). A few representative examples from the NW European wheat growing centers are as follows:

To begin with, Scott (1978) underlines the complementary character of host resistance and chemical control practices. He has found that the winter wheat acreage treated in 1977 with MBC-fungicides in the United Kingdom and in W. Germany is inversely proportional to the partial resistance of wheat cultivars to eyespot. An eyespot prognosis published by Effland (1973), taking into consideration more crop and weather criteria, has been widely used in W. Germany. The features

preceding crop, sowing date of winter wheat, crop density, climatic conditions and regional aspects are assessed on a point scale for causing slight, moderate or severe infections.

In France disease complexes are generally observed and broad-spectrum fungicides are used for their control. As a rule diseases of the stem base and lower leaves are the most dangerous and diseases on the upper leaves and ears occur less frequently. Poussard et al. (1977) have analysed nearly 1000 experiments for the profitability of spraying at the 2-node and the heading stage. In an over-all assessment after two sprays only 35-40% of these trials gave economic benefit. However, a 100% level of profitability was obtained after early sowing of winter wheat, the preceding crops being leguminous crops, cereals, potatoes or rape and on deep soils. Due to different climatic conditions in Belgium, the Netherlands and Denmark diseases of the upper leaves and the ears prevail. With the exception of the extremely dry year of 1976, ear treatments with broad-spectrum fungicides have nearly always been profitable. Thus this control measure is standard in intensive wheat cultivation in these countries. Additional sprays may be carried out against eyespot powdery mildew or yellow rust.

Once again a static wheat protection system is practiced in the north of Germany (Schleswig-Holstein) (Teuteberg, 1978). High yields are the result of proper fertilization and protective treatments for every wheat cultivar. The most important characteristics of each cultivar are represented in a cultivar passport, where specific weak points are pointed out. The treatments which are likely to be economically useful are carried out at standard growth stages.

Concluding this description of static, empirical predictive systems, we must remember that cropping conditions are the major criteria for the decision as to wheather to spray or not. In general no observation of disease development is done, so there is no flexibility concerning the time of application. In practical farming this is a first useful step towards abandoning routine sprays.

3.1.2. Dynamic, Partly Inductive Predictive Systems. A survey of the methods of chemical control in the countries under consideration reveals that single pathogens often occur epidemically. To overcome the risk of attack different practices are used, specific observations of the disease occurrence and the disease progress being essential.

As first example eyespot control will be analysed. In the northern countries winter infections of *Pseudocercosporella* prevail. At the beginning of stem elongation the infection can be seen with the naked eye. An infection level of 20-30% is considered to be the critical threshold for a control spray with MBC-generators. In W. Germany eyespot infection in spring is often important. For a decision whether or not to spray visual assessments are too late, as the pathogen may already have invaded the stem. Here a regression model from Schroedter and Fehrmann (1971) has been usefully employed since 1973. At 90 official meteorological stations temperature and air humidity are recorded at 1-hour intervals, and with the aid of a central computer the probability of infection is calculated. As the model does not include any inoculum assessments, local experience on infection pressure has to be incorporated. Based on the investigations of Schroedter and Fehrmann (1971), Rapilly et al. (1979) have adapted this model for winter eyespot infections. They accept the submodel for eyespot infections and add another one for the penetration of the fungus through the five leaf sheaths of the young plant. Treatment is necessary when on 20-25% of the tillers the fungus has invaded the last (inner) and the last-but-one leaf sheath. Control of powdery mildew epidemics in most countries is based on visual assessment. The critical threshold and the time of fungicide application are achieved when 1% of the leaf area is attacked and when the weather conditions are likely to favor further development of the disease. In Belgium the decision criteria are defined more precisely: a mildew treatment is necessary when 25 pustules are visible on the upper 3 leaves. Yellow rust generally is controlled in a similar way. A simplified model for yellow rust has been in use since 1978 in the Netherlands, which is built up on weather data, initial inoculum observations from farmers and major cultivation criteria. In the first year it gave correct advice in more than 99% of commercial fields under observation.

Septoria glume blotch may be a major disease in some years. Recently several simple models have been presented which use weather data or laboratory assessments for the inoculum evaluation. Rapilly and Jolivet (1976) have published a detailed simulation model for the vertical spread of *Septoria nodorum* in wheat leaves, but, it has only been used for testing horizontal resistance of wheat varieties.

Losses due to the grain aphid are extremely variable from year to year and so farmers are recommended to spray against aphids only upon advice of the warning service or on their own aphid observations. The provisional threshold for grain aphid control is 15 aphids per ear, when a further increase in the aphid population is likely. A population model of *Sitobion avenae* will be presented by Dr. Rabbinge. Grain aphids are also being investigated by an IOBC working group on

"Integrated Control in Cereals", which began in 1973. In addition it may be noted that in 1977 on the same international level a working group on "Use of Models in Integrated Crop Protection" was established. The first aim is to make a catalog of the existing disease and pest models. These then have to be tested and improved.

From this, one can see that dynamic predictive systems are already in operation against some of the major wheat parasites. They are more effective than static methods in reducing chemical control, as they give decisions as to whether and when to spray. These forecast systems, however, are still incomplete.

3.2. Ecologically Selective Compounds and Methods of Application

The impact of pesticides on the ecosystem can be minimized also by the choice of less harmful pesticides and by selective ways of application. In 1974 and IOBC working group on "Pesticides and Beneficial Arthropods" was established. It is its objective to work out methods for testing pesticides for their effect on 31 species of beneficial arthropods. Since 1975 in W. Germany test methods for the side-effects on seven species have been made available to chemical companies. According to our present knowledge the new fungicides are neutral to most parts of the beneficial fauna. In a completely different way ecological selectivity is achieved with compounds for seed treatment against seed-borne and exogenous pathogens. Although a seed treatment is always preventive, it is less damaging to the environment because of its localized as opposed to generalized application and low dosage. An outstanding example is triadimenol, which is effective against several seed-borne pathogens including loose smut, and against some exogenous pathogens, such as powdery mildew and rust.

4. Final Remarks

This short presentation on chemical control practices in intensive wheat cultivation in NW Europe reveals that the control of diseases and insect pests is carried out according to several concepts which do indeed aim to reduce the impact of the chemical component of crop protection. However, it must be admitted that crop protection is still far from the ideal of complete integrated control.

Within the scientific field the methods of disease and pest prognosis have to be improved and levels of damage defined. The farmer, too, has to alter his outlook and to learn again. He should conceive the wheat crop as a multitude of biological interactions which may be severely deranged by chemical manipulation. For this an intensive education and training are necessary. The farmer has to be supported by an efficient advisory service, which quickly transforms scientific results to a practical level. A close network of pest monitoring and warning services is required where excessive demands are placed on the farmer alone in finding the right decision.

Acknowledgements

Many colleagues have kindly helped with valuable information and advice. In particular the author wishes to thank: Dr. J. C. Zadoks, Wageningen, The Netherlands; Dr. P. R. Scott, Cambridge, England; Dr. H. Scheinpflug, Leverkusen, Dr. H. Sturm, Dr. W. Zwick, Limbergerhof, and Mr. W. Teuteberg, Kiel, W. Germany. Dr. C. J. Mappes, Limburgerhof, W. Germany, kindly corrected the English text.

Selected References

Dekker, J., 1977: The fungicide-resistance problem. Neth. J. Plant Pathol. 83, Auppl. 1, 159-167.

Effland, H., 1973: Bekämpfung von Pilzkrankheiten im intensiven Getreidebau durch Sprossbehandlungen. Mitt. Biol. Bundesanst. Land- Forstwirtsch. Berlin-Dahlem H. 151. 75-117.

Fraselle, J. and Martens, P. H., 1973: Le problème des residus en relation avec les traitements fongicides appliqués contre les maladies tardives du froment d'hivers. Mededel. Fac. Landbouwwetenschap. Rijksuniv. Gent 38, 865-874.

Obst, A., 1975: Present situation of disease control in North Western Europe (excluding Gt. Britain). Proceedings 8th British Insecticide and Fungicide Conference, 953-966.

Poussard, C., Maumene, J., Goulet, J. P., and Lescar, L., 1977: Les traitements fongicides du blé tendre d'hiver en cours de végétation. Compte rendu de la journée d'étude organisée par l' I.T.C.F. "La lutte contre les maladies et les ravageurs des céréales", 169-177.

Rapilly, F., and Jolivet, E., 1976: Construction d'un modèle (EPISEPT) permettant la simulation d'une épidemie de *Septoria nodorum* sur blé. Revue de Statistique Appliquée 24, 31-60.

Rapilly, F., Eschenbrenner, P., Choisnel, E., and Lacroze, F., 1979: La prévision du piétin verse sur blé d'hiver. Perspectives Agricoles 23, 30-40.

Schrödter, H., and Fehrmann, H., 1971: Okologische Untersuchungen zur Epidemiologie von *Cercosporella herpotrichoides*. III. Die relative Bedeutung der meteorologischen Parameter und die komplexe Wirkung ihrer Konstellationen auf den Infektionserfolg. Phytopathol. Z. 71, 203-222.

Scott, P. R., 1978: Physiologic specialization in the eyespot pathogen, in the context of the introduction into wheat of resistance from *Aegilops ventricosa*. Paper presented at the 1st Meeting of the EPPO Study Group on Assessment of Eyespot of Cereals (*Pseudocercosporella herpotrichoides*), Munich, 15th August 1978.

Teuteberg, W., 1978: Winterweizen 1978/79—Sortenpass für Schleswig-Holstein. Landwirtschaftskammer Schleswig-Holstein (editor), 72 pp.

Vereijken, P., 1976: Integrated control of insects in cereals. Faculté des Sciences Agronomiques de l'Etat et Centre de Recherches Agronomiques, Gembloux. Semaine d'Etude Céréaliculture, Compte-Rendu des Séances, 399-410.

Zimmerman, G., 1978: Laborversuche über den Einfluss systemischer Fungizide auf Verpilzung und Konidienbildung durch aphiden-pathogene Entomophthoraceen (Zygomycetes) bei Getreideblattlausen. Z. Pflanzendrankh. Pflanzensch. 85, 513-524.

30.1 Overview of Rice Pest Problems

N. C. Brady

The International Rice Research Institute
P. O. Box 933, Manila, Philippines

Rice is the staple diet of about 60% of the world's population. The crop provides at least one quarter of the diet for 90% of the world's low-income people and is a critical source of food in areas with the highest populations.

In monsoon Asia where 86% of the total rice hectarage is found, the traditional rices are low yielding and do not respond to chemical inputs. They are leafy, tall, low tillering, and sensitive to day length. Because of their weak stems, they lodge easily; hence, they cannot make effective use of fertilizers or of natural soil fertility.

In the early 1950s and 1960s scientists in China and India and (in the sixties) at the International Rice Research Institute (IRRI) developed cultivars of a new plant type, which was semidwarf, stiff-strawed, high tillering, and relatively insensitive to changes in day length. Cultivars like IR8 responded to added fertilizers and produced yields of 6-10 t/ha on irrigated lands. These cultivars helped start the *Green Revolution* in rice.

But soon after the high yielding semidwarf rice cultivars were released, attention was called to yield constraints stemming from rice pests (insects, diseases, and weeds). The availability and use of the short-statured high yielding rices may have aggravated already serious pest problems. Even though the pest resistance of these modern rices is as good as, or better than, that of the average traditional cultivars they replaced, in some areas their production may have encouraged the pest enemies of the rice plant.

Three primary reasons have been given for the increased pest pressures associated with the modern cultivars. First, many traditional cultivars with a wide range of host resistance were replaced by a few modern cultivars with good but nevertheless limited host resistance. That, in effect, narrowed the germplasm base for resistance in the field. Second, the cultural conditions that assured high yields provided relatively ideal conditions for some of the rice pests. Heavy fertilization and short stature encouraged weed competition. Dense planting, high tillering, and the vigorous crop growth stimulated by fertilizers led to a crop canopy environment that proved ideal for some diseases and insect pests.

Because the potential yield level of modern cultivars is high, the yield-depressing effect of pests became more noticeable than it had been with traditional cultivars. A 20% reduction of a yield of 6 t/ha is much more noticeable than a comparable reduction in a 2-t crop.

The third reason for increased pest management problems is increasing cropping intensities. Greatly expanded irrigation schemes permitted two or three rice crops a year in some areas where only one had been grown previously. The development and release of modern high yielding cultivars requiring a growth period of only 105 to 120 days (compared to 150-170 for most native cultivars) encouraged double and triple cropping. The extent to which rice monoculture is followed in crop intensification definitely influences opportunities for pest buildup.

The most widespread diseases affecting rice are those caused by viruses (tungro, grassy stunt, and ragged stunt), by fungi (hoja blanca, blast, and sheath blight), and by bacteria (bacterial blight and bacterial leaf streak). Major insect pests include planthoppers, leafhoppers, stem borers, gall midge, seedling maggots, caseworm, leaf folder, armyworm, and rice seed bugs.

Weeds are a universal problem whether rice is grown as a dryland crop, in wetland fields, or as floating rice in deep water. With nonflooded conditions, weeds are as great a constraint as, or greater than, insects and diseases. Of particular importance are weeds of the genera *Echinochloa, Cyperus, Eleusine, Fimbristylis, Ischaemum, Monochoria,* and *Sphenochlea.*

Absolute yield losses from insect, disease, and weed pests are hard to assess because they are dependent on the type of rice culture and the yield potential, and are highly location and time specific. Cultural practices developed by farmers include a number of methods that reduce pest populations. Many scientists believe that the paddy system of transplanting rice on puddled soil and keeping the fields flooded evolved primarily to solve the weed problem. Out of necessity, farmers over the centuries have selected rices that are tolerant of diseases such as blast and that can compete with weeds.

Several epidemics, including tungro virus transmitted by the green leafhopper and hopperburn caused by brown planthopper, occurred in various parts of Asia in the 1970s. Blast reduced the yields of the 1978 rice crop in Korea where large-scale adoption of superior cultivars had resulted in the world's highest national yields. Heavy usage of pesticides in Japan has resulted in pest resistance to insecticides and a buildup of tolerant weeds. Resurgence of the brown planthopper following repeated insecticide usage has been demonstrated by scientists at the International Rice Research Institute (IRRI).

To be acceptable, pest management practices in the tropics must be cheap and simple to apply. Rice farmers in the region have limited land and financial resources. The average farm size is less than 2 hectares and per capita annual incomes are mostly under $300. Recognizing such resource constraints, IRRI and its cooperators in national programs have emphasized the development of host resistance to rice pests.

At IRRI, research on host resistance is done through an interdisciplinary genetics evaluation and utilization (GEU) program. Teams of crop protection scientists and plant breeders systematically screen the large germplasm collection (currently about 50,000 accessions) for tolerance or resistance to major pathogens and insect pests. Resistant cultivars are included in the crossing programs and the resulting progeny are screened to determine if genetic resistance has been incorporated. Tens of thousands of cultivars and breeding lines are screened each year. This extensive breeding effort has produced numerous lines and cultivars with good resistance to six major diseases and insect pests.

But the work at IRRI headquarters is only a fraction of that being done worldwide to fully utilize the genetic potential of the rice plant. Early generation progeny of crosses made at IRRI are made available for testing in national programs. Through the IRRI-sponsored International Rice Testing Program (IRTP), nearly 600 nursery tests are run each year by scientists in national programs. Rices submitted for testing by these researchers are included in the IRTP nurseries. The IRTP insures the testing under widely varying ecological conditions of genetic materials from throughout the world. The nurseries concerned with resistance to pests are among the most important in the IRTP program.

Natural selection of insect biotypes and pathogen races has complicated the GEU approach to pest management. Although rice scientists are currently managing most pest outbreaks by anticipating the development of new races and biotypes, it is evident that not all pest problems can be solved through plant resistance.

Pest management systems that incorporate other control measures such as the timeliness of pesticide application, conservation of natural enemies of insect pests, and cultural controls are being developed or fully exploited for the various environments in which rice is grown. There is now significant interest in the development and implementation of rice integrated pest management (IPM) programs in countries throughout tropical Asia.

In Asia FAO has recently formulated a project to incorporate the known methods of rice pest control into viable pest management systems. Much of the rice pest management technology in Asia has been developed or fostered by IRRI. IRRI is cooperating with FAO in developing this IPM project and is continuing collaboration with national programs to help prepare them to implement IPM at the local level. The fruition of pest management systems can be achieved only by the strength of the research and extension capabilities of national programs.

Inputs from three primary sources are necessary to develop and implement successful integrated pest management programs. International organizations such as FAO and IRRI can be the sources of background information and can help encourage an international focus on pest management. Pest management scientists and extension specialists from the more developed countries can provide research data and organizational assistance. But the third source of inputs, the scientists, extension specialists, and farmers in the developing countries will ultimately determine the success or failure of integrated pest management programs. Coordinated and concerted efforts must be made to help them do their job.

30.2   Strategies for Rice Disease Control

V. T. John, K. C. Ling, and M. C. Rush

All India Coordinated Rice Improvement Project, Hyderabad 500030, India; The International Rice Research Institute, P. O. Box 933, Manila, Philippines; and Department of Plant Pathology, Louisiana State University, Baton Rouge, Louisiana 70808, USA, respectively.

Although rice diseases have been recorded for several centuries, epidemics were rare until the extensive cultivation of high-yielding, semi-dwarf cultivars began during the mid-sixties of this century. Concomitant with the use of these cultivars was the associated use of high nitrogen inputs, more intensive management practices, and the increased use of continuous cropping, which gave significantly increased rice yields in most parts of the world. However, the use of these cultivars and cultivation practices also increased the prevalence and severity of diseases, leading at times to epiphytotics which caused large scale losses. Consequently, it became necessary to intensify efforts directed toward controlling plant diseases.

Major Diseases of Rice

More than 70 rice diseases caused by viruses, bacteria, fungi, and nematodes have been recorded. Some are distributed worldwide, while others are localized in occurrence. The diseases vary greatly in economic importance. Diseases that have caused significant losses are discussed below:

Fungal diseases

The rice blast disease, caused by *Pyricularia oryzae,* continues to be a problem in many countries, particularly in Africa, South America, and the hilly and temperate areas of the Himalayas. In 1978 the disease occurred in epidemic form in Korea, causing heavy losses [1]. The blast disease has been thoroughly studied and many aspects of its etiology and epidemiology are well documented in the literature. The disease causes severe damage to seedlings in nurseries and on infected foliage, nodes, and panicles in the field. Many pathogenic races of *P. oryzae* have been reported.

The sheath blight disease, caused by *Thanatephorus cucumeris,* the perfect state of *Rhizoctonia solani,* received scant attention until recently when it became a major disease of high-yielding cultivars. The disease has been reported to cause serious losses in India, Japan, Philippines, the United States, and other rice-producing countries. The fungus survives between crops as sclerotia, as mycelium in plant debris in soil, and on weeds and other crop plants. The fungus also causes a serious disease on soybean in the USA. Varieties and lines with moderate levels of resistance have been identified, however, resistance varies at different test locations and is very sensitive to environmental influences.

Various other diseases caused by fungi are occasionally epidemic or locally serious. The most important of these diseases are brown spot, caused by *Cochliobolus miyabeanus* (*Helminthosporium oryzae*), stem rot, caused by *Magnaporthe salvinii* (*Sclerotium oryzae*) [2]; narrow brown spot, caused by *Cercospora oryzae;* sheath rot, caused by *Acrocylindrium oryzae,* and leaf scald, caused by *Rhynchosporium oryzae.*

Bacterial diseases

Many improved cultivars are susceptible to the vascular disease caused by *Xanthomonas oryzae,* which causes the symptoms of "kresek" (wilting) and leaf blighting of the foliage. Bacterial leaf blight is widespread during the wet season in many countries in Asia. Recently the disease was also found in Africa [3], Australia [4], Mexico, and Central America [5]. Isolates of *X. oryzae* differ in their pathogenicity and virulence [6,7].

Virus diseases

A total of 15 virus and virus-like diseases have been identified since rice dwarf was first recorded in Japan in 1883. The yellow mottle disease is distributed in Africa, the giallume disease is found only in Italy, and hoja blanca is unique to the Americas. Black streaked dwarf, dwarf, necrosis mosaic, stripe, transitory yellowing and waika occur in temperate Asia. Grassy stunt, orange leaf, ragged stunt, and tungro are distributed in tropical Asia. The necrosis mosaic and yellow mottle viruses are mechanically transmissible and are transmitted through soil or by chewing insects, respectively. The giallume virus is transmitted by aphids. The other rice viruses are transmitted by a total of 13 species of leafhoppers. Variation in the incidence of the insect-transmitted virus diseases is a function of: (1) populations and activity of the vector, (ii) quantity and quality of the virus source, (iii) susceptibility of the rice cultivars to the pathogen and the vector, (iv) biotic and abiotic environmental conditions, and (v) time. Forecasting the incidence of rice virus diseases is difficult, however, if any one of the factors required for an epidemic is absent or unfavorable, there will not be an outbreak of disease [8].

Nematode diseases

Several diseases caused by nematodes have caused significant losses. The white tip nematode *Aphelenchoides besseyi,* is most common, followed by the stem nematode *Ditylenchus angustus,* root nematode *Hirschmanniella oryzae,* and root knot nematodes *Meloidogyne* spp.

Control Strategies

The concepts of "integrated pest control" and "integrated disease control" have recently been defined and redefined. It is clear that disease control techniques must first be developed and integrated before disease control can be integrated with the other pest control measures. The major factors being integrated are host resistance, the use of pesticides, cultural practices, biological control, physical control methods, and legislative or regulatory control measures.

Recently the term "management" has been used in reference to plant diseases to refer to the maintaining of a pathogen population, and the resulting disease, at levels below a predetermined economic threshold. This term should be restricted to those control methods which would limit the pathogen population, either throughout the season or during some critical growth stage where disease development will restrict yield, but which do not apply selection pressure to the pathogen population that might lead to the selection of new races or strains. It should be possible to divide the major disease control measures into two categories based on the above premise. One category would include eradicative-control measures, such as, protective or eradicative pesticides with biocidal activity, including soil fumigants; physical control methods, and vertical or monogenic resistance. The second category would include management-control measures, such as, pesticides with biostatic activity, horizontal or polygenic resistance, the equivalent of "slow rusting", multiline cultivars, biological control, cultural practices, and regulatory activities such as certification programs.

An integrated disease control program would then be one that combines all possible control measures to lower or restrict the pathogen population to the maximum. An integrated disease management program would combine the management-type control measures to restrict pathogen populations, and disease, below some predetermined economic threshold. The selection of program would depend primarily on the variability of the pathogen and the success or failure of earlier control measures.

The following disease control measures have been used in rice production:

Host resistance

Disease control through the use of resistant cultivars has historically been the most economic and environmentally safe method. For several decades emphasis has been placed on the development of disease resistant rice cultivars [9]. Methods have been developed for screening and testing varietal resistance to major diseases by natural infection in the field or by artificial inoculation in the field or greenhouse [10, 11].

Recently, the International Rice Testing Program, coordinated by the International Rice Research Institute in more than 70 countries, have organized nurseries for testing resistance to several major diseases. The objectives of these programs are to find cultivars with stable resistance to the various diseases and the preliminary determination of pathogen variability on selected cultivars [12].

Numerous sources for resistance to rice diseases have been identified through screening programs throughout the world. Rice breeders have been actively incorporating resistance into commercial cultivars. In certain cases stable resistance has been transferred to commercially acceptable cultivars. A gene for resistance to grassy stunt was transferred from *Oryza nivara* into cultivated varieties of *O. sativa* [13]. The incidence of grassy stunt has been reduced to a negligible level in most of tropical Asia since 1974 due to the increased use of cultivars with grassy stunt resistance. In the United States, the cultivar Saturn was released in 1964, primarily because of its resistance to blast. Although a race of *P. oryzae* has developed which can infect Saturn, natural infections remain rare and no significant yield losses to blast have been recorded on this cultivar [14].

Several major problems have been encountered in programs with primary emphasis on inherent resistance for disease control. Sources of stable resistance have not been found for several important diseases. Often cultivars with resistance to one important disease were susceptible to one or more other major diseases. The incorporation of multiple resistance into commercial varieties has been a complex, long term undertaking and the so called "pyramiding of resistance genes" has been successful only to a limited extent. There have been difficulties in combining genes from some resistance sources, particularly between cross-sterile materials. The most important drawback has been the breakdown of resistance due to the development of new variants of the pathogens.

Disease resistance has historically been a front-line defense against disease losses. Breeding programs are being adjusted to compensate for or overcome these inate problems through the use of horizontal or multigenic resistance, the equivalent of "slow rusting", multiline cultivars, tissue culture, and similar techniques.

Chemical control

The use of pesticides for disease control in rice has been increasing. Chemical control agents, such as fungicides, bactericides, insecticides, and nematicides, often can be located and made available in a comparatively short time. Pesticides may be specific for a single pathogen species, for instance tricyclazole and *P. oryzae,* or they may be broad spectrum and give control of several diseases. Benomyl, for example, controls the blast, narrow brown spot, sheath blight, and stem rot diseases to some extent. Pesticide use can be restricted to areas threatened by or having epidemic conditions. These factors make pesticides an ideal component of disease control or management systems. Disadvantages of pesticides include their high cost, costs of application, unfavorable effects on nontarget organisms, residues, environmental pollution, development of pathogen resistance, and the delay in obtaining approval for the use of efficacious pesticides by government agencies charged with protecting human life and environment. A solution to these problems is to integrate pesticide usage with the other control measures into a continuously changing disease management program, which is dynamic, responding to changes affecting each control practice, and minimizing the development of epidemic conditions.

Because chemicals have effectively reduced disease and increased yields, their use has been increasing in some countries, particularly in Japan and the United States [15, 16, 17].

Fungicides developed in Japan and used for blast control include Blasticidin S, Kasugamycin, Hinosan, and Kitazin. Fungicides used for sheath blight control in Japan include MAF (iron methanearsonate), Polyoxin, and Validamycin. In the United States, benomyl is registered for use on foliar diseases of rice. Since the registration of Benlate 50W in the United States in 1976, the acreage sprayed has increased yearly to about 125,000 hectares in 1978.

Although there are no viricides available, chemicals are applied to control insect vectors to reduce the spread of virus diseases. Organophosphates, carbamates, phenyl-substituted carbamates, pyridaphenthion, and prophaphos have been effective for controlling vector leafhoppers and planthoppers [18].

The present pesticide screening programs and development activities offer only the minimum underpinning of a system for integrating the use of pesticides into disease management programs. Essential elements of a program to develop pesticides for disease management systems must at a minimum include screening tests to determine efficacy against pathogens, development of residue and toxicity information, monitoring for development of resistance by pathogens, testing and monitoring of application equipment and techniques, determination of the effects of cultural practices on pesticide applications, determination of the effects of pesticides on microorganisms antagonistic to or competitive with rice pathogens, development of scouting and prediction methods (forecasting) to minimize the frequency of pesticide applications, development of disease thresholds for economic loss, and the continual monitoring of disease incidence and status.

Cultural, physical, and regulatory control methods

Cultural practices for disease control have often been integrated with other cultural practices for so long that their disease control function is no longer recognized. Almost all aspects of the cultural practices used in rice production have some effect on the development of disease. Rice is unique among the field crops in that it is most commonly grown under flooded conditions. Water management often has a critical impact on disease development as it affects seeding, transplanting, retention of fertilizer in the soil, survival of soilborne pathogens, movement of inoculum, various stresses associated with drought and flooding, and many other factors. Production of seedlings in seedbeds and transplanting helps to minimize seedling diseases and irregularities in stand. Timing of planting and harvesting, seed storage, fertilizer rates, rotation crops, and weed control measures often have an effect on disease control. These and other cultural practices can be manipulated to minimize inoculum production, survival, and dissemination, infection, and disease development.

Physical control measures have not been used extensively for disease control in rice. Heat treatment of seeds has been used to a limited extent for controlling the white tip nematode and other seedborne pathogens. Regulatory methods have also been mainly directed to the control of seedborne pathogens through quarantine and inspection systems.

Conclusion

The identification of pathogens affecting rice and their causal relationships to disease have been extensively studied. The definition of the management unit, the agroecosystem, has been rudimentary and requires intensive research efforts in the future. The interacting aspects of pathogen and host ecosystems must be identified, conceptualized and systematized so that the effectiveness of the human element of the interaction is maximized. Self-buffering managment systems can then be devised that utilize biological processes combined with a minimum imput of non-system practices (pesticides, etc.) to minimize the epidemic potential of the agroecosystem.

Literature Cited

1. Second Korea-IRRI Cooperative Rice Research Plan. Off. Rural Dev., Suweon, Korea, 1979, Feb. 21-23.
2. Krause, R. A.; Webster, R. K. Mycologia, 1972, 64, 103-114.
3. Buddenhagen, I. W.; Vuong, H. H.; Ba, D. D. IRRN, 1979, 4(1), 11.
4. Aldrick, S. J.; Buddenhagen, I. W.; Reddy, A. P. K. Aust. J. Agri. Res., 1973, 24, 219-227.
5. Lozano, J. C. Plant Dis. Reptr., 1977, 61, 644-648.
6. Ezuka, A.; Sakaguchi, S. Rev. Plant Prot. Res., 1978, 11, 93-118.
7. Mew, T. W.; Vera Cruz, C. M. Phytopathology, 1979, 69, 152-155.
8. Ling, K. C. "Rice Virus and Virus-like Diseases"; Int. Rice Res. Inst., Los Banos, Philippines, 1979; 16 p.
9. International Rice Research Institute. "Rice Breeding", Los Banos, Phillippines, 1972, pp. 1038.
10. Jennings, P. R.; Coffman, W. R.; Kauffman, H. E. "Rice Improvement", International Rice Research Institute, Los Banos, Philippines, 1979, pp. 186.
11. Hoff, B. J.; Rush, M. C.; McIlrath, W. O.; Morgan, A. 68th Prog. Rept. LSU Rice Expt. Sta., Crowley, Louisiana. 1976, 68, 142-192.
12. International Rice Research Institute. "Report of 1978 International Rice Testing Program Activities", IRRI, Los Banos, Philippines, 1978, pp. 21.
13. Khush, G. S.; Ling, K. C.; Aquino, R. C.; Aguiero, V. M. Proc. 3rd Int. Cong. SABRAO, Canberra Australia, 1977, 1, 4(b), 3-9.
14. Marchetti, M. A.; Rush, M. C.; Hunter, W. E. Plant Dis. Reptr. 1976, 60, 721-725.
15. Bhakthavatsalem, G.; Reddy, A. P. K.; John, V. T. Pesticides, 1977, 11, 13-16.
16. Rush, M. C.; Lindberg, G. D.; Whitam, H. K. Louisiana Agriculture, 1977, 20(4), 10-11.
17. Yamaguchi, T. Plant Prot. Japan, 1976, Agri. Asia, Special Issue 10, 108-121.
18. Iwata, T. Plant Prot. Japan, 1976, Agri. Asia, Special Issue 10, 122-131.

30.3   Weed Control Practices in Rice

Roy J. Smith, Jr.

Res. Agron., Agric. Res., Sci. and Ed. Admin.
U.S. Dep. Agric., and Univ. Arkansas, Stuttgart, AR, 72160 USA

and Keith Moody

Agron., Int. Rice Res. Inst., Manila, Philippines.

Rice (Oryza sative L.) is a major world crop that millions of people depend on as a staple food. About 85 countries produce 374 million metric tons of rice on approximately 144 million hectares. Weeds reduce the yield and quality of rice on approximately 144 million hectares. Weeds reduce the yield and quality of rice and require the use of herbicides, and cultural and mechanical practices for their control. In the USA, the total estimated direct losses from weeds and the cost of their control represented 28% of the value of the crop annually during 1975-1977. Barnyardgrass [Echinochloa crusgalli (L.) Beauv.] in Taiwan reduced yields of transplanted rice more than 80%. In Australia, it reduced yields of direct-seeded rice 75%.

Barnyardgrass is the most troublesome weed of rice in the world. Other rice field weeds of world importance are junglerice [Echinochloa colonum (L.) Link], Cyperus difformis L., C. rotundus L., C. iria L., Eleusine indica (L.) Gaertn., Fimbristylis *miliacea* (L.) Vahl, *Ischaemum rugosum* Salisb., *Monochoria vaginalis* (Burm. f.) Presl, and *Sphenoclea zeylanica* Gaertn.

The best approach to controlling weeds in transplanted and direct-seeded rise is to use an integrated system that combines preventive, cultural, mechanical, chemical and biological practices. The system that omits any one of these components is often inadequate.

Weed Control in Transplanted Rice

Rice is transplanted by setting plants, that have been grown in nurseries, into wet paddy fields. Special weed problems and weed control technologies are associated with transplanted rice.

Cultural-Mechanical Methods

Land preparation, plant spacing, cultivar selection, age of seedlings, water and fertilizer management, hand and mechnical weeding, and crop rotations are important components of integrated weed control systems for transplanted rice.

Preplant tillage is divided into primary and secondary tillage operations. During primary tillage, rice stubble and weeds are incorporated into the soil by plowing 10 to 15 cm deep. Secondary tillage or puddling involves several harrowings of wet soil to destroy weeds missed during primary tillage and to promote germination of weed seed that can be killed by subsequent harrowing. With deep plowing of the soil, weeds such as barnyardgrass are reduced. In Japan, the practice of transplanting rice by machine requires both an increased interval between puddling and transplanting to permit soil to become firm and the use of smaller rice seedlings suitable for machine planting. These practices increase weed infestation and the need for herbicide treatments.

When rice is transplanted in narrow rows, weed competition is reduced. Yield losses from weeds in the Philippines averaged 52, 29 and 19% for transplanted spacings of 25 × 25, 20 × 20, and 15 × 15 cm, respectively.

Short-statured, erect rice cultivars are less competitive with weeds than tall, drooping cultivars. In the Philippines, Ir 442-2-58, an intermediate-statured cultivar, competed better with weeds than IR 20, a semidwarf cultivar. The cultivar being grown also affected the type of weeds that grew in the rice field. Weeds that grew in association with BR3, a short, erect cultivar, were sedges, such as *Fimbristylis miliacea*, *Cyperus difformis* and *C. iria*. However, barnyardgrass, rather than sedges, was associated with the traditional, tall, drooping Dharial cultivar.

The age and size of rice seedlings at time of transplanting affect weed competition. Weeds in Japan reduced yields from 27 to 59% when seedlings were 23 cm tall at time of transplanting, compared with 80 to 89% when 14-cm-tall seedlings were transplanted.

Unavailability of irrigation water and poor water management increase problems with weeds. In the Philippines, a continuous flood of as little as 1- to 2-cm reduced infestations of many grasses and sedges; a continuous flood 15-cm deep eliminated all weeds. Perennial weeds are less sensitive to water depth than are annual weeds. In the Philippines, perennial weeds such as knotgrass (*Paspalum distichum* L.) and *Scirpus maritimus* L. were not controlled by normal flooding depths. Unfortunately, controlled water management is practical on only about 20 % of the rice culture in Asia; the other 80% is rain-fed culture where supplies of water are unreliable.

Nitrogen management (rates and timing of application) affects weed competition in rice. In Taiwan, barnyardgrass and *Cyperus difformis* reduced yields of rice that received high rates of nitrogen more than *Monochoria vaginalis* or *Spirodela polyrhiza* (L.) Schleid., but the latter two weeds reduced yields of rice that received low-nitrogen levels more than the former two weeds.

Manual weeding by hand or with the aid of tools is the prevalent method of weed control in transplanted rice in Asia. This is because hand labor is readily available and wage rates are low. Where land has been well prepared for transplanting, irrigation is managed well, and a competitive cultivar is transplanted at optimum spacings, one properly timed manual weeding usually controls weeds well enough to prevent yield and quality losses.

Mechnical weeding is feasible in rice transplanted in straight rows or checked-rows. However, farmers prefer random transplanting because less time is required to plant the crop. The best mechnical device for weed control in transplanted rice is the hand-pushed, rotary weeder which can reduce the time of weeding by 50%.

Crop rotations that include upland crops grown in a system with rice reduce problems with certain weeds. In the Philippines when transplanted rice was grown continuously, *Scirpus maritimus* infestations were at constantly high levels for a 3-year period, but *C. rotundus* infestations increased. When upland crops were rotated with rice, infestations of *S. maritimus* reduced rice yields less than in the continuous rice-cropping system.

Herbicides

In Asia, herbicides are most commonly used where manual labor is unavailable or its cost is high.

Countries that use large amounts of herbicides include Japan, Korea and Taiwan. Rice, that was grown in Japan in 1976, received an average of more than two herbicide treatments per crop season. About half of the rice is treated with herbicides in Korea and Taiwan. Relatively inexpensive herbicides such as 2,4-D and MCPA are used to control broadleaf and sedge weeds in rice grown in developing nations of tropical Asia. However, most of the rice in the developing countries of tropical Asia receives no herbicide application.

The principal herbicides used in rice in Japan in 1976 were thiobencarb + simetryn, CNP and oxadiazon. Butachlor, thiobencarb + simetryn, nitrofen and 2,4-D are important herbicides used in weed control programs in Korea. In these two countries, high usage of herbicides has changed the weed flora in rice fields. Herbicides control the annual weeds, but ecological shifts have increased problems with perennial weeds such as *Cyperus serotinus* Rottb., *Sagittaria* spp., *Eleocharis kuroguwai* Ohwi and *Scirpus juncoides* Roxb. New technology is being developed to control many of these hard-to-kill perennial weeds. Perfluidone applied 5 days after transplanting (DAT) or bentazon applied 15 DAT control these perennial weeds. Rotating herbicides or combining two or more herbicides can prevent buildup of tolerant weed species.

Integrated Control Systems

The development of integrated weed control systems that include cultural, mechanical, ecological, biological, and chemical practices is essential to rice production. Herbicides in combination with good water management control weeds more effectively than either practice alone. In the Philippines, barnyardgrass was controlled better when herbicides were combined with continuous flooding than when they were used with alternate flooding and draining. In addition, a greater number of treatments controlled barnyardgrass effectively when rice was continuously flooded than when it was alternately flooded and drained. Most herbicides give better weed control when combined with other herbicides or manual weeding than when used alone. The use of propanil and MCPA, combined in a weed control program with mechanical weeding, controlled weeds in transplanted rice in India better than when either herbicide was used alone or when only herbicides or mechanical weeding were used. Weed control and grain yields were superior when herbicides were combined with hadweeding than when either practice was used alone.

Fertilizer management, combined with herbicide treatments produces highest yields. Rice yielded better in Bangladesh when herbicides and fertilizers were applied together than when they were applied alone. Control of weeds at the time of applying fertilizer also improved efficiency of the fertilizer. Spacing of rice seedlings in the field influences the efficacy of herbicides. In the Philippines, a greater number of herbicide treatments controlled weeds in rice transplanted at 15 × 15 cm spacing than when transplanted at 25 × 25 cm spacing.

Weed Control in Direct-Seeded Rice

Rice is direct-seeded by drilling or broadcasting into moist soil or by broadcasting dry or sprouted seed into the floodwater. Drill- or broadcast-seeded rice may be flooded after emergence and grown as paddy rice or it may be grown as a rain-fed upland crop.

Cultural-Mechanical Methods

Crop rotation, seedbed preparation, seeding method, water and fertilizer management and cultivar selection are important components of integrated weed control systems for direct-seeded rice.

Rotations of upland crops with rice is the only means of controlling many hard-to-kill weeds such as red rice (*Oryza sativa* L.). Because red rice is the same species as white rice, it is difficult to control in the rice crop. Rotating upland crops with rice permits the use of herbicides that control red rice effectively in soybeans [*Glycine max* (L.) Merr.]. If red rice is controlled for 2 years in the upland crops, rice can be grown in the third year without serious competition from red rice.

Thorough seedbed preparation helps to control weeds that infest rice fields. The goal is to eliminate all weed growth up to the time of seeding rice. Weeds that germinate and emerge with the rice crop are easier to control with herbicides.

Seeding method influences weed populations in rice. Barnyardgrass is more prevalent in dry-seeded than in water-seeded rice. However, aquatic weeds, such as ducksalad (*Heteranthera* spp.), redstem (*Ammannia* spp.) and waterhyssop [*Bacopa rotundifolia* (Michx. Wettst.)] are more troublesome in water-seeded rice.

Water management influences weed growth in rice fields. Floodwater applied 1 to 2 days after treatment with propanil prevents barnyardgrass from reinfesting rice fields. Maintenance of the floodwater on rice fields for 2 to 3 weeks after applying molinate is essential to obtaining control of tillering barnyardgrass.

Some rice cultivars are more competitive with weeds than others. In the USA, all-season competition of barnyardgrass reduced grain yields of

Starbonnet and Bluebelle cultivars 40% and 64%, respectively. It was believed that competitiveness was associated with the time required for maturity of the cultivars.

Although cultural-mechanical methods of weed control for rice are beneficial, nonchemical methods alone seldom control weeds adequately. They control weeds more effectively when combined with herbicide treatments.

Herbicides

Use of herbicides for control of weeds in direct-seeded rice is an important component of weed control systems. In Australia, the principal herbicides used for weed control in rice are propanil, MCPA and 2,4-D. In Italy, important herbicides for weed control in rice include propanil, molinate, thiobencarb and MCPA. Herbicides used in rice in South America include propanil, molinate, thiobencarb, butachlor and MCPA. Those used in the USA are propanil, molinate, bifenox, oxadiazon and phenoxys.

Propanil

Propanil is applied postemergence when grass weeds are in the 1- to 4-leaf stages of growth. It may be applied aerially or by ground equipment. Sequential applications are frequently made to about 70% of the rice in the southern USA. The activity of propanil is influenced by rainfall and termperature. At least 8 hours without rain after propanil treatment are required for effective weed control. Best control of weeds by propanil occurs when daily maximum air temperatures range from 21 to 32 C and daily minimums are above 16 C. Irrigation water must be managed properly to obtain satisfactory weed control with propanil. If the soil is dry, irrigating the rice field 2 to 5 days before propanil treatment increases weed control. Flooding rice 8 to 10 cm deep within 1 to 5 days after treatment increases weed control. Flooding rice 8 to 10 cm deep within 1 to 5 days after treatment prevents germination of more grass weeds. Also, floodwater must be lowered to expose weeds to propanil.

Molinate

Molinate is applied as a postemergence treatment into the floodwater when barnyardgrass ranges from the 4-leaf to jointing stages of growth. Rice is tolerant to molinate if the plants have grown above the flood water. Water management is critical for effective weed control when using molinate. As grass weeds get larger, longer flood periods are required to kill or suppress them. Granular molinate is usually applied aerially to large commercial fields where aerial equipment is available. However, it may be applied by hand-operated rotary spreaders to small fields where aerial equipment is unavailable.

Recently in Australia, emulsifiable molinate has been metered into the floodwater to control barnyardgrass and other weeds selectively in direct-seeded rice. In the USA molinate metered into the floodwater controls barnyardgrass effectively in direct-seeded rice, but farmers have not adopted this practice.

Phenoxy Herbicides

Phenoxy herbicides such as 2,4-D, MCPA, 2,4,5,-T and silvex are applied as postemergence sprays. The stage of growth greatly influences the response of rice plants to phenoxy herbicides. Rice in the late-tillering to early-jointing stages is not injured by phenoxy herbicides. The tolerant stage can be positively identified when the basal internodes begin to elongate from 0.6 to 1.3 cm. Rice cultivars vary in the length of time required to reach this tolerant growth stage. Also, environmental and cultural practices influence the period required for rice to reach the tolerant stage.

Environment and nitrogen fertilization may influence the activity of phenoxy herbicides on rice and weeds. Phenoxy herbicides control weeds effectively if rain occurs no sooner than 6 hours after treatment. Even when rice is treated during the tolerant stages, high temperatures (above 34 C) may increase rice injury by phenoxy herbicides. Temperatures below 15 C during the week before treatment may slow weed growth and reduce control. Because phenoxy herbicides may cause some injury to the crop even when applied in the tolerant stage of growth, nitrogen applied within 5 days after treatment helps the rice to recover from the herbicide injury.

Other Herbicides

Many new herbicides are being used in or developed for direct-seeded rice. They control many species of weeds that standard herbicides fail to control. Many also possess better pre-emergence or residual activity than standard herbicides. New herbicides recently registered for use in direct-seeded rice include thiobencarb, bifenox, oxadiazon, bentazon and butachlor. One or more of these are now used in Australia, Italy, South America and the USA. These herbicides frequently give better weed control when combined in tank mixtures with propanil.

Integrated Control Systems

The integration of weed control technologies into systems for rice is important. Effective weed control systems for direct-seeded rice combine preventive, cultural, mechanical, biological and chemical methods. Components of integrated control of barnyardgrass in the USA are (a) preplant tillage to kill all weed growth at the time of seeding rice, (b) seeding rice in such a way as to obtain stands of fast-growing plants that compete with and shade the barnyardgrass plants, (c) the use of timely, sequential treatments of propanil, molinate or other effective herbicides to kill the weed plants during the early season before they compete with the crop, and (d) timely flooding or proper water management before and after herbicide treatments.

Integrated weed control systems for hard-to-kill weeds, such as red rice, are more complex than those for barnyardgrass. Present technology for control of red rice in the rice crop is very limited. Therefore, cropping-herbicide-cultivation systems are essential for control of red rice. A well planned program for control of red rice includes (a) use of weed-free crop seed, irrigation water free of red rice in all crops; (c) mechanical cultivation; (d) careful crop and water management; and (e) herbicides.

Summary and Conclusions

Integration of all available weed-control technologies is required to reduce losses in yield and quality of transplanted and direct-seeded rice caused by weed and to minimize potential environmental damage. An effective weed management system for rice integrates preventive practices, crop rotation, soil and water management practices, cutivation, management of the rice crop to minimize weed competition, natural enemies, and herbicides. Use of herbicides, combined with other control practices to reduce losses by weeds, permits rice farmers to abandon more traditional, inefficient agronomic practices. By the use of herbicides, rice may be grown under a minimum tillage regime that permits more efficient use of floodwater, fertilizer and energy. Integrated systems of weed management must be compatible with management of other pests, other practices to increase production, and with a quality environment.

The future of weed science in rice will bring continuing and new challenges. Rice culture will be confronted with shifts in weed flora as susceptible weeds are removed from the rice monoculture and tolerant weeds take their place. New herbicides and new uses for standard herbicides will be required to control new weed problems. New herbicides are being combined with standard herbicides, such as propanil and molinate, to increase the spectrum of weeds controlled. Presently, standard herbicides are being applied via the irrigation water, rather than by air, to reduce labor for and cost of application. Herbicides that control hard-to-control weeds, but may be toxic to the rice crop, can be made safe on rice by use of herbicide antidotes (crop protectants) applied as dressings to the planting seed. Plant breeders may be able to develop new rice cultivars that are more competitive with weeds and all new cultivars must be tolerant to extablished herbicide programs. New production technologies will require simultaneous use of new weed control technologies. For example, rice irrigated by sprinkler systems, may require only half as much water as rice grown with flood irrigation. Because the weed ecology in a sprinkler irriation culture is different from that of a flood culture, new weed control technology will be required to control these new species selectively. Natural enemies, such as weed-eating insects, fish or shrimp, or plant pathogens or nematodes, will increase in importance as components of weed control systems.

References

Aria, M. and S. Matsunaka. 1968. Japan Agric. Res. Qr. 1:5-9.

Chang, W.L. 1970. J. Taiwan Agric. Res. 19(4):18-24.

De Datta, S. K. and R. Barker. 1977. pp. 205-228. *In* Integrated Control of Weeds, Ed. J. D. Fryer and S. Matsunaka. Univ. Tokyo Press, Tokyo, Japan.

Floresca, E. T., F. B. Calora and S. R. Obien. 1978 9th Ann. Conf. Pest Control Counc. Philippines. Manila.

Matsunaka, S. 1974. pp. 443-456. *In* Rice in Asia. Univ. Tokyo Press, Tokyo, Japan.

Moody, K. and S. K. De Datta. 1977. 6th Asian-Pacific Weed Sci. Soc. Conf. Jakarta, Indonesia. 17 p.

Noda, K. 1977. pp 17-46. *In* Integrated Control of Weeds. Ed. J. D. Fryer and S. Matsunaka. University Tokyo Press. Tokyo, Japan.

Roa, L. L., D. C. Navarez and K. Moody. 1977. 8th Ann. Conf. Pest Control Counc. Philippines, Bacolod City.

Sahu, B. N. and P. Das. 1969. Indian J. Agron. 14:200-204.

Smith, R. J. Jr., W. T. Flinchum, and D. E. Seaman. 1977. U.S. Dep. Agric. Handb. 497. U.S. Gov. Printing Office, Washington, D.C. 78 pp.

30.4 Development of Rice Insect Pest Management Systems for the Tropics

E. A. Heinrichs, V. A. Dyck, R. C. Saxena and J. A. Litsinger

*The International Rice Research Institute, P. O. Box 933
Manila, Philippines*

There is an urgent need to develop and implement insect pest management systems for rice in the tropics where insects are a major constraint to rice production. Yield losses have been estimated to be about 30% on Asian farms [1]. Insect control often increases yields by 0.5 to 1.0 t/ha.

The yellow stem borer *Tryporyza incertulas* and the gall midge *Orseolia oryzae* are recurring pests causing extensive yield losses. Some insects are more cyclic in appearance and in recent years have increased in importance. The brown planthopper *Nilaparvata lugens,* once a minor pest, since 1973 has caused losses estimated to be $100 million in Indonesia alone [2].

Insecticide use is the major method of controlling rice insects. In some countries the frequent use of insecticides has led to the development of resistance, resurgence of target pests, and cases of human and fish poisoning.

In the last several years there have been significant advances in the development of insect pest management components for rice [3]. This paper discusses the status of exploratory pest management programs in tropical Asia and the components that have been developed for more effective and economic rice insect control.

Components for Rice Insect Pest Management

Decision Making. Prior to the 1970's pest control recommendations emphasized prophylactic treatments. But rice farmers traditionally apply control measures when pest damage becomes visible. The economic injury levels for several major pests have now been determined, and tentative economic or action thresholds are included in pest control recommendations issued to farmers. For example, thresholds are available for *Nilaparvata lugens* (1 per tiller), rice bug *Leptocorisa* spp. (4 per sq m), *Orseolia oryzae* (5% galls) and early damage by stem borers (10% damaged tillers). Only a few pests still require prophylactic control measures, such as the whorl maggot *Hydrellia* sp.

Surveillance programs are operated in many rice-growing regions, but often the scouting is done by teams of government technicians for purposes of large-scale, short-term forecasting of pest outbreaks; and not by farmers for immediate pest control purposes. As knowledge of pest behavior, population dynamics and population models increases, surveillance techniques are being developed for forecasting purposes, to detect selection for *N. lugens* biotypes and to improve the use of economic thresholds [4].

Practical sampling methods are available for the major pests and natural enemies, but improvements in efficiency are needed [5]. Even remote sensing for insect damage is being investigated. Light traps are widely used to monitor insect populations. Net and pheromone traps are being evaluated.

Varietal Resistance. Cultivation of insect-resistant rice cultivars has become the most valuable component of rice pest management in the tropics. Resistant cultivars are especially valuable to the Asian farmer whose plantings are small, and whose economic constraints and inadequate knowledge of pesticides limit their proper use.

Resistance to almost all major rice insect pests has been found in the large and diverse rice germplasm [6]. Within the last decade several thousand rice cultivars have been screened for insect resistance both at the International Rice Research Institute (IRRI) and in other Asian national rice improvement programs. Valuable sources of resistance have been found primarily among traditional *indica* rices. Earlier concerns that resistance is frequently associated with poor grain quality and low yield have been overcome largely by breeding resistance into improved rice cultivars. Such cultivars are utilized in insect pest management programs in various Asian countries although high levels of resistance are limited to only a few major insect pests.

Resistance to rice stem borers is moderate and it has been difficult to increase this level. The resistance level to rice planthoppers is high. The first *N. lugens* resistant cultivar, IR26, was released in 1973 and brought about a significant decline in the insect's abundance in the Philippines. An upsurge in population in 1976 was likely due to the selection of a biotype which was capable of surviving on IR26 and sibling cultivars. Since then cultivation of new rice cultivars such as IR36, IR38, and IR42, carrying a different source of resistance, has effectively curbed *N. lugens* incidence in the Philippines and Indonesia, and rendered substantial savings in the use of insecticides. Now resistant cultivars are being planted in over 90% of the irrigated rice area in the Philippines. Entomologists and breeders are evaluating horizontal resistance to cope with the biotype problem.

Orseolia oryzae, a serious pest in several Asian countries, is difficult to control with insecticides. High levels of resistance to this insect have been found and incorporated into cultivars with good yield potential. Cultivars resistant in one region, however, are not necessarily resistant in another, suggesting the occurrence of biotypes.

Because rice has many insect pests, efforts have also been made to combine resistance to more than one pest in one cultivar, e.g. IR36 is resistant to three major insect pests.

Chemical Control. Insecticides are a major component in the management system for rice insects. IRRI, along with national research programs, has been instrumental in identifying effective insecticides. Coded and commercial compounds are first evaluated in the laboratory. Those insecticides found effective in the laboratory are then field evaluated. The major target pests are the green leafhopper, *Hephotettix viresçens*; *N. lugens*; white plant hopper, *Sogatella furcifera*; striped stem borer, *Chilo suppressalis*; leaf folder *Cnaphalocrocis medinalis* and *Leptocorisa oratorius*.

Some insecticides have an ovicidal action against eggs of rice pests. Insecticides with fumigant activity have been identified. This is an important attribute in the control of *N. lugens* which feeds at the base of the plant and is difficult to contact directly with a sprayable formulation after the crop canopy is closed.

Recently it has become evident that potential rice insecticides also must be evaluated for their activity as resurgence-causing chemicals. Insecticides that significantly increase the reproductive rate of *N. lugens* have been identified [7]. Because of this characteristic, methyl parathion use has been banned on rice in the Philippines.

Pheromones have potential in the control of *Chilo suppressalis*. Recent field studies have shown their potential as mating disruptants.

Antifeedants are currently under evaluation to maintain pest populations at sub-economic injury levels without directly harming natural enemies. Oil from the seed of the neem tree, *Azadirachta indica*,

has antifeedant activity against *N. lugens* and *Cnaphalocrocis medinalis* in IRRI studies.

Steam distillate extracts of an insect-resistant rice cultivar have been shown to be biologically active against insect pests in IRRI laboratory studies. More studies are under way to determine their activity when applied as field sprays.

Based on light trap catches and field monitoring, the best time to spray *N. lugens* is about 2 weeks after peak trap catches, provided field sampling shows the economic threshold has been reached. One insecticide application directed at the third and fourth instars of the second generation is more effective than several calendar-based applications.

New application methods are being evaluated with the aim of increasing the cost-benefit ratio of insecticides. Hand-held, battery-operated, controlled droplet application (CDA) sprayers are being tested at IRRI with the objective of determining optimal droplet sizes which provide effective canopy penetration. CDA applicators utilize ultra-low volumes of spray solution, thus limiting the laborious task of carrying large amounts of water through the paddy field that is common in high-volume spraying.

Residual activity of both granules broadcast into the paddy water and foliar sprays is short in the tropics because their degradation is rapid. By placing a systemic insecticide into the soil, termed *root-zone application*, a residual activity can be greatly lengthened. One application of carbofuran in the root zone at transplanting provides insect control and yields equal or superior to those obtained with four broadcast or foliar spray applications. Root-zone application is compatible with a system of rice-fish culture, which is rapidly expanding in the Philippines.

Biological Control. The important parasites and predators of most rice insect pests have been identified. Recent evidence shows that several natural enemies play a critical role in pest control. Because little information is known on the exact number of natural enemies needed to control pests, measuring parasite or especially predator density is so far only occasionally useful for making practical control decisions.

Conservation of natural enemies is important. Commercial insecticides with some selectivity have been identified. Some parasites and predators appear not to be easily killed by insecticide application, but selective chemicals are needed for other predators. Insecticides having low toxicity to a predator of *N. lugens*, the spider *Lycosa pseudoannulata*, and yet effective against *N. lugens*, have been reported [8].

Direct use of biotic control agents has been limited to treatment with *Bacillus thuringiensis* and mass release of trichogrammatid parasites against lepidopterous pests, and herding ducks in rice fields for *N. lugens* control [9]. Except for one case in Hawaii years ago, the few introductions of exotic parasites of rice stem borers evidently have not had an impact on pest suppression.

Cultural Control. Cultural methods to control rice insects have been developed by generations of farmers through years of experience. Those practices fit both the agronomic and anthropological definition of the term "culture." They are agronomic in that they utilize resources and methods primarily intended for crop culture, not directly for insect control. They are anthropological in that they are patterns of behavior representing indigenous technical knowledge handed down from generation to generation. Cultural control methods can be classified as those which are effective within one farm only and those which suppress pest popluations over a farming community.

The use of a seedbed for transplanting a hectare of rice concentrates the crop within a few hundred square meters during the seedling stage to allow easy inspection of developing pest problems. If insecticide treatment is warranted, the area to be protected is minimal. The close spacing within a wet seedbed reduced oviposition by *Hydrellia* sp. Dry seedbeds control *Hydrellia* sp. and the caseworm *Nymphula depunctalis* which require standing water. The seedbed often serves as a trap crop for stem borers as egg masses are removed manually before transplanting. The trap crop concept is currently being tested for *N. lugens* control.

Flooded paddy fields eliminate many soil-borne insects—white grubs, root aphids, and termites—that occur in nonflooded rice cultures. Draining fields is practiced to control *Hydrellia* sp. and *N. lugens.*

The trend toward early maturing cultivars (100 vs 150 days) reduces the threat of late season pests such as the planthoppers and stem borers. Other escape mechanisms are ratooning and transplanting of older seedlings. Planting early or late, depending on the location, results in escape from highly seasonal pests such as *O. oryzae.*

In many areas farmers synchronize the planting time within large contiguous land units, and thus reduce the period of host availability. Rotation with nonrice crops breaks pest cycles, and tillage destroys volunteer rice and other weed hosts.

Selective removal of alternative insect weed hosts of leafhopper virus vectors, area wide at the beginning of the growing season, is practiced in China. Small-scale farmers can utilize labor and indigenous materials to replace cash-intensive methods of insect control. Hand removal of large insects, use of baits or traps, use of plant parts as repellants, or botanical or other indigenous insecticides are methods of choice in many areas.

Harvest methods affect stem borer larvae and pupae harbored in stems. Cutting the plant close to ground surface and burning the straw kills *Chilo suppressalis* and the pink stem borer *Sesamia inferens* larvae, which prefer the stem portions. Plowing under the stubble or prolonged flooding destroys *T. incertulas* larvae, which pupate at the soil surface.

Implementation

It is encouraging that today there is a good appreciation and understanding of the need to implement rice insect pest management programs in the tropics. The status of the development varies among countries, but there is now sufficient technology available to implement programs on at least a pilot basis. Pilot programs are already operating in India, Indonesia, Philippines, Taiwan, and Thailand.

The FAO-sponsored project for rice pest management in tropical Asia can provide the needed impetus to accelerate program implementation throughout Asia. It will require cooperation between national programs and the various international agencies. More important it will require strengthening of farmer organizations and closer interdisciplinary cooperation among biological and social scientists in national programs.

References Cited

1. Cramer, H. H. 1967. Plant protection and world food production. Plf.˙ Schutz-Naehr Bayer AG, Leverkusen, Germany. 524 p.
2. Dyck, V. A. 1979. The brown planthopper problem. p. 3-17 *In:* International Rice Research Institute, The brown planthopper: threat to rice production in Asia. 369 p.
3. Kiritani, K. 1979. Pest management in rice. Annu. Rev. Entomol. 24:279-312.
4. Oka, I. N. 1978. Quick method for identifying brown planthopper biotypes in the field. International Rice Research Newsletter 3(6):11-12.
5. Chen, C. N. 1977. Guidelines on rice crop surveillance methods for arthropod pests. p. 27-36 *In:* H. G. Zandstra (ed.) Proceedings of a Workshop on Crop Surveillance, East-West Food Institute, Honolulu.
6. Pathak, M. D., and R. C. Saxena. Breeding approaches in rice. *In:* Breeding plants for resistance to insects. Wiley Interscience, New York (in press).

7. Chelliah, S., and E. A. Heinrichs. 1979. Identification of insecticides that induce brown planthopper resurgence when applied as a foliar spray. International Rice Research Newsletter 4(1):15.

8. Dyck, V. A., and G. C. Orlido. 1977. Control of the brown planthopper (*Nilaparvata lugens*) by natural enemies and timely application of narrow-spectrum insecticides p. 58-72 *In:* The Rice Brown Planthopper, Food and Fertilizer Technology Center, Taipei. 258 p.

9. N. A. S. 1977. Insect control in the People's Republic of China. CSCPRC Report No. 2, National Academy of Sciences, Washington. 218 p.

30.5   Integrated Rice Insect Control in China

Shin-Foon Chiu

Department of Plant Protection
South-China Agricultural College
Kwangchow, China

Integrated plant protection for rice requires an interdisciplinary approach in research and in implementation. Rice pest management includes noxious insects, pathogens, and weeds. However, this paper reviews the work on integrated control of selected insect pests of rice in China.

There are more than 200 species of insects injurious to rice in China, but only about 20 species are common and of major significance. Insect pests are particularly abundant in central and southern China where two or three rice crops are grown each year. In southern China the most important species are the yellow stem borer *Tryporyza incertulas,* the striped stem borer *Chilo suppressalis,* the leaf folder *Cnaphalocrocis medinalis,* the brown planthopper *Nilaparvata lugens,* the green leafhopper *Nephotettix* spp., the gall midge *Orseolia oryzae,* the armyworm *Mythimna separata* and the rice thrips *Baliothrips biformis.* In recent years enough evidence has been gathered to show that the brown planthopper, the leaf folders, and armyworms are migratory over long distances.

The Government of the People's Republic of China has for many years laid down the principles of plant protection as "prevention combined with integrated control". The emphasis on prevention is implemented by a well organized pest forecasting system set up at four levels: the county, the commune, the production brigade and the production team. In every production team certain members are chosen as plant protectors. Thus a pest forecasting network—with the county's station as the center, the commune forecast group as the backbone, and plant protectors in the teams as the base—has been formed. The county forecasting station advises the commune and brigade when and where to carry out control operations. The data upon which forecasting is based are gathered by monitoring insect populations in the field. Records accumulated during the previous years are analyzed. Decisions on control, however, are based mainly on local conditions: the status of insect pests, the rice plant and the weather (particularly rainfall). Rice pest management in China has been developed based on cultural practices, insecticides, biological control, light traps, varietal resistance, and other control methods.

Cultural Control

Sound cultural practices are the foundation of integrated insect control. The cropping system determines the pest status and the number of pest species as well as their relative abundance. In the past several years in Kwangtung Province double-and triple-cropping of rice in the same locality has provided conditions favorable for the outbreaks of gall midge, yellow stem borer and brown planthopper. Since the introduction of the double and triple cropping, the peak of abundance of the borer has shifted to the fifth generation, and the population density has increased greatly. Under such conditions, a successful method of controlling borers is to maintain a short interval between the two rice crops, thus breaking the pest cycle. Another effective measure is to separate the double-cropped fields from those that are triple-cropped. This is particularly effective in controlling the outbreaks of the gall midge which migrates only a limited distance.

Flooding rice fields early in the spring (in regions around Kwangchow usually prior to March 6 each year) is very effective in controlling the yellow stem borer. There is a critical period for killing hibernating larvae by submergence. If the rice fields are flooded in winter or in autumn shortly after harvest when the larvae have already moved to the lower part of the stubble inside the stem and are ready to overwinter, they are very resistant to the effect of water and it takes more than 1 month to kill them. But in early spring when the overwintering larvae begin to pupate, they are very susceptible to flooding and it takes only 7 days to reach 95-100% mortality. Technique of flooding the rice field in early spring began to be used over an extended area in the Pearl River Delta about 20 years ago when the double-cropping system was adopted. This strategy has proven to be very successful, and is still practiced in certain localities. For example, at Tahsia Commune in the County Si-hui, Kwangtung where the integrated control progrm for rice pests has been carried out for a number of years, 80% of the rice fields are immersed in several centimeters of water prior to March 6 each year as one of the most effective methods of control. Even in fields where green manure is placed in the winter, flooding for 3 days kills the insects during the pupation period without damaging the crop.

Proper arrangement of the dates of sowing and transplanting is also an important cultural practice. Planting early maturing cultivars may avoid the damage caused by the second generation of the yellow stem borer.

Other cultural practices including weeding, trapping insects and draining rice fields are also adopted in certain districts. Meticulous removal of weeds particularly the weed *Leersia hexandra* which is the main overwintering host of the gall midge, is considered to be an effective control measure. Drainage suppresses the population of the brown planthopper, and at the same time is a good method for controlling sheath blight and bacterial blight.

Host Plant Resistance

Although some studies have been made in Chekiang and Hunan Provinces on cultivars of rice resistant to the brown planthopper, breeding for resistance to rice insect pests is in its infancy. But the importance of host resistance is recognized, and recently research projects have been started in Kwangtung on breeding cultivars of rice resistant to the gall midge and the brown planthopper.

Chemical Control

Insecticides constitute an important aspect of the integrated control program. Annually a large quantity of insecticides is used on rice. Insecticide treatments are applied on the basis of insect population data and the stage of growth and development of the rice plant, and are usually confined to a part of the total acreage, to individual fields, or even to infested spots within a field (e.g. an area occupying 5-8 hills of rice wherein one egg mass of the yellow stem borer had been deposited). This tactic not only reduces the overall use of insecticides and labor, but also minimizes the adverse effects on the ecosystem, thus promoting natural control of rice insect by natural enemies.

The correct timing of insecticides application is very important. For controlling the yellow stem borer and leaf folder, foliar sprays of contact insecticides should be applied during the peak time of egg hatching and not during the peak of adult emergence. Newly hatched larvae are very susceptible to insecticides; therefore in applying insecticides to the infested spots in the field, treatments must be done in the early stage of "deadhearts", i.e. when the shoots just begin to show signs of withering. In controlling the second generation of the yellow stem borer, the proper time for insecticide treatment is during the peak of egg hatching which coincides with the first appearance of the panicles. Spraying is usually more effective than dusting.

The selection of the proper insecticide and correct formulation is not only important to the efficacy of control but also in conservation of natural enermies. Broadcasting granules or soil particles impregnated with insecticide is better than conventional spraying or dusting. Mixed formulations have been recommended particularly for the control of insecticide-resistant strains of insects. For example, mixtures of kitazine and dimethoate or malathion are effective against planthoppers and leafhoppers resistant to BHC or DDT.

Finally the method of insecticide application holds great potential for planning an efficient integrated control program. Seedling dips with trichlofron, dimethoate, or chlordimeform are effective against the yellow stem borer and the gall midge. ULV ground sprays of formulation with 25% dimethoate, 25% octanol and 50% mixture of polyalkyl-benzene and polynaphthalene were effective against leaf folders and armyworms. A 25% trichlorfon ULV formulation has been used on a large scale in Szechuan Province. Results of studies in South China have shown that root zone application of systemic insecticides is a breakthrough in integrating the biological and chemical methods of control. For example, one root zone application of the new product "Pyrimioxythion" (O,O-diethyl-O)2-methoxy-r-methyl-pyrimidyl-6)-phosphorothionate) at a rate of 15 kg a.i./ha gave very effective control of the gall midge for the entire crop, and had no adverse effects on the natural enemies of rice insect pests.

Among the insecticides, a large tonnage of mixtures of BHC-parathion and BHC-methyl parathion has been used over an extended area of rice in China. In recent years chlordimeform, padan, carbofuran, acephate, parathion, malathion, phosphamidon, dimethoate, fenitrothion, cyanophas, carbaryl, and MIPC have been used in various provinces. Recently it was found that dimilin is very effective against armyworms. From the standpoint of insect control, the successful application of cordimeform has given spectacular improvements in the control of stem borer and leaf folders. Even under heavy infestations the application of chlordimeform 1-2 times at a rate of 0.75-1.13 kg a.i./ha gave very encouraging results. In Nan-Ao County, near Swatow, Kwangtung Province, chlordimeform has been applied successfully for 5 years in about 300 hectares of rice fields for the control of hibernating yellow stem borer. The average density of borers became as low as 78 borers per *mu* (1 mu = 0.066 ha or 0.1647 acre). Incidentally this will afford favorable conditions for the pioneering experiments on the sterile male release technique for the eradication of the yellow stem borer, a basic study of which has already been conducted in China. In short, our results during the last decade strongly suggest that a breakthrough in the chemical control of rice insect pests is feasible and compatible with the ecological approach.

Biological Control

Heavy emphasis has been placed on the biological control of rice insects in China. Basic studies on natural enemies particularly arthropod predators and parasites have been conducted in several provinces. Altogether 276 species of natural enemies (parasites and predators including spiders) have been described. This information provides guidelines for the introduction of natural enemies from foreign countries. Biological control in the integrated control of rice pests includes the following measures:

1) *Spiders*—There are 39 species of spiders belonging to 12 families in the rice fields of South China. Spiders play an important role in suppressing the populations of leafhoppers and planthoppers. A large scale experiment on the protection of spiders covering an area of about 186 hectares was conducted in 1977 in Xiang-Yin county, Hunan Province. In protected fields wherein no insecticide was applied the population density of spiders was 7-8 times higher than in fields with insecticide treatments. In the early crop the predominant species was *Erigonidium graminicolum* and if the equilibrium density (spiders: hoppers) was in the ratio of 1:4, no outbreaks of green leafhoppers and planthoppers occurred. In the late crop the predominant species were *Lycosa* spp. and *Pirata* spp., and if the equilibrium density was in the ratio of 1:8-9, again no outbreaks occurred. Under such conditions insecticide treatments were not necessary. But spiders are of little value for controlling rice borers and leaf folders and when infestations are heavy the application of insecticides becomes inevitable. The decrease in the number of insecticide treatments and the application of selective insecticides are important for the protection of spiders. Some of the carbamate compounds such as MIPC and BPMC and organophosphorus compounds such as pyridafenthion are toxic to hoppers and other insects but practically have no adverse effects on spiders.

2) *Trichogramma wasps*—Among some 12 species of *Trichogramma* reported from China, four (*T. dendrolimi, T. confusum, T. ostrineae,* and *T. japonicum*) are found in rice fields. Usually only *T. dendrolimi* and *T. confusum* are mass produced in special laboratories and in some communes. In South China they are released chiefly for control of rice leaf folders. Depending on the density of leaf folder eggs, *Trichogramma* are released at about 30,000 per *mu* at definite points several times during a single crop. Usually a parasitization of about 80% is obtained, which is adequate for control. Use of *Trichogramma* for the control of rice leaf folder is a major success in some places. But this method has not been popular because two problems remain to be solved: one is the supply of a sufficient amount of host insect eggs, and the other is that it is necessary to apply insecticides when an outbreak of borers, thrips or planthoppers occurs. *Trichogramma* wasps are very susceptible to chemicals, and any spraying or dusting kills the parasites. Under such conditions the integration of chemical control with the release of

parasites presents great difficulties, and often the rice farmers adopt the practice of spraying insecticides because they can expect higher yields.

3) *Pathogens*—Basic studies and field experiments have been done on entomogenous bacteria—mainly forms of *Bacillus thuringiensis* (B.t.)—for the control of leaf folders, skippers, and borers in rice fields. The bacteria are particularly effective against the skippers, and 90% control can be obtained with only two treatments. More than 50 strains have been identified at various institutes which differ considerably in their virulence against various insects. Among them, B. *thuringiensis* var. *galleriae* is the most common. In some counties in Kwangtung Province special production facilities for B.t. were set up. The effectiveness of B.t. is greatly increased if used in combination with a low-dosage (1/5-1/10 of the ordinary amount) application of a conventional insecticide such as dimethoate or trichlorfon. Generally it is important to have a highly virulent form of B.t., and it is necessary to have proper conditions of temperature, moisture and time of fermentation, making the product rather expensive for general use. For this and other reasons, during the last 2 years, B.t. has not been widely used in integrated rice insect control programs in Kwangtung Province.

4) *Ducks*—Ducks have been used by Chinese farmers for the control of rice insects for more than half a century, but the development of ducks as important biological control agents is recent. Results of extensive field trials indicated that ducks, when properly "herded" through paddies, control planthoppers and, to some extent, leaf folders and armyworms. Ducks are used at an average of 2-3 per *mu*. Due to the use of ducks and the protection of spiders and frogs in the rice fields, the application of insecticides has been greatly reduced in areas where planthoppers are the main potential pests, greatly improving environmental quality. Ducks however also destroy some beneficial insects, and they are of no use in controlling borers, gall midge and thrips. Besides, in some regions such as the county Hua, Kwangtung Province, where to control sheath blight the rice fields for the most part of the growing season are kept only moist, conditions for duck "herding" are difficult.

The tropical and subtropical climate in South China is advantageous to the utilization of natural enemies. The use of biological agents in integrated control shows promise. But just like other control measures it has its limitations. At present, yellow stem borers and thrips cannot be effectively controlled by the release of parasites and predators, so the sophisticated use of insecticides is necessary.

Also in subtropical ricefields usually more than one species of potential insect pest occurs at the same time and generations are more or less overlapping, making the utilization of biological methods more difficult. Our experiences in South China strongly suggest that with the rational use of pesticides and the protection of natural enemies over a wide area, the implementaiton of biological control in an integrated probram will be much more feasible.

Mechanical Control

Blacklight traps have been used widely as a control technique in some regions. While suppressing significantly the populations of some insects, for others such as stem borers the light traps only serve as as supplementary factor. Because of the large numbers of beneficial insects being trapped and the high cost of the traps the use of light traps has been greatly reduced during the last few years.

Examples of Successful Integrated Program

Integrated control of rice insects has been carried out extensively in nearly all the rice-growing regions of China. Regarding the management of the key pests, the most successful cases are the yellow stem borer, the brown planthopper and the gall midge.

An example of the integrated control program of the gall midge is as follows:

1) Cultural practices—Weeding, especially of *Leersia hexandra*, to eradicate overwintering larvae, proper timing of sowing and transplanting, separation of areas for the two-rice cropping and the three-rice cropping systems, and the cultivation of rice cultivars having comparatively high tolerance to the insect's attack.

2) Chemical control—Spraying or dusting is done during the peak of adult emergence during the tillering stage of rice and the most effective insecticides are "pyrimioxythion", carbofuran and pyridafenthion. Other treatments are seedling dip with 0.1% trichlorfon and root-zone application of systemic insecticide either in the seedbed or in the field.

3) The use of light and pheromone traps for monitoring and for direct control is a recent development. Results of field experiments showed that gall midges are very sensitive to light of the wave lengths around 400 μm (blue color), and that males are strongly attracted by females or dichloromethane extracts of females.

Summary

The integrated control of rice insects in China is based on cultural practices, chemical control and the use of biological control agents. Cultural practices are the backbone, and the judicious use of insecticides when infestation is above the economic threshold is a very important technique.

Investigations on the breeding of resistant rice cultivars have been done only in recent years. This in fact is a missing link in the integrated control of rice insects in China. As to biological control, the conservation of indigenous natural enemies, particularly through the rational use of insecticides, is of basic importance. This may be much more effective than the mass rearing and liberation of parasites and predators. Any unilateral approach, either chemical or biological, does not contribute to the development of an efficient program of integrated control. Chemical control must be integrated with biological control, and some effective measures have been developed.

Since the foundation of integrated control is a detailed knowledge of the crop ecosystem, great emphasis should be put on studies of the stability and instability of the rice agroecosystem. The population dynamics of the key pests may be studied with the help of empirical models of predator (parasite)—pest interacting systems. The model should be developed on the basis of data obtained mainly under natural conditions. The paddy agroecosystem, a man-modified environment, is an integrated water-dependent system which includes man, rice, crops other than rice, weeds, animals and microorganisms. Since rice grows along water routes, the wildlife damage from pesticides that drift downstream cannot be overlooked. Therefore intensive investigations on the rational use of pesticides and field and environmental toxicology with an ecological approach, should be made. Finally it must be pointed out that the efficiency of any integrated control program can be fully realized only when it is used over an extended area. Thus integrated programs call for a new type of large-scale cooperative grower organizations for which the socialist system of the People's Republic of China, particularly the unique organization of communes, possesses distinguished superiority. We believe that the development of integrated control of rice insects in China will be greatly accelerated during the course of the "Four Modernization" program.

31.1   Role of Certification Programs in Integrated Plant Protection

Orville T. Page and James E. Bryan

International Potato Center
Arpatado 5969, Lima 100, Peru

Historical

One of the many aspects of integrated plant protection are those of a regulatory nature involving the containment, suppression and eradication of diseases and pests. This is particularly true for the potato (*Solanum* spp.) which may be affected by any of more than 300 pests and diseases (Monro, J. 1978. Seed potato improvement in Canada. Can. Plant. Dis. Sur. 58 (2):26-28).

The need for potato seed tuber certification was first recognized in Europe in the early 1900s and was begun through the efforts of Dr. Otto Appel in Germany (Appel, O. 1934. Vitality and vitality determination in potatoes. Phytopathology 24:482-494). The certification of potato seed is the oldest and most extensive plant certification program and has provided the stimulus for certification schemes for other crops (Orton, W. A. 1914. Improvement of potato seed stocks through official inspection and certification. Proc. 1st Ann. Meeting Nat. Potato Assoc. Am., Philadelphia 37-43). Through the influence of Appel, Dr. W. A. Orton of the United States Department of Agriculture presented a plan for a potato seed certification program to the First Annual Meeting of the Potato Association of America in 1914. From these early beginning potato certification programs have been established in many countries in recognition of the need to provide healthy potato seed tuber stocks.

Terminology

The objectives of seed certification has been succinctly stated by Eastman (Eastman, P.J. 1974. Present seed production technology in the U.S. and Canada *in* Report of the Planning Conference on Seed Production Technology, International Potato Center, pages 32-41): "The overall goal of certification of potato seed tubers is to produce a product uniform as to variety and free enough of seed-borne diseases and pests so that the commercial crop produced will not be damaged as to yield or quality". This general aim of seed certification has been embodied in various legal and para-legal regulations which detail certification standards and their implementation. Terms such as: basic; foundation; elite; premier; certified; virus tested; virus-tested cutting stocks; elite I, II, III; S, SE, E, A, B, C; A- clones, B- clones; and so on are in general use. Numerical disease limitations, usually expressed as percent maximum allowable tolerances, vary widely among countries and even states reflecting differences in recognized grade standards. Two widely used terms are "basic seed" and "certified seed". For example, according to rules of the UN-Economics Commission for Europe (Economic Commission for Europe. 1961. Revised European Standard recommended by the Working Party of Standardization of Perishable Foodstuffs of the Economic Commission for Europe. UN-ECE Standard No. 19/Rev. 2) and accepted as minimal European Common Market standards, basic seed tubers are produced in accordance with strict requirements and conservative selection with respect to variety and healthiness for the subsequent production of certified seed. Certified seed is the direct offspring from basic seed, from certified seed, or from an older generation than basic seed which at an official inspection meet the conditions of basic seed (Keller, E. R. 1974. Present seed production technology in Europe *in* Report of the Planning Conference on Seed Production Technology, International Potato Center, pages 42-49).

Parameters for Basic Seed

According to general Common Market Standards, basic material may not exceed 2.09% blackleg (*Erwinia carotovora* var *atroseptica* in official field inspections. In the direct offspring the percentage of atypical plants may not exceed 0.25%, the percentage of other varieties may not exceed 0.1% and the percentage of plants which show symptoms of serious or mild mosaic diseases must not exceed 4.0%.

The standards for basic seed in individual Common Market countries are usually more stringent than the above mentioned. For example, Great Britian has a maximum tolerance of 0.05% for atypical types, 0.25% for blackleg and zero tolerance for severe mosaic during the second or final field inspection. The Netherlands require 100% purity of type. If blackleg caused by *Erwinia* is present or rogued, crops cannot be accepted as basic seed (Keller, E. R. 1974. Present seed production technology in Europe *in* Report of the Planning Conference on Seed Production Technology, International Potato Center, pages 42-49).

Parameters for Certified Seed

The quality standards for certified seed derived from basic seed are generally less stringent than those for basic seed. For example, standards set by the UN-Economics Commission for Europe set limits for blackleg, atypical plant and virus infection for certified seed at twice the permissible level for basic material (Keller, E. R. 1974. Present seed production technology in Europe *in* Report of the Planning Conference on Seed Production Technology, International Potato Center, pages 42-49). Field inspection may involve a number of factors including determining the suitability of isolation of seed fields for certification, border row requirements, roguing during the growing season, ensuring insecticide application and top killing to minimize aphid and virus infection, timing of harvest to ensure firm skin, freedom from specific diseases, tolerance for blemishes and freedom from nematodes.

In developed countries seed certification may be controlled by a variety of agencies. For example, in Oregon seed certification is administered by the Dean of the School of Agriculture of Oregon State University. In Montana the Montana Potato Improvement Association certifies seed. In Britian and Canada the Department of Agriculture is responsible for certification while in Switzerland the Swiss Experiment Stations for Agriculture certifies seed. Thus a variety of agencies are responsible for the enforcement of certification regulations, conducting field and storage inspections, tagging, labeling and sometimes shipping point inspections.

Seed certification in developed countries has been reasonably effective as part of intergrated protection for potato production. The continuing need for disease control through certification is apparent when yield losses due to leaf roll and severe strains of PVY and PVA can be as high as 50% (Reestman, A.J. 1970. Importance of the degree of virus infection for the production of ware potatoes. Potato Res. 13:248-268). Since these and other viruses, as well as bacterial and fungal pathogens, pests, and physiological disorders are also prevelant in developing countries, what is the role of seed certification in integrated protection of potatoes in these countries?

Seed Certification in Developing Countries

The countries in this category generally lie between 35°N Lat. and 25°S Lat. Potato production per capita is low being about one-ninth world production (van Der Zaag, D. E. 1976. Potato production and utilization in the world. Potato Res. 19:37-72). Factors contributing to the low production of approximately 9 kg per person are many, including the lack of good quality seed. Europe has been the principal source of certified seed for developing countries since no substantial amounts of good quality seed is produced except some limited production in India and several South American countries.

To organize a basic seed program in developing countries is analogous to putting a roof on a house. It is the last part built. For most developing countries a basic seed scheme is also the last program that should be structured. The logical course, then, instead of setting up a basic seed program, would be a one-generation multiplication of imported certified seed. If this is successful, attempt a second multiplication and if it is successful a third multiplication might be attempted if the incidence of virus is not serious. Inherent in this concept is the renouncement of perfection as a first step in the practical implementation of a certification program.

Close liaison between growers and government personnel is essential for the effective operation of a seed potato improvement program. The integration of these human resources, essential to the integration of seed certification as a component of successful production methodology is, at best, nebulous in most developing countries. A progressive grower, unable to obtain a steady supply of costly imported seed and without the assistance of honest, well trained inspectors, might rely on his own negative selection program. This consists of removing all plants with disease symptoms or those not of desirable type. This system works best in controlling the more destructive virus diseases, but is generally not effective with the so called "latent" viruses. Alternately, in some situations, it is useful to stake healthy appearing plants and to save tubers from these plants for seed. That the potato has been cultivated as a principal food source for a thousand years in the Andes where the small farmer grows a mixture of clones suggests that there is reduced risk through a multi-line approach.

The Use of True Seed

True seed, produced in potato berries derived through sexual processes, offers an alternative approach to the usual propagation of potatoes by seed tubers. True seed is a natural filter for pests and most pathogens except for a viroid and two minor viruses. The use of true seed to propagate a potato crop is an alternative to certification as presently practiced in developed countries. In developing countries where food production is of paramount importance, cosmetic attributes of the potato are of little significance.

The International Potato Center in Peru has initiated a breeding and agronomic management program to develop practical methods to propagate potatoes by means of true seed. We are encouraged in this approach by first hand observations on the successful use of true seed in the Peoples Republic of China. While each plant arising from a true seed is genetically different, relative phenotypic homogeneity is relatively easy to obtain while preserving heterozygosity in the plant population. Whether or not such background heterozygosity is a useful hedge against plastic pathogens such as the late blight fungus remains unknown.

Changing to the use of true seed in developing countries were sufficient technical infrastructure is usualy lacking to supervise a certification program presents a challenge to the developed countries traditionally involved in exporting certified seed tubers. Instead of shipping two metric tons of tubers required to plant a single hectare only 250 g or half a pound of true seed, a five-fold plant equivalent, would ensure an adequate stand. At present, several European countries have initiated research with a long-term objective of developing monoploid lines for the ultimate production of controlled hybrids.

Finally, while much reliance is placed on the use of certified seed, chemical control and agronomic practices to produce a healthy crop in developed countries, relatively little emphasis is being placed on the development of potato cultivars resistant ro specific pests and diseases. Public demand for cosmetically perfect potatoes runs counter to the decreasing availability of inexpensive energy supplies.

31.2 Potential for Plant Resistance and Biological Control to Manage Insect Pests

Ward M. Tingey

Department of Entomology, Cornell University
Ithaca, New York, USA 14853

The potato, man's fourth most important food plant, is subject to losses by arthropod pests virtually everywhere in the world it is grown. In recent years, the drawbacks associated with dependence on chemical insecticides for pest control have become increasingly apparent, and alternate control strategies have gained greater attention and appeal for consumers, farmers, and legislators alike. This paper reviews the current situation and potential for management of potato-infesting arthropods by use of two non-chemical control strategies, varietal resistance and biological control.

Plant Resistance

Sources and Nature of Resistance

Tuber-bearing species of the genus *Solanum* have been systematically assessed for arthropod resistance over the past 40 years, particularly in North America and Europe. A major portion of this effort has been directed toward identification of resistant germplasm, with less emphasis placed on the nature and inheritance of resistance. The following discussion provides a summary of the current situation in breeding for resistance to extensively studied pests, i.e. green peach aphid, *Myzus persicae* (Sulzer); Colorado potato beetle, *Leptinotarsa decemlineata* (Say); and potato leafhopper, *Empoasca fabae* (Harris).

Green peach aphid. Clones and cultivars of Group Tuberosum (*Solanum tuberosum tuberosum, S. t. andigena*) vary in their response to aphid infestation, but levels of genetic resistance are low (1, 15, 50) and the expression of resistance fluctuates with stage of plant growth (49). The greatest genetic diversity for aphid resistance is in wild, tuber-bearing *Solanum* species. Radcliffe and co-workers (22-25) have screened a major portion of the U.S. holdings of this germplasm, and reported excellent levels of resistance in accessions of various species, including *S. brachistotrichum, S. bulbocastanum, S. cardiophyllum, S. michoacanum, S. sanctae-rosae, S. stenophyllidium,* and *S. stoloniferum.* Many of these species are difficult to exploit by traditional plant-breeding methods because of their genetic incompatibility with Gr. Tuberosum, but Sams *et al.* (29) have reported encouraging progress in incorporating aphid resistance of *S. stoloniferum* and *S. sanctae-rosae* in tetraploid clones. The plant factors conferring aphid resistance in these species are unknown. In contrast, aphid resistance of the wild species, *S. berthaultii, S. tarijense* and *S. polyadenium,* is conditioned by glandular pubescence. Trichome secretions of these species appear to confer resistance primarily by entrapment and immobilization of aphids (6). As a male parent, *S. berthaultii* can be directly crossed to Gr. Tuberosum, yielding some fertile tetraploid progeny with glandular trichomes and significant levels of aphid resistance (7).

Potato leafhopper. Clones and cultivars of Gr. Tuberosum vary dramatically in infestation and yield response to the potato leafhopper (32, 34, 35, 43, 50) but plant age and clonal differences in time of tuberization or foliage maturity tend to confound the expression of genetic resistance (30). The resistant cultivar 'Sequoia' is unique as the only named potato variety for which insect resistance is considered a major feature (5). Excellent levels of resistance to this pest are also found in accessions of wild species such as *S. berthaultii, S. chacoense, S. commersonii, S. demissum, S. hougasii, S. kurtzianum, S. pinnatisectum, S. polyadenium, S. polytrichon, S. spegazzini* (26, 44, 51, 52), and in Gr. Tuberosum stocks with *S. chacoense* and *S. demissum* ancestry (30, 32).

At least 2 different mechanisms are associated with potato leafhopper resistance in wild *Solanum* species. First, resistance of *S. berthaultii* and *S. polyadenium* is conferred, in part, by glandular

trichomes, secretions of which entrap adults and nymphs, or prevent feeding by occlusion of the labium (51). Secondly, resistance is correlated with levels of leaf glycoalkaloids. Foliage of highly resistant species such as *S. brachycarpum, S. chacoense, S. kurtzianum, S. medians,* and *S. polyadenium* contains 7- to 20-fold greater levels of these steroidal glycosides than that of susceptible Gr. Tuberosum cultivars (52). These compounds significantly limit survival and feeding activity (in vitro) of the potato leafhopper (3, 26).

Colorado potato beetle. Gr. Tuberosum cultivars are uniformly susceptible to *L. decemlineata,* although some clones differ slightly in their ability to tolerate and compensate for defoliation (36). The highest levels of resistance are present in accessions of wild species such as *S. polyadenium, S. chacoense,* and *S. demissum* (46). Resistance in these species to the Colorado potato beetle is associated largely with levels and types of foliar glycoalkaloids (20, 21, 37, 46). Individual glycoalkaloids differ in their activity against *L. decemlineata;* tomatine, chaconine, demissine, and the water-soluble leptines appear to be the most toxic or repellent of the *Solanum* glycosides (14, 42). Two other potential but largely unexplored defense mechanisms against the Colorado potato beetle are glandular trichomes (8) and proteinase inhibitors (10).

Inheritance of Resistance

Despite considerable progress in identifying sources of arthropod resistance in potatoes, relatively little is known about inheritance mechanisms. In part, this may be due to the scarcity of information on specific resistance mechanisms, and the complex genetics of the genus Solanum. Nevertheless, significant progress toward an understanding of heritability has been made in several cases. Resistance to the green peach aphid is partially dominant and relatively well-buffered against environmental influence in certain wild, tuber-bearing *Solanum* species and progenies of their interspecific crosses with Gr. Tuberosum (29). Heritability estimates of 50-70% were obtained for these populations. The heritability of resistance to the potato leafhopper, potato fleabeetle (*Epitrix cucumeris*), and tobacco fleabeetle (*Epitrix hirtipennis*) in a population of tetraploid clones with *S. chacoense* and *S. demissum* ancestry has also been demonstrated (30-32).

The inheritance of glandular trichomes and steroid glycoalkaloids has also been investigated. According to R. W. Gibson (personal communication), a single dominant gene controls the presence of Type **B** glandular trichomes (9) on foliage of *S. berthaultii* and *S. tarijense*. In addition, at least 1 set of recessive genes is necessary for full expression of this character in interspecific hybrids of these species with *S. phureja* and Gr. Tuberosum. As for glycoalkaloids, most studies of inheritance have focused on pooled glycoalkaloid content of tubers. Tuber glycoalkaloid content of Gr. Tuberosum is inherited in a polygenic manner (33). It has recently been proposed (17) that the presence of solanine and commersonine in tubers of *S. chacoense* is determined by alternative codominant alleles at 1 locus, whereas that for chaconine is controlled by a major gene that segregates independently of the solanine-commersonine locus. In addition, a recessive gene linked with the conmersonine allele is epistatic to the chaconine gene, causing production of β-chaconine rather than α-chaconine.

Considerations in Development and Implementation of Resistance

Other pests of regional and world-wide importance. The major effort to develop arthropod resistance in potato has been directed toward relatively few pests, e.g. *Empoasca fabae, Leptinotarsa decemlineata, Macrosiphum euphorbiae* (Thomas), and *Myzus persicae*. Although these are among the most important species limiting potato production in North America and Europe, more attention should be paid to the development of varietal resistance for other major pests of either regional or cosmopolitan importance, such as the potato tuberworm, *Phthorimaea operculella* (Zeller), the Andean weevil complex, and soil insects such as cutworms, wireworms, and white grubs.

Genetic diversity and germplasm resources. In view of the importance of the potato as food for man and animals, it is surprising that the development of varietal resistance to arthropods has lagged behind that for field and forage crops. In part, this may be due to the reluctance of potato breeders and entomologists to look beyond Gr. Tuberosum for sources of resistance. Far greater genetic diversity is available in wild, unadapted species than in the cultivated species of *Solanum,* but the use of wild, mostly diploid species poses a considerable challenge in breeding programs because of the complex genetics of the genus *Solanum.* Considerable progress has been made in recent years, however, in the development of methods for easing crossability barriers between normally incompatible species (11, 28), More knowledge of *Solanum* genetics and of the inheritance of specific resistance mechanisms may also help breeders more effectively incorporate

desirable traits in populations of interspecific hybrids while eliminating genes for undesirable characteristics.

Selection and exploitation of resistance mechanisms. Some highly effective arthropod resistance factors may confer undesirable plant characteristics. An excellent example is that of leaf glycoalkaloids associated with resistance to *Empoasca fabae* and *Leptinotarsa decemlineata.* In general, levels of these broadly-toxic steroidal glycosides in foliage and tubers are highly correlated. Although it may be possible to select clones with pest-active levels of these compounds in foliage and safe levels in tubers (38), current methodology for glycoalkaloid determination is too time-consuming and complex for routine adoption in most potato breeding programs (16). Until there are economical and streamlined analytical methods adaptable for use in screening large populations, breeding for glycoalkaloid-based resistance is not feasible. Glandular pubescence, on the other hand, appears to be relatively free of serious limitations as a practical arthropod resistance mechanism: Glandular trichomes are active against a wide range of small, soft-bodied pests (9); the glandular species, *S. berthaultii,* can be readily hybridized with *S. tuberosum,* and the former species appears to be relatively free of undesirably high levels of glycoalkaloids (26, 52); and pests are likely to counter-adapt relatively slowly to the physical entrapment aspect of glandular trichome resistance.

Finally, more effort should be directed toward analysis of specific defense mechanisms of resistant germplasm. It is remarkable, for example, that 30 years passed before the significance of glandular trichomes in resistance of *S. polyadenium* was fully appreciated (6, 44). Knowledge of specific defense mechanisms may help clarify the genetics of resistance, thus providing a more quantitative basis for selection and breeding.

Biological Control by Predators, Parasites, and Pathogens

The potential for biological control of potato-infesting arthropods by use of predators, parasites, and pathogens has been best studied for potato tuberworm, Colorado potato beetle, and aphids.

Predators and Parasites

Several species of parasitic Hymenoptera, including *Orgilus lepidus* Muesebeck, have been introduced into Australia, India, South Africa, New Zealand, and the western U.S. for control of the potato tuberworm (19). Success varies from region to region, but in California combined parasitism from introduced and native species may exceed 50% (18). The potential for control of the Colorado potato beetle by the predacious pentatomid, *Perillus bioculatus* (F.), and the parasitic tachinid, *Doryphorophaga doryphoryae* (Riley), has been studied in the U.S. and Europe (4, 12, 13, 41, 47, 48). A parasitic nematode has also been evaluated for control of *L. decemlineata* (55). The potato-infesting aphids, *Myzus persicae, Macrosiphum eurphorbiae, Aulacorthum solani* (Kaltenbach), and *Aphis nasturtii* Kaltenbach, are subject to mortality by a wide variety of parasitic Hymenoptera and predators such as coccinellids (39). In general, however, the natural level of control is insufficient to retard initial infestation of the crop and subsequent loss in yield and quality by aphid-borne viruses (40).

Fungal and Viral Pathogens

By far the greatest interest in recent years for biological control of potato insect pests has been centered on the use of entomogeneous microorganisms. A major advantage of this strategy is that viral and fungal pathogens can be mass cultured and applied to the target pest as necessary. In most instances, formulations of entomopathogenic microorganisms can be applied to the crop with the same equipment and methods used for chemical insecticides. The entomogeneous fungi *Beauveria bassiana* and *Entomophthora thaxteriana* have shown considerable potential for control of the Colorado potato beetle and green peach aphid, respectively, when used in this manner (45). Granulosis viruses pathogenic to the potato tuberworm have also been studied in Australia and South Africa (2, 27).

Summary

We must be realistic and recognize that neither varietal resistance nor biological control is likely to provide the near complete pest suppression achieved with chemical insecticides. However, we should not be discouraged from development and implementation of either of these strategies, because one of their major advantages is compatibility with other control measures. The integrated use of resistant cultivars with biological control, chemical insecticides, or other control measures may provide additive or synergistic control sufficient to reduce pest population levels below economic injury thresholds.

References

1. Bradley, R. H. E. & R. Y. Ganong. 1951. Can. J. Zool. 29:329-38.
2. Broodryk, S. W. & L. M. Pretorius. 1974. J. Ent. Soc. Sth. Afr. 37:125-8.
3. Dahlman, D. L. & E. T. Hibbs. 1967. Ann. Ent. Soc. Amer. 60:732-40.
4. Franz, J. M. & A. Szmidt. 1960. Entomophaga 5(2):87-100.
5. Gardner, M. E., R. Schmidt & F. J. Stevenson. 1945. Am. Potato J. 22:97-103.
6. Gibson, R. W. 1971. Ann. Appl. Biol. 68:113-9.
7. Gibson, R. W. 1974. Potato Res. 17:152-4.
8. Gibson, R. W. 1976. Ann. Appl. Biol. 82:147-50.
9. Gibson, R. W. and R. H. Turner. 1977. PANS 22(3):272-7.
10. Green, T. R. & C. A. Ryan. 1972. Science 175:776-8.
11. Hermsen, J. G. Th. & M. S. Ramanna. 1973. Euphytica 22:457-66.
12. Jasic, J. 1964. Praha, 1964, pp. 47-49.
13. Jermy, T. 1962. Folia Entomol. Hung. (N.S.) 15(2):17-23.
14. Kuhn, R. & J.|Löw. 1955. p. 122-32. In: Origins of Resistance to Toxic Agents. M. G. Sevag, R. D. Reid, O. E. Reynolds (eds.). Academic Press.
15. Landis, B. J., D. M. Powell & L. Fox. 1972. Am. Potato J. 49:63-9.
16. Mackenzie, J. D. & P. Gregory. 1979. Am. Potato J. 56:27-33.
17. McCollum, G. D. & S. L. Sinden. 1979. Am. Potato J. 56:95-113.
19. Oatman, E. R., G. R. Platner & P. D. Greary. 1969. Ann. Ent. Soc. Am. 62:1407-14.
20. Pierzchalski, T. & E. Werner. 1958. Hodowla. Rosl. Aklim. Nasienn. 2:157-80.
21. Prokoshev, S. M., E. I. Petrochenko & V. Z. Baranova. 1952. Doklady Akad Nauk (USSR) 82:955-8.
22. Radcliffe, E. B. & F. I. Lauer. 1970. J. Econ. Ent. 63:110-4.
23. Radcliffe, E. B. & F. I. Lauer. 1971. Univ. Minn. Agric. Exp. Stn. Techn. Bull. 286. 22 pp.
24. Radcliffe, E. B. & F. I. Lauer, 1971. J. Econ. Ent. 64:1260-6.
25. Radcliffe, E. B., F. I. Lauer & R. E. Stucker. 1974. Environ. Ent. 3:1022-6.
26. Raman, K. V., W. M. Tingey & P. Gregory. 1979. J. Econ. Ent. 72:337-41.
27. Reed, E. M. 1971. Bull. Ent. Res. 61:207-22.
28. Sams, D. W., P. D. Ascher & F. I. Lauer. 1977. Am. Potato J. 54:355-64.
29. Sams, D. W., F. I. Lauer & E. B. Radcliffe, 1976. Am. Potato J. 53:23-9.
30. Sanford, L. L. & J. P. Sleesman. 1970. Am. Potato J. 47:19-34.
31. Sanford, L. L. & R. D. Peel. 1970. Am. Potato J. 47:169-75.
32. Sanford, L. O., O. V. Carlson & E. T. Hibbs. 1972. Am. Potato J. 49:98-108.
33. Sanford, L. L. & S. L. Sinden. 1972. Am. Potato J. 49:209-17.
34. Sanford, L. L. & J. P. Sleesman. 1974. Am. Potato J. 51:44-50.
35. Sanford, L. L. & R. E. Webb. 1977. Am. Potato J. 54:581-86.
36. Schalk, J. M., R. L. Plaisted & L. L. Sanford. 1975. Am. Potato J. 52:175-7.

37. Schreiber, K. 1957. Zuchter 27:289-90.
38. Schwarze, P. 1962. Zuchter 32:155-60.
39. Shands, W. A., G. W. Simpson & R. H. Storch. 1972. J. Econ. Ent. 65:799-809.
40. Shands, W. A., G. W. Simpson, C. F. W. Muesebeck & H. E. Wave. 1965. Maine Agric. Exp. Stn. Tech. Bull. T19.
41. Sikura, A. I. 1963. Zashch. Rast. 8(2):45-6.
42. Sinden, S. L., J. M. Schalk & A. K. Stoner. 1978. J. Amer. Sco. Hort. 103:596-600.
43. Sleesman, J. P. & J. Bushnell. 1937. Am. Potato J. 14:242-5.
44. Sleesman, J. P. 1940. Am. Potato J. 17:9-12.
45. Soper, R. S., F. R. Holbrook, I. Majchrowicz & C. C. Gordon. 1975. Maine Life Sci. Agric. Exp. Stn. Tech. Bull. 76.

46. Swiniarski, E., E. Werner & Z. Mierzwa. 1958. Hodowla Rosl Aklim Nasienn 2:623-31.
47. Szmidt, A. & W. Wegorek. 1967. Entomophaga 12(4):403-8.
48. Tamaki, G. & B. A. Butt. 1978. USDA Tech. Bull. 1581, 11 p.
49. Taylor, C. E. 1962. Eur. Potato J. 5:204-19.
50. Tingey, W. M. & R. L. Plaisted. 1976. J. Econ. Ent. 69:673-6.
51. Tingey, W. M. & R. W. Gibson. 1978. J. Econ. Ent. 71:856-8.
52. Tingey, W. M., J. D. Mackenzie & P. Gregory. 1978. Am. Potato J. 55:577-85.
53. Tremblay, E. & N. Zouliamis. 1968. Boll. Lab. Ent. Agric. Filippo Silvestri 26:99-122.
54. Wegorek, W. & A. Szmidt. 1962. Biul. Inst. Ochr. Rosl. 17:7-27.
55. Welch, H. E. & L. J. Briand. 1961. Can. Ent. 93:759-63.

31.3   Establishing Action Thresholds for Insect Pests

E. B. Radcliffe, R. E. Cancelado, and W. S. Cranshaw,

University of Minnesota, St. Paul, MN 55108

Insects cause direct yield losses estimated at 6.5% of the world's potential production of potato, *Solanum tuberosum* L., and indirectly cause losses of 24.3% through disease transmission (*1*). Current estimates, cited by Thurston (*2*), place U. S. potato yield and quality losses at 10-12% compared to 22% prior to the advent of modern synthetic insecticides.

U.S. potato growers use ca. 1300 metric tons of insecticide per year (2.5 kg AI/treated ha) (*3*). The benefit/cost return to potato growers from their insecticide investments is conservatively 5:1. Insecticides provide such reliable and economical means of preventing losses from insects that, to the present, pest management approaches have not been widely adopted by potato growers. Indeed, much insecticide use on potatoes is viewed as crop insurance and is applied with little regard to pest pressure.

The major motivation for grower interest in pest management approaches is the increasing frequency of insecticide resistance in pest species. Potato flea beetle (PFB), *Epitrix cucumeris* (Harris), was perhaps the first crop pest to develop resistance to DDT (*4*). DDT resistance was observed in Colorado potato beetle (CP), *Leptinotarsa decemlineata* (Say), in 1952 [5]. Introduction of organophosphate and carbamate insecticides subsequently reduced CPB to minor economic status for many years. With development of resistance to many of these insecticides, CPB again ranks as the major insect pest

of potatoes in some parts of the eastern U.S. (*6*). But, it is the aphids, especially green peach aphid (GPA), *Myzus persicae* (Sulzer), in which insecticide resistance presents the greatest threat due to their tremendous reproductive potential and role as virus vectors. GPA developed resistance to DDT and certain organophosphate insecticides in the mid-1950's, to carbaryl in the arly 1960's, and is now resistant to most insecticides registered on potatoes (*7, 8*).

A major obstacle to application of insect pest management approaches for potato is the difficulty of satisfactorily documenting economic thresholds. This problem stems from the complex interactions occurring between various insect pests, diseases, cultural practices, cultivars, weather, and difficulties in obtaining precision in yield estimates. For some species, there exists an essentially linear relationship between insect numbers and yield reductions, but the relationship between vector numbers and crop injury is confounded by a varying incidence of vector infectivity.

Minnesota-North Dakota Potato Insect Problem

In Minn-ND potatoes (79,700 ha in 1978) potato leafhopper (PL), *Empoasca fabae* (Harris), is the insect species most often responsible for yield losses. Nymphal PL numbers are closely correlated with percent hopperburn and even low densities can cause substantial yield reductions (9, 10). PL does not have natural enemies of consequence so insecticides are routinely applied for its control. Insecticide resistance has not been observed, probably because PL migrates northward each season from the Gulf States where it is of no economic importance (11).

GPS is almost invariably present in Minn-ND potato fields. Usually, GPA is a low density pest and does not cause direct yield losses except when outbreaks are induced by insecticidal treatments selectively detrimental to its natural enemies. However, it is not uncommon for potato growers to inadvertently induce GPA outbreaks as a result of their insecticidal spray programs.

In seed production (27,100 ha in 1978) GPA control is regarded as essential since the perpetuation and spread of virus diseases is governed by the abundance and activity of aphid vectors (12). Most growers apply systemic insecticides at planting and spray on a regular schedule. Although insecticides protect potatoes against colonization, transmission may occur before migrant vectors are killed (13).

Defoliators such as PFB, CPB, various noctuids, and grasshoppers usually are not abundant in Minn-ND potatoes. However, growers often apply insecticides unnecessarily because defoliators are conspicuous and their injury evident. Soil insects are of localized importance, but are not discussed here.

Methods

Based on surveys of pest densities typical of well managed Minn-ND potato fields (14), we selected a range of action thresholds for evaluation, i.e., pest densities at which to apply insecticidal sprays. We evaluated these action thresholds in terms of the number of spray applications required to maintain levels below prescribed densities and in terms of comparative pest numbers over sampling dates (15, 16). In the defoliation studies, leaves were removed with a razor blade to simulate insect injury (17).

For GPA experiments the plots were seeded with aphids from an organophosphate resistant culture. Weekly applications of azinphosmethyl were applied throughout the season to enhance GPA populations by elimination of natural enemies and competing phytophagous species. Plots were monitored at 10-day intervals and presumed effective aphicides, pirimicarb in 1977 and fenvalerate in 1978, were applied whenever counts of apterae exceeded prescribed thresholds.

PL populations were enhanced by interplanting 6 m wide strips of alfalfa, *Medicago sativa* L., the preferred host, between every 8-12 rows of potatoes.

Many PL adults produced in the alfalfa moved into the potatoes to oviposit and predators harbored in the alfalfa also spilled into the potatoes virtually eliminating aphids. Methamidophos was applied whenever PL densities exceeded prescribed action thresholds.

Results and Discussion

GPA Experiments

Treatments in which the action threshold for GPA was 30 apterae/105 leaves (3 replications of 35 leaves) required up to 3 less spray applications than did treatments with action thresholds of 1-10 (Tables 1-3). Through July, treatments with action thresholds of 1-10 apterae/105 leaves showed no consistent differences in GPA numbers and were not appreciably different from treatments in which the threshold was 30. Application of phorate at planting provided initial protection against GPA colonization, but did not reduce the number of sprays required to maintain GPA densities below prescribed action thresholds. In fact, mid to late-season GPA numbers tended to be higher where the systemic had been used.

Table 1. Green Peach Aphid, 1977.[a] Insecticidal Sprays Required With Various Action Thresholds.

Apterae/ 105 Leaves	Spray Dates[b]						
	6/30	7/10	7/20	7/30	8/9	8/19	8/29
1	S	S	S	S	S	S	S
3		S	S		S	S	S
10			S	S	S	S	S
30				S	S	S	S
100					S	S	S
Control	145	12	11	21	26	71	259

[a] Phorate, 3.36 kg AI/ha, applied in-furrow at planting, 5/18, and azinphosmethyl, 0.56 kg AI/ha, sprays applied weekly except to control.
[b] Application of pirimicarb, 0.28 kg AI/ha, sprays indicated by S.

Table 2. Green Peach Aphid, 1977.[a] Insecticidal Sprays Required With Various Action Thresholds.

Apterae/ 105 Leaves	Spray Dates[b]						
	6/30	7/10	7/20	7/30	8/9	8/19	8/29
1		S		S	S	S	S
3	S		S	S	S	S	S
10		S			S	S	S
30					S	S	S
100					S	S	S
Control	150	25	12	33	59	93	105

[a] Planted 5/18, azinphosmethyl, 0.56 kg AI/ha, sprays applied weekly except to control.
[b] Application of pirimicarb, 0.28 kg AI/ha, sprays indicated by S.

Table 3. Green Peach Aphid, 1978.[a] Insecticidal Sprays Required With Various Action Thresholds.

Apterae/105 Leaves	Spray Dates[b]					
	7/21	7/31	8/10	8/20	8/30	9/9
1	S		S	S	M	S
3	S	S	S	S	M	S
10		S		S	M	
30			S	S	M	
100				S		M
Phorate	3	79	922	4288	15209	14388
No Phorate	8	43	379	4023	17043	18546
No Insecticide[c]	40	116	49	11	1	2

[a] Phorate, 3.36 kg AI/ha, applied in-furrow to sub-units of each treatment at planting, 5/22-23, and azinphosmethyl, 0.56 kg AI/ha, applied weekly. Controls (phorate and no phorate) also sprayed weekly with azinphosmethyl.

[b] Application of fenvalerate, 0.224 kg AI/ha, sprays indicated by S, application of methamidophos, 0.84 kg AI/ha. sprays indicated by M.

[c] Control of separate, but adjacent experiment.

Applications of both pirimicarb and fenvalerate proved ineffective from mid-season on in suppressing GPA densities below prescribed action thresholds. Although severe GPA pressure was produced as the result of the weekly azinphosmethyl sprays, mid to late-season control difficulties appeared to result primarily from increased insecticide tolerance in the target population. In another experiment, initial late-season applications of pirimicarb and fenvalerate proved highly effective against a GPA population that had been sprayed repeatedly with azinphosmethyl. Methamidophos is now the only insecticide presently registered on potatoes that has never failed to give us a high level of GPA control regardless of previous treatment history.

Growers producing potatoes for fresh market and processing should be able to achieve adequate GPA control, using foliar sprays without the need for systemics at planting. We concluded that foliar sprays should be applied only when GPA densities approach 30 apterae/105 leaves. The stricter standards of vector control required by the seed industry justifies use of systemics at planting with aldicarb being the material of choice. Foliar insecticides should be applied when GPA flight activity is noted because migrants are more likely to be viruliferous. If only because zero aphids is an unobtainable goal, it follows that there must be some tolerable density for GPA apterae even in seed potatoes. Certainly the risk of inducing an outbreak outweighs potential benefits to be derived from applying insecticidal sprays when GPA densities are already at acceptable levels. Our data suggests an action threshold of ca. 10 apterae/105 leaves is appropriate for seed fields.

PL Experiments

For PL, 1-5 spray applications were required depending upon the prescribed action threshold (Tables 4-5). Treatments with threshold of 3-10 nymphs/105 leaves did not suffer appreciably in total PL numbers over sampling dates from that where the action threshold was 1, but the former required 2-3 fewer spray applications. When phorate was applied at the standard rate, sprays were required only where the action threshold was 1. The .25× rate of phorate gave complete control of 1st generation PL nymphs.

Nymphs are much more injurious than adults (9) so with usual Minn-ND PL population levels spray applications can be delayed until nymphal eclosion. Well timed insecticidal spray applications eliminated the bulk of a given PL generaion. Sprays applied too early miss some nymphs and if applied too late may not prevent injury from occurring. Timing is less critical with more persistent insecticides; but these chemicals may be more disruptive of GPA natural enemies. Our data suggest that ca. 10 nymphs/105 leaves is the appropriate threshold for the application of insecticidal sprays against PL; not more than 2 sprays per season should be required for Minn-ND potatoes.

Table 4. Potato Leafhopper, 1977. Insecticidal Sprays Required With Various Action Thresholds.

Nymphs/105 Leaves	Spray Dates[a]							
	6/30	7/10	7/20	7/30	8/9	8/19	8/29	9/8
1	S	S			S	S		S
3		S			S			
10		S			S			
30		S				S		
100					S			
Phorate[b]	0	0	0	0	0	1	2	0
Control	5	43	17	64	136	178	196	70

[a] Application of methamidophos, 0.84 kg AI/ha, sprays indicated by S.

[b] Phorate, 3.36 kg AI/ha, applied in-furrow, at planting 5/18.

Table 5. Potato Leafhopper, 1978. Insecticidal Sprays Required With Various Action Thresholds.

Nymphs/105 Leaves	Spray Dates[a]						
	7/11	7/21	7/31	8/10	8/20	8/30	9/9
3	S			S			
10	S			S			
30				S			
100					S		
Phorate (1X)[b]	0	0	0	1	4	0	0
Phorate (.25X)[b]	1	0	0	16	58	48	91
Control	25	15	15	67	105	94	59

[a] Application of methamidophos, 0.84 kg AI/ha, sprays indicated by S.

[b] Phorate, 3.36 (1X) kg AI/ha or 0.84 (.25X) kg AI/ha, applied in-furrow at planting, 5/23.

Defoliation Experiments

In the simulated insect defoliation studies potatoes generally recovered from lower rates of injury (10-33% defoliation) inflicted early in the season and sometimes showed yield increases. Heavier defoliation (67%) resulted in slight yield reductions. Injury at mid-season resulted in substantial yield reductions. Defoliation after mid-season had gradually less effect on yield. Removal of top leaves was most detrimental to yields and removal of middle leaves was least detrimental. Removal of lower leaves reduced yields only during mid-season. Comparison of equal amounts of defoliation applied uniformly (equally to each plant) or nonuniformly (more severe injury, but with plants skipped) indicated that nondefoliated plants do not fully compensate for injured neigbors. Late maturing cultivars recover more completely than do early maturing cultivars.

Our studies suggest potatoes can sustain much more defoliation before yield losses occur than is generally tolerated by growers. However, the feeding habits of defoliators must be considered in estimating the effects their feeding will have on crop yield. Insects that are nonuniformly distributed or which concentrate their attack on upper portions of the plant would have greatest impact. Paper no. 10,912, Sci. J. Series, MAES, IAFHE, U. Minn.

Literature Cited

1. Cramer, H. H., *Pflanzenschutz-Nachrichten*, (1967), *20*, 1-524.
2. Thurston, H. D., "Pest Control Strategies", Academic Press, (1978), 117-36.
3. Andrilenas, P. A., *Econ. Res. Serv., U.S. Dept. Agric., Agric. Econ. Rept.*, (1974), *252*, 1-56.
4. Palm, C. E., *Adv. in Chem. Ser.*, (1950), *1*, 218-22.
5. Cutkomp, L. K.; A. G. Peterson; and P. E. Hunter, *J. Econ. Entomol.*, (1958), *51*, 828-31.
6. Tingey, W. M., (personal communication).
7. Chapman, R. K., *Misc. Publ. Entomol. Soc. Am.*, (1960), *2*, 27-39.
8. Radcliffe, E. B., *Proc. N. Cent. Branch Entomol. Soc. Am.*, (1973), *28*,100-2.
9. Sanford, L. L.; and R. E. Webb, *Am. Potato J.*, (1977) *54*, 581-6.
10. Peterson, A. G.; and A. A. Granovsky, *J. Econ. Entomol.* (1950), *43*, 484-7.
11. Peterson, A. G.; J. D. Bates; and R. S. Saini, *J. Minn. Academy Sci.*, (1969), *35*, 98-102.
12. Bishop, G. W.; and J. W. Guthrie, *Am. Potato J.* (1964), *41*, 28-34.
13. Bacon, O. G.; V. E. Burton; and J. A. Wyman, *Calif. Agric.*, (1978), *32*, 26-7.
14. Robinson, D. P., Univ Minn M. S. Thesis, (1978).
15. Cancelado, R. E.; and E. B. Radcliffe, *J. Econ. Entomol.* (1979a), 72, 566-9.
16. Cancelado, R. E.; and E. B. Radcliffe, *J. Econ. Entomol.* (1979b), 72, 606-9.
17. Cranshaw, W. S.; and E. B. Radcliffe, *J. Econ. Entomol.* (1980), 73 (in press).

31.4 Management of Potato Nematodes

F. G. W. Jones

Rothamsted Experimental Station, Harpenden, England

Nematodes injurious to the potato crop are in Table 1. By far the most important are the cyst-nematodes. Much is known about their population dynamics and, for main crop cultivars, mathematical models now make it possible to simulate the effects on population densities of almost any desired rotation or practice. Management of species other than round cyst-nematodes remains empirical because adequate information on their population dynamics is lacking.

Control methods are reviewed by Winslow and Willis (1972) and background information is in Evans and Trudgill (1978). Because nematode populations are relatively immobile, damaging

Table 1. Nematodes Injurious to Potatoes

Endoparasitic	
Cyst	*Globodera rostochiensis, G. pallida*
Root-knot[1]	*Meloidogyne* spp
False root-knot[2]	*Nacobbus aberrans*
Tuber-rot	*Ditylenchus destructor*
Root-lesion	*Pratylenchus*

Ectoparasitic	
Stubby-root[3]	*Trichodorus, Paratrichodorus*
Needle nematodes	*Longidorus*

[1] Mainly in warm, sub-tropical soils
[2] Potential pest
[3] Vectors of tobacco rattle virus causing spraing disease of tubers

numbers are generated by preceding crops. Dispersal is slow (Jones, 1979a) and, except during the colonisation of new territory, immigration and emigration can be discounted. Attempts to prevent ingress to areas thought to be uninfested by quarantine regulations or to limit spread after the first foci are found, are liable to ultimate failure (Jones, 1969), and any success in delaying the spread of *G. rostochiensis* from Long Island, to mainland USA probably owes more to the fact that Long Island is an Island and metropolitan New York a desert than to the drastic measures taken.

General Methods

Methods of management other than the use of nematicides are applicalbe to all nematode pests. They rely greatly on the avoidance of fields known to harbour heavy infestations, on choosing immune crops and on rotations that do not encourage the development of dense populations. The suppression of alternative weed hosts is practised. Unfortunately all these methods are only palliative. Circumstances force growers in developing countries to plant potatoes in unsuitable fields and long rotations are economically unacceptable in developed countries. Furthermore, the wide host ranges of root-knot nematodes, root-lesion nematodes, stubby-root nematodes and other ectoparasitic species render the choice of alternative crops difficult. Even the tuber-rot nematode, which has few hosts in potato fields, finds these among stoloniferous weeds (e.g. *Mentha* spp) which are difficult to kill with herbicides.

Most countries try to produce clean seed tubers, more to control tuber-borne pathogens, especially viruses, than to control nematodes, by raising seed in cool, hilly, windswept terrain away from the ware crop. Here, long rotations are followed that discourage tuber-borne nematodes (root-knot, false root-knot, and root-lesion nematodes) or soil-borne ones likely to be transmitted on or in association with seed tubers.

Many attempts have been made to disinfect tubers by hot-water treatment, dry heat, fumigating with SO_2 from suphur candles (Fenwick, 1942) washing under high pressure water jets (Mabbott, 1960) and, recently, by steeping in hypochlorite solution to destroy eelworm cysts (Wood & Foot, 1977). Although tuber disinfestation is a useful quarantine measure against cyst-nematodes, it is hardly worthwhile in countries where they are well established because the numbers transmitted with seed are trivial compared with existing field populations.

It is sometimes claimed that large amounts of farmyard manure, green manures, crop residues and organic amendments seem to be effective against root-knot nematodes. Whether this is because they change the soil environment, liberate toxins during breakdown, or encourage enemies, is unknown. Enemies of nematodes certainly exist (Jones, 1974; Jones, 1979b) but those of pest species are patently ineffective. Fungal parasites of round-cyst nematodes seem fewer than those of species with lemon-shaped cysts (*Heterodera* spp) (Tribe, 1979). An Oomycetous fungus, *Nematophilus gynophila* (Kerry & Crump, 1979), common in British cereal fields, controls *Heterodera avenae* except in dry soils. This and associated fungi are suppressed by formalin but seem unaffected by fumigant, oxime and carbamate and organophosphorus nematicides.

Although there is evidence that some cultivars resist or tolerate nematode attack better than others, only against round-cyst nematodes have really effective resistant ones been bred and exploited. Attacks can sometimes be alleviated by varying planting dates, providing additional fertilizer, especially nitrogen (Jones, 1977), and by irrigation (Tables 2 and 3).

Nematicides

Control of nematodes in soil is difficult because of its bulk, 2500 tonnes/hectare to plough depth. The

Table 2. Effects of Additional N Fertilizer on Land Infested with *G. rostochiensis*, Total Tubers, tonnes/ha

	Farmyard Manure t/ha				
	0	12.5	25	50	Means
Infested site					
No nitrogen	15.7	17.5	20.8	29.6	20.9 .
Nitrogen[1]	31.1	35.2	37.4	*38.7*	35.6
Three uninfested sites					
No nitrogen	28.0	30.7	33.9	35.6	32.1
Nitrogen	31.2	34.0	36.0	*37.5*	34.7

[1]113 kg/ha as ammonium sulphate

Note: extra P and K had little effect. 12.5 t of farmyard manure contains 90 kg N, 15 kg P and 115 kg K.

Table 3. Effects of 'Blue Print' Farming, Irrigation and Nematicide on Land Infested with *G. rostochiensis* in a Drought Year, Ware Tubers, tonnes/ha

	Nematicide		No nematicide	
	Blue print	Standard	Blue print	Standard
Farming Irrigation				
Full	71.2	54.0	37.1	20.0
None	34.8	23.6	7.4	8.2

Nematicide = aldicarb, 'blue print' farming = additional N, more seed.

481

earliest nematicides, used mainly against potato cyst-nematodes, were fumigants such as DD (dichloro-propane-dichloropropene mixtures). They must be applied to trash free soil in seed bed condition and time allowed for the toxic vapour to disperse before planting. In the cold, wet soils of N.W. Europe this means autumn injection which is agronomically difficult. Fumigants work best in sandy soils and at best kill no more than 70% of cyst-nematodes, more of other species. The powder nematicide dazomet is somewhat more effective especially on soils containing sand and peat but has the same disadvantages. Oxime carbamates and a few organophosphates are nematostatic; they kill indirectly by preventing host finding and feeding. Granules can be distributed at planting and are more effective than fumigants in a greater range of soils (Whitehead, 1975, 1978). The cost of applying nematicides is considerable, a factor greatly limiting their use. Toxic hazards and lack of suitable machines are additional limitations in developing countries.

Management of Cyst-Nematode Populations

Historical

Until 1970 it was thought that there was only one species of potato cyst-nematode. Now it is known that there are two (G. rostochiensis and G. pallida) that do not interbreed freely and that both exist as a series of pathotypes distinguished by their behaviour on potato plants with genes for resistance (Kort, Ross, Rumpenhorst & Stone, 1977). Traditionally, in Europe, control was and still is by crop rotation supplemented by pre-planting estimates of population density and reinforced by legislation especially in the Netherlands. In Great Britain from the 1940s, and somewhat later in the Netherlands and other mainland countries, soil sampling systems and field rating were devised and applied on a large scale: these have been outstandingly successful.

The advent of resistant varieties and of oxime carbamate and organophosphate nematicides have changed the picture. In the starch-potato areas of the Netherlands a scheme integrating rotations, fumigation and resistant cultivars has been operating for some years (Nollen & Mulder, 1969). In areas of Great Britian where G. rostochiensis is dominant, the cultivar Maris Piper bearing gene H_1 from S. tuberosum spp. andigena is widely grown.

Most fields infested with G. rostochiensis are challenged by G. pallida which is likely to replace the former after five or six crops of Maris Piper. Cultivars resistant to a greater range of pathotypes of both species are being bred and the first of these have been released by Dutch breeders.

Within both species the gene-for-gene relationship exists between pathotypes and resistant potato plants as in rust fungi and rust-resistant cereals (Jones, 1975, Parrott, 1979). This information facilitates modelling of population genetics.

Monoculture of Early Maturing Cultivars

These make appreciable root growth below 10°C whereas cyst-nematodes do not hatch freely until 10°C is exceeded (Chitwood, 1951). Moreover, these cultivars are harvested within 100 to 120 days which curtails the egg production of females and limits post harvest population densities (van den Brande and D'Herde, 1964). In part of Britian, Belgium, Cyprus (Jones, 1976) and elsewhere, early planting and early lifting permits continuous or nearly continuous cropping, but not without yield penalty which can be demonstrated by applying nematicides. In Cyprus monoculture is assisted by extra nitrogen and irrigation.

Mathematical Models Simulating Cyst-Nematode Population Dynamics and Their Use

These were introduced by Jones, Parrott and Ross (1967), revised by Jones, Kempton and Perry (1978), Jones and Perry (1978) and extended by Perry (1979) making it possible to include items listed in Fig. 1. Although the models have limitations, they simulate well the effects of main-crop potatoes grown in fields with well-established infestations. The following symbols are used: S susceptible, R resistant potatoes, O other crops and fallow, N fumigant, X oxime carbamate nematicide, G16 and G4 ground-keepers, $16/m^2$ and $4/m^2$ respectively. Symbols may be placed in any desired order except that X must precede and G14 or G4 must follow the symbols S or R. The computer reads the sequence and uses the appro-

Table 4. Simulated Effects of Ground-Keepers

Crop Sequence/Population Density					
–	0.37	–	0.37	–	0.37
S	0.52	S	0.52	S	0.52
S	0.50	O	0.34	SG16	0.34
S	0.50	O	0.22	SG4	0.22
S	0.50	O	0.14	SG4	0.15

Crop Sequence/Population Density/ Frequency of New Pathotype								
–	0.37	0.01	–	0.37		–	0.37	
R	0.13	0.01	R	0.13		R	0.13	
R	0.05	0.03	O	0.08	0.01	RG16	0.08	0.01
R	0.03	0.08	O	0.05		RG 4	0.05	
R	0.03	0.24	O	0.03		RG 4	0.03	

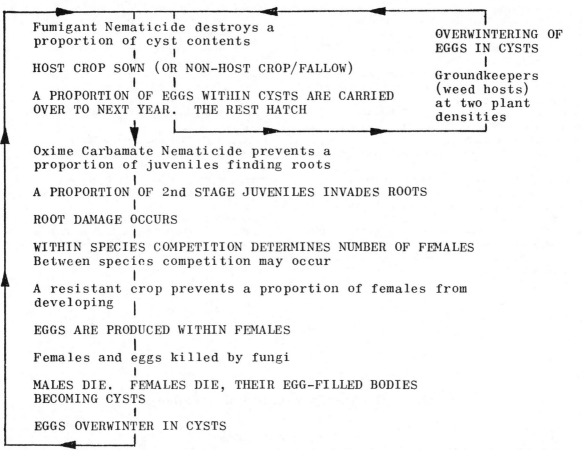

Figure 1. Events in the life-cycle of cyst-nematodes that can be simulated. Those in upper case are in the main model. Those in lower case are optional.

priate equation to calculate P_f, the population density after harvest from P_i, that at planting. For convenience, densities are expressed as fractions of E_f, the equilibrium density that would be established by undamaged root systems. Model parameters and other details are in Jones and Perry (1978). Approximate values for British and some NW European conditions are known and the effect of a range of values can be explored. The models promise to be useful tools in planning management strategies, especially as rotational experiments may last longer than the research span of individual workers and

Table 6. Simulated Effect of Rotations

Non-Host Crops	Stable Population Density[1] Beginning of Cycle	End of Cycle	Non-Host Crops	Stable Population Density[1] Beginning of Cycle	End of Cycle
0 (continuous)	0.20	0.20	4	0.06	0.36
			5	0.04	0.36
1	0.15	0.24	6	0.02	0.32
2	0.12	0.28	7	0.01	0.20
3	0.09	0.33	8	Extinction	

[1]Developed after one or more cycles depending on starting density.

require a commitment of men and resources difficult to sustain. Space permits only three examples of abbreviated output from the models (Tables 4, 5, 6).

References

Brande, J. van den and D'Herde, J. (1964) *Nematologica,* 10, 25-28.

Chitwood, B. G. (1951) *U.S. Dept. Agric., Washington,* Circular 875, pp. 48.

Table 5. Simulation of Selection of Mutant or Immigrant Gene (N or n) in Nematode Population

Crop Sequence	Years to reach 50% (90%) frequency, Initial Frequency 1 in 100,000 Dominant (NN + Nn)	Recessive (nn)
RRR	7 (322)	50 (70)
RORO	13 (574)	135 (161)
ROOROO	19 (829)	160 (295)
ROOOROOO	29 (1145)	237 (521)
ROOOOROOOO	36 (1656)	246 (986)

Evans, K. and Trudgill, D. L. (1978) In *The potato crop*. Ed. P. M. Harris, London. Chapman & Hall, pp. 440-469.

Fenwick, D. W. (1942) *J. Helminth.*, 20, 41-50.

Jones, F. G. W. (1969) In *Nematodes of tropical crops*. Ed. J. E. Peachey. Farnham Royal, Commonw. Agric. Bur., pp. 67-80.

Jones, F. G. W. (1974) In *Biology in pest and disease control*. Eds. D. Price Jones and M. E. Solomon. London, Blackwell Scientific Publications, pp. 247-268.

Jones, F. G. W. (1975) *Nematologica*, 20, 437-444.

Jones, F. G. W. (1976) *Report on the golden nematode in Cyprus*. Rome, F. A. O., pp. 21.

Jones, F. G. W. (1977) *Proc. 12th Int. Potash Inst. Izmir, Turkey, 1976*, pp. 233-258.

Jones, F. G. W. (1979a) In *Comparative epidemiology*. Ed. J. Palti. Amsterdam, PUDOC Press (in the press).

Jones, F. G. W. (1979) *Rep. Rothamsted expl. Stn. for 1978* Pt 1, 176-177

Jones, F. G. W., Kempton, R. A. and Perry, J. N. (1978) *Nematropica*, 8, 36-56.

Jones, F. G. W., Parrott, D. M. and Ross, G. J. G. (1967) *Ann. appl. Biol.*, 60, 151-171.

Jones, F. G. W. and Perry, J. N. (1978) *J. appl. Ecol.*, 15, 349-371.

Kerry, B. R. and Crump, D. H. (1979) *Trans. Brit. mycol. Soc.*, (in the press).

Kort, J., Ross, H., Rumpenhorst, H. J. and Stone, A. R. (1977) *Nematologica*, 23, 333-d359.

Mabbott, T. W. (1960) *Eur. Potato J.*, 3, 236-244.

Nollen, H. M. and Mulder, A. (1969) *5th Brit. Insectic. Fungic. Conf. Brighton*, 3, 671-674.

Parrott, D. M. (1979) *Rep. Rothamsted expl. Stn. for 1978*, Pt 1, p. 182.

Perry, J. N. (1979) *J. appl. Ecol.*, 15, 781-787.

Tribe, H. T. (1979) *Ann. appl. Biol.*, 92, 61-72.

Whitehead, A. G. W. (1975) *ARC Research Rev.*, 1, 17-23.

Whitehead, A. G. W. (1978) In *Plant nematology*. Ed. J. F. Southey. MAFF ADAS Pub. GD1, London, HMSO, pp. 283-296.

Winslow, R. D. and Willis, R. J. (1972) In *Economic nematology* Ed. J. M. Webster. London, Academic Press, pp. 17-48

Wood, F. H. and Foot, M. A. (1977) *NZ J. expl. Agric.*, 5, 315-319.

31.5   The Management of Foliar Pathogens

W. E. Fry and R. I. Bruck

Plant Pathology, Cornell University, Ithaca, NY

As with any group of pathogens the amount of disease, *D* induced by those which affect potato foliage is determined by several factors. These factors including those pertaining to the pathogen, *p*, the host, *h*, and the environment, *e*, during time (Eq. 1). Pathogen factors include the size, aggressiveness, and compatibility of the pathogen population. Host

$$D_t = \sum_{i=0}^{t} f(h_i, p_i, e_i), \qquad (1)$$

factors include tissue susceptibility, and influences on the microenvironment. Environmental factors include the weather, and aspects of disease management technology. The time during which these factors interact influences the amount of disease. If other things are equal, a greater amount of disease will occur when these factors interact for a longer time than when they interact for a shorter time. For many different diseases the individual influences of some factors have been identified, but they have not been integrated to permit an accurate prediction of epidemic development.

The necessary intensity of disease management depends on the interaction of all factors. For example, use of high quality potato seed tubers is helpful to reduce the initial inoculum of several virus-induced diseases, and when these are used there is less need to apply other techniques to suppress virus-induced disease. Especially susceptible cultivars require a greater intensity of disease management than do more resistant cultivars. The environment experienced by the pathogen is influenced by cultural techniques such as irrigation, and planting density; when these favor disease development greater intensity of disease management is required. The length of interaction influences the needed intensity of disease management and longer seasoned cultivars may require more attention than shorter seasoned cultivars.

One of the goals in disease management is to employ technology efficiently to suppress disease to tolerable levels: the level of management activity should be adjusted so that greater activity is employed when a larger amount of disease is probable, and less activity is employed when a small amount of disease is probable. General guidelines evolving from an understanding of one or few of the influencing factors have been developed. For example, disease forecasts to identify the need for fungicide sprays for suppression of potato late blight have been developed all over the world [9, 11]. Most of these forecasts use weather data as criteria for determining whether or not disease is likely to develop. Although weather favorable to *Phytophthora infestans* is mandatory for development of potato late blight [2, 3], weather is not the only influential factor. When the pathogen population is present in a very low amount, disease forecasts based on weather have indicated premature fungicide application; in other cases when the initial inoculum was quite high, disease forecasts predicted disease which was less severe than that which actually occurred. Thus for enhanced efficiency disease forecasts should employ the amount of initial inoculum. Most disease forecasts have also neglected the effect of variation among cultivars in their susceptibility to the pathogen. However, late blight is more likely to occur sooner in susceptible varieties than it is in the more resistant varieties.

An important goal of our research has been to identify probable intensities of disease development so that the appropriate intensity of disease management can be applied. We know that for maximum efficiency we need to quantify host susceptibility, pathogen amount and probable aggressiveness. Potato growers in northeastern USA and in southeastern Canada have observed differences among cultivars and some growers have exploited these differences without well defined guidelines. For example, growers on Prince Edward Island, Canada, have, on the average, applied fewer sprays to the more resistant cultivar Sebago than they have applied to the more susceptible cultivar Green Mountain [8]. The next section will describe attempts to integrate several factors (host resistance, fungicide and climate) to suppress late blight development and then to describe potential integrations to suppress leaf roll and early blight.

Late Blight

Because climate, host resistance, and fungicides influence pathogen development in different ways it was important to identify a technique for comparing the influences quantitatively. The unifying measurement was the effect of each of these factors on epidemic development. Host resistance is manifested in several ways, including effects on infection efficiency, lesion development and sporulation [7,

12, 13]. However, the integral of these effects is to influence epidemic development [4]. Climate not only influences infection efficiency, lesion development, and sporulation, but also influences dispersal and survival of the pathogen at any of several stages of development. Again, however, the integral effect is to influence epidemic development. Most fungicides applied to potato foliage primarily prevent penetration of the tissue. However, because of the low probability of covering all plant tissue with adequate amounts of fungicide, the integral effect of applying fungicides again is to suppress epidemic development [4]. With a given amount of initial inoculum and for a specified time, these factors interact to influence the amount of disease development.

We have used apparent infection rates [14] and areas under the disease progress curve to quantify the effect of host resistance and fungicide. Both plant resistance (race non-specific) and regular application of fungicides reduce apparent infection rates [4, 5] and the areas under disease progress curves [6]. Where epidemics occur under different climatic conditions at different times of the year, the area under the disease progress curve has been a more useful measurement with which to make comparisons [6]. Using these two techniques we have now defined the relative resistance of the most popular Northeast potato cultivars and have identified the manner in which this resistance can complement fungicide applications. For example, by observing the effect of resistance and fungicide on the rate of epidemic development, we identified that the greater degree of resistance in the cultivar Sebago relative to the more susceptible cultivar Russet Rural was equivalent to about 0.67 kg of mancozeb/acre applied weekly to Russet Rural [4].

Host resistance can be integrated with fungicide either by adjusting the dosage of fungicide per application or by adjusting (within limits) the interval between regular application [6]. For example, when fungicide dosage was altered by adjusting the amount per application the effects on disease development were similar to those observed when fungicide dosage was altered by adjusting the interval between applications. In both cases the greater resistance in Sebago relative to that in Hudson was equivalent to the effect of about 0.06 kg mancozeb/ha/day applied to Hudson.

Fungicide is used most efficiently when the amount applied is a function both of climate and host resistance. In one comparison when a reduced dosage of fungicide was applied to a resistant variety as dictated by a forecasting scheme, disease suppression was equivalent to a standard dosage of fungicide applied weekly to a more susceptible cultivar [5]. However, nearly twice as much fungicide was used when the standard amount of fungicide was applied weekly to the more susceptible cultivar.

It is likely that disease forecasting techniques can be modified to take advantage of host resistance. For example, Blitecast [9] recommends that no fungicide application is necessary, a fungicide application every 7 days is necessary, or an application every 5 days is necessary as the climate becomes increasingly favorable for late blight [9]. If these various application schedules are sufficient for adequate disease suppression on a susceptible cultivar then a resistant cultivar should require less fungicide for equivalent disease suppression. For example, if application on a 5-day schedule is sufficient to suppress disease in a susceptible cultivar such as Hudson then a 6-day schedule should be sufficient to suppress disease in a resistant cultivar such as Sebago. Similarly, a 7-day schedule on Hudson should be equivalent to a 9-day schedule on Sebago. Thus, the frequency of fungicide applications could be determined from a consideration of the effect of weather on disease development in a susceptible cultivar and then modified on the basis of plant susceptibility.

Early Blight

Integration of plant resistance and fungicide application should be possible with early blight as well as with late blight. Some potato cultivars are more susceptible to early blight than are other potato cultivars, e.g. Hudson appeared to be somewhat more susceptible than did Russet Burbank [1]. Additionally, plant vigor affects host susceptibility. In one experiment fungicides in combination with high nitrogen fertilization (175 lbs/acre) were more effective in suppressing early blight than when combined with low nitrogen (100 lbs/acre). Thus, for management of potato early blight, genetic resistance, as influenced by plant nutrition, could influence the amount of fungicide needed for adequate suppression of disease.

Leaf Rool

Finally, it seems possible to integrate the effects of resistance to potato leaf roll with insecticides to suppress populations of the aphid vector. Again, potato cultivars differ in their susceptibility to potato leaf roll (Bruck, et al., unpublished results; Cetas, unpublished results). Whether susceptibility is to the virus or to the aphid has not been definitely determined. However, by observation and experi-

ment a cultivar such as Chippewa appears more susceptible to leaf roll than does a cultivar such as Abnaki. Epidemics of potato leaf roll in plots of Abnaki developed more slowly than did epidemics in plots of Chippewa (Fig. 1). A systemic insecticide (Aldicarb) suppressed leaf roll epidemic development in each cultivar but the epidemic in Abnaki was less intense than the epidemic in Chippewa. The insecticide reduced aphid populations on each cultivar, but aphid populations on Abnaki were less than those on Chippewa. From these results it seems that insecticide would be used most efficiently if adjusted to complement host resistance to leaf roll. More susceptible cultivar should receive larger amounts of insecticide than should more resistant cultivars.

These three examples illustrate how host resistance can be integrated with pesticides for more efficient disease management. This approach is only a first step in using all of the important factors which influence disease development to enhance disease management efficiency. The influence of several factors yet remains to be quantified. For example, quantifications of the effects of canopy density and soil texture on microclimate have not been done. These and other influences need to be quantified and integrated to determine the most efficient and effective uses of disease management technology. In a very well defined system one ought to be able to estimate the influence of pathogen population and aggressiveness, host resistance, climate, and disease

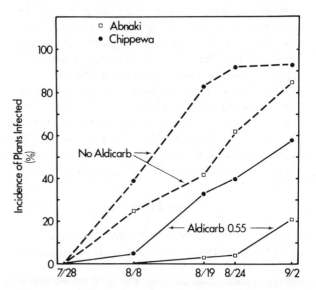

Figure 1. Epidemics of leaf roll induced by potato leaf roll virus (and transmitted by the green peach aphid) as influenced by host resistance and insecticide. Aldicarb was applied as a granule to the soil at planting. The cultivar Chippewa is more susceptible to leaf roll than is Abnaki.

management technology during time, to develop the most efficient approach to suppress disease to a tolerable level.

As pest management specialists, we are at the threshold of supressing pest populations efficiently. Our next step is to integrate the diverse information in a systems approach to avoid harmful interactions and to exploit beneficial ones.

Literature Cited

1. Abdel-Rahman, M. 1977. Evaluation of the interaction between cultivars of potatoes and different fungicides and programs of application on early blight control. Plant Dis. Reptr. 61:473-476.
2. Beaumont, A. 1947. The dependence on the weather of the dates of outbreak of potato epidemics. *Brit. Mycol. Soc. Trans.* 31:45-53.
3. de Weille, G. A. 1963. Laboratory results regarding potato blight and their significance in the epidemiology of blight. *Eur. Potato J.* 6:121-130.
4. Fry, W. E. 1975. Integrated effects of polygenic resistance and a protective fungicide on development of potato late blight. *Phytopathology* 65:908-911.
5. Fry, W. E. 1977. Integrated control of potato late blight—Effects of polygenic resistance and techniques of timing fungicide applications. *Phytopathology* 67:415-420.
6. Fry, W. E. 1978. Quantification of general resistance of potato cultivars and fungicide effects for integrated control of potato late blight. *Phytopathology* 68:1650-1655.
7. Guzman-N., J. 1964. Nature of partial resistance of certain clones of three Solanum species to Phytophthora infestans. *Phytopathology* 54:1398-1404.
8. James, W. C., C. S. Shih, and L. C. Callbeck. 1973. Survey of fungicide spraying practice for potato late blight in Prince Edward Island. 1972. *Can. Plant Dis. Surv.* 53:161-166.
9. Krause, R. A., and L. B. Massie. 1975. Predictive systems: Modern approaches to disease control. *Annu. Rev. Phytopathol.* 13:31-47.
10. Krause, R. A., L. B. Massie, and R. A. Hyre. 1975. Blitecast: a computerized forecast of potato late blight. Plant Dis. Reptr. 59:95-98.
11. Shrum, R. D. 1978. Forecasting of epidemics. In *Plant Disease. An Advanced Treatise.* Vol. 2. How disease develops in populations. (J. G. Horsfall and E. B. Cowling, Eds.). Academic Press, NY. pp. 223-238.
12. Thurston, H. D. 1971. Relationship of general resistance: Late blight of potato. *Phytopathology* 61:620-626.
13. Umaerus, V., and D. Lihnell. 1976. A laboratory method for measuring the degree of attack by Phytophthora infestans. *Potato Res.* 19:91-107.
14. Van der Plank, J. E. 1963. *Plant Diseases: Epidemics and Control.* Academic Press, NY. 349 pp.

31.6   The Potential for Integrated Management of Potato Pests in Developing Countries

J. Franco

International Potato Center, Apartado 5969, Lima, Peru

The mid-1970 population of developing countries (DCs) was 1.7 billion people. This population grew by an average of 30% during the 1970s. More than 30 developing countries record population growth between 3.0 and 3.5% a year. Latin America, the Mid East and Africa each have 10 of these countries; the remainder are in Southeast Asia.

Food production in the DCs has stayed ahead of population growth in the 1970s. During the last quarter century food production in DCs has been enough to feed growing populations and to add slightly to the average diet although wide variations in performance are evident for the more than 100 countries involved. Projecting on the basis of recent growth rates, the average hectare of crop-land in the DCs that now provides food for three persons must provide food for one additional person by 1985 and for two more persons by 2000. These averages conceal dramatic differences between countries including some reflecting nearly insolvable food outlook situations.

Developing countries often suffer more from losses by pests, diseases and environmental stress because adequate technology is not available due to lack of communications, poor advisory service to farmers or high costs. Thus joint efforts of national research programs and international institutions must have high priority for successful integrated pest management. Here we are defining integrated pest management as a systems approach that encompasses not only immediate objectives of preventing pest losses but also considers long term objectives with regard to economics, society, environment and health protection.

Joint efforts of this type have already been initiated by national and international institutions to spearhead agricultural research, training and extension activities and to improve agricultural productivity as well as ecological, social and economic well being. The International Potato Center (CIP) headquarters in Lima, Peru, as a single crop center, focusing on the tuber-bearing species of *Solanum*, is one international organization using this approach.

The Center's two basic objectives are to increase potato yields and ability, stability and efficiency of production in the DCs where potatoes are grown, and to extend the potato range of adaptation to new territories, including the lowland tropics. To accomplish these objectives, breeders, agronomists, physiologists, pathologists, entolomogists and nematologists work on the various identified cultural and disease problems. CIP economists, sociologists and anthropologists work on other problems of potato production such as marketing, farmer and consumer acceptance and where the potato fits into the farming system.

As a result of this joint effort during the earlier years of CIP, some traditional potato research has been conducted in the DCs to build up an essential base of knowledge on problems of potato production in tropical highland and lowland areas.

Concurrently, all major components and/or tactics utilized in the Integrated Management of Potato Pests (IMPP) are being investigated. These include reducing damage caused by the most common environmental stresses, pests and diseases of the potato crop as well as agronomic factors to increase farmer's yields, within their present natural and socio-economic limitations. Although the factors to be considered for implementing an IMPP are various and depend on the problems encountered, the major component at CIP is to build clones with multi-resistances.

The most common pests, diseases and enviromental stresses identified for the hot and humid lowland tropics and cool highland tropics are summarized in Fig. 1.

Although methods of plant breeding are directed toward population improvement to obtain germplasm with an enhanced frequency of genes for yield, resistance to common pests and diseases, increased dry matter and protein contents, and crop adaptation (i.e. to hot and humid lowland tropics),

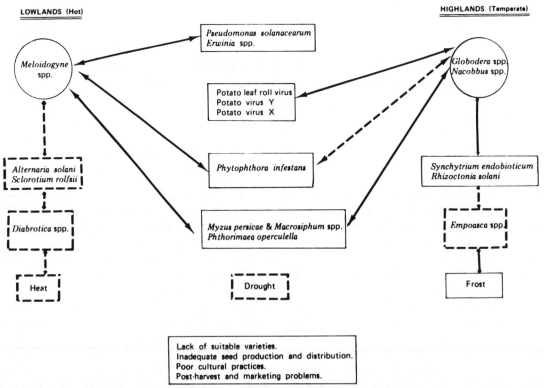

Figure 1. Principal constraints to potato production in developing countries. The most common pests, diseases and environmental stresses identified for the hot and humid lowland tropics and cool highland tropics. Both are indicated by full lines (actual breeding) and broken lines (to be considered in future breeding).

488

the utilization of pesticides has not been banned, although they have not been promoted either.

The potential for establishing Integrated Pest Management of potatoes in developing countries is enormous and offers one way of helping to solve problems of providing sufficient food for these countries. However, the application of sound management practices used in developed countries is difficult for developing countries due to the great range of agricultural, ecological, socioeconomical, cultural and political structures.

After this general view and as a nematologist I will now cover some aspects of the research in this field directed towards solving the most important identified problems and their complex interrelationships.

Nematologists and other plant scientists are developing multidisciplinary strategies (i.e. plant resistance, rotations, fertilizer effects, tolerance and biological control) in an effort to manage nematode population complexes in temperate and tropical areas where the potato cyst-nematode *Globodera* spp. and root-knot nematode *Meloidogyne* spp. are widely distributed. The strategy for management of these nematodes varies because of specific ecological associations. For instance, the need for combined resistance to both root-knot nematode and bacterial wilt (*Pseudomonas solanacearum*) is in contrast to the need to integrate control of potato cyst-nematodes with frost tolerance, black wart (*Synchytrium endobioticum*) and late blight (*Phytophthora infestans*) resistances. A similar situation occurs with certain insect pests. Furthermore, management of potato cyst-nematodes in developed countries involves working with low nematode densities, while high densities of nematodes of wider variability in the Andes require different managment techniques. Because of this, tolerance to these nematodes is an important part of integrated management programs.

Therefore, to accomplish this program, research on the population behavior of potato cyst-nematodes and their management under natural conditions has been done. This research has covered three general aspects: (a) potato cyst-nematode populations, (b) screening and breeding for resistance, and (c) management.

Collection, identification, distribution and classification of pathotypes of potato cystnematode in the Andes and other countries is being done. The populations in CIP's collection are maintained and used for further taxonomic and behavioral studies. As result, a map of the distribution of nematode species in Latin America reveals that the two species of potato cyst-nematodes occupy different zones in the Andes. The demarcation line separating the two species is near latitude 15.6°S. To understand why the species are distributed in this way and their variability, a series of studies were conducted with selected populations. Data on the effects of certain factors, such as temperature, day-length, host and soil were obtained upon rates of development of these species.

The classification and determination of pathotypes in the Andean region by the use of differential plants has also been studied. This is of twofold importance. First, the Andean region is where the potato cyst-nematodes evolved and, therefore, one expects to find in these regions a wider spectrum of pathotypes than those found in the potato-growing areas of Europe and other countries. The second reason why classification is crucial is that it provides a basis for resistance, and also dictates the areas where these resistant genes will be useful in controlling the nematode. The frequency distribution of pathotypes has been established with Andean populations and it is clear that P_4A and P_5A are the predominant pathoptypes (Table 1).

The last line of research on population studies has been on the host-parasite relationships to learn more about the host range of potato cyst-nematodes species in solanaceous and non-solanaceous plants.

On the second aspect of screening and breeding for resistance to the potato cyst-nematodes suitable and efficient techniques have been developed to screen either seedlings or tuber material form CIP's world germplasm collection and other programs. In the early years four potato cyst-nematode populations were used in these tests but only two are now being utilized. As a result, two promising *Solanum andigena* clones have been selected as resistance sources (CIP 702535 and 702698). These resistant clones tested with Andean populations of potato cyst-nematodes showed promise as parental material. However, this research emphasizes the

Table 1. Frequency of Pathotype Distribution in Latin Countries According to Their Reaction on Different Clones

Pathotype Identification	Resistant Source of Solanum	Frequency Number	%
P_1 B (*)	S. multidissectum	1	1.9
P_2 A (*)	S. kurtzianum	6	11.5
P_3 A (*)	S. vernei GLKS	1	1.9
P_4 A (P_a2)	S. vernei $(VT^n)^2$ 62.33.3	13	25.0
P_5 A (P_a3)	None	20	38.5
R_1 A ($R_0$1)	S. tub. ssp. andigena	1	1.9
R_1 B ($R_0$4)	S. tub. ssp. andigena	1	1.9
R_2 A ($R_0$2)	S. kurtzianum	3	5.8
R_3 A ($R_0$3)	S. vernei GLKS	6	11.5
	TOTAL:	52	100.0

*No equivalence in European classification.

necessity for a permanent process of breeding for resistance through search for and use of new sources of resistance. Also, material resistant to other diseases has been tested to select multiple resistant lines. When material resistant to one pest is found to be susceptible to other pests, intercrosses have been made such as : frost × potato cyst-nematode, late blight × potato cyst-nematode, wart × potato cyst-nematode, false root-knot × potato cyst-nematode and bacterial wilt × potato cyst-nematode. Finally, the most promising clones are selected, screened against other diseases, multiplied and tested in international field trials.

Under the third aspect of research other tactics useful in pest management and which can be integrated into a practical viable strategy have been investigated. These include cultural control, biological control, interaction with other organisms, pesticides and others. Furthermore, because inter-relationships between nematode populations and plant metabolic processes are not well known, water relationships, nutrient uptake and photosynthesis have been investigated in susceptible and resistant plants to determine the factors limiting growth and eventually yield in potato cyst-nematode infested plants.

Research shows that resistant varieties grown in nematode infested soil used water more efficiently than susceptible ones but photosynthesis and uptake of N and K were affected by nematodes in both types of plants. However, it also seems that tolerance to potato cyst-nematode damage is related to the water efficiency used by that potato plant since some degree of drought tolerance is also shown. Field tests of native and commercial varieties as well as advanced material from other programs show that certain clones possess a degree of tolerance to potato cyst-nematode when compared with the most common cultivars. However, because large nematode multiplication occurs on this material its use should be limited to areas where the nematode is widespread or in situations where a strategy of integrated management is being conducted. Other ways to improve potato yields in potato cyst-nematode infested soils are to use P-fertilizer and pre-cropping with other Andean crops (*Lupinus sp., Oxalis tuberosum, Chenopodium quinoa*). These crops suppress nematode reproduction. Another consideration is the interaction of nematodes with other organisms. Fungi such as *Phoma exigua, P. andinum* and *Colletotrichum coccodes* suppress the number of newly formed cyst of *Globodera* spp. although *P. exugia* became a root parasite instead of foliage parasite as it is usually known. On the other hand, *Pseudomonas solanacearum* resistant plants become susceptible when root-knot nematode *Meloidogyne* spp. is present.

Regarding biological controls using "natural enemies," larvae emergence of potato cyst-nematode has been suppressed by *Ulocladium botrytis* and *Drechslera halodes* which were isolated from naturally infected eggs of *G. pallida*. Furthermore, a fungus isolated from root-knot nematode eggs and identified as *Paecilomyces lilacinus* has shown potential for destroying young females and eggs of *Meloidogyne* spp. and to a lesser degree the cyst-nematodes.

Finally, although my approach to this subject has been from a nematological point of view, I believe that multidisciplinary cooperation across many regions is essential. This would help accomplish an effective integrated management program to minimize losses due to environment stress, pests and diseases. Additionally, it would optimize those agronomic components to reduce risks of developing country farmers and aid in solving the food and other problems that face the developing world population.

31.7 Integrated Plant Protection for Potatoes—Concluding Remarks

V. Umaerus

Sweden

Integrated plant protection for potatoes has the same elements of control methods as in most other crops.

Legislative measures are rather demanding in a vegetatively propagated crop like the potato, both concerning quarantine regulations and seed certification programs. The latter has been dealt with in the first paper by Dr. Page and his associate James Bryan. They conclude that seed certification in developing countries, in CIP's experience, is not very successful. The various ingredients necessary to maintain certification standards are lacking. CIP has initiated a breeding and agronomic management program to develop practical methods to propagate potatoes by means of *true seed* thus avoiding many of the drawbacks of tuber propagation.

Another obvious ingredient of an integrated pest management program is the use of *resistant cultivars* with a good example in the case of resistance to the cyst nematodes (*Globodera rostochiensis* and *G. pallida*) as presented by Dr. Jones and Dr. Franco. Many other examples can be mentioned: resistance to wart disease, late blight, virus diseases and insects, and all speakers emphasis the importance of breeding for resistance to pests and diseases.

Biological control has some interesting implications in control of nematodes by fungi as indicated both by Dr. Jones and Dr. Franco, but prospects are perhaps greater in the management of insect pests as presented by Dr. Tingey.

Cultural Control or General Farming Practices—Crop rotation, fertilization, irrigation, weed control, time of planting, time of harvest—are all important ingredients. Unfortunately, these methods are only "palliative" as Jones put it in relation to nematode control. "Circumstances force growers in developing countries to plant potatoes in unsuitable fields and long rotations are economically unacceptable in developed countries." This is also true for other pests and diseases.

The *use of chemicals* is unavoidable in modern potato cultivation for economical reasons in developed countries but of less interest in developing countries for obvious other reasons, e.g. lack of application methods. Forecasting and warning systems are valuable tools to tailor the amount of chemicals needed to disease or pest pressure. Considerable difficulties exist to establish threshold values in insect control as demonstrated by Radcliffe. Also foliar diseases like late blight and early blight raise problems in establishing good and reliable forecast systems. Fry and Bruck point out the need of methods to establish the amount of initial inoculum present in the fields, and the influence of variability in resistance of cultivars used.

These are the cornerstones of most integrated pest management programs. What are the prospects of the potato? Potato production is very conservative in many aspects:

1. *Breeding and the use of cultivars* is very much concerned with "cosmetically perfect potatoes" as Dr. Page has put it. A long row of cultivars with different combinations of resistance factors has been produced, but for some reason the old cultivars continue to be used.

    The breeder's right system has put in an element of commercial interest and there are breeders who do not utilize the possibilities to breed for virus resistance to stimulate the turnover rate of basic seed.

2. *Seed production* in Europe has entered a vicious circle.

    To avoid virus infection basic seed has to be produced in cool areas at the cost of increase of latent fungal and bacterial diseases (*Phoma*, ringrot, black leg).

    This has forced the seed trade to consider more sophisticated methods of growing prebasic seed through meristem culture and stem cutting techniques, which obviously will raise the cost of production.

3. *Chemicals* are used as crop insurance with little regard to pest pressure.

    Forecasting and warning service is offered in several countries but receive little attention from the growers for several reasons: (a) too large acreage to spray within the risk period of warning, (b) contracted aerial spray not flexible in time, and (c) cost of spraying is low at present.

4. *Cultural control methods* are not popular—the farmer often requires a simple nonsophisticated method to solve a disease problem.

I think, however, that we are close to a breakthrough in the implementation of an integrated pest management program for potato for several reasons: (a) the energy shortage, (b) the respect for a clean environment, and (c) the concern of producing risk-free food.

In many ways the problems in the developed and the developing countries will be much the same. We have therefore a mutual interest in developing such a program and see it used in practice but the applications are different according to each country's need.

32.1 Worldwide Sugarbeet Production Trends

Raymond Hull

Little Sacham, Bury St. Edmunds
Suffolk, England

The sugarbeet crop demands skillful husbandry and its culture has undergone more revolutionary developments during the last four decades than any other arable crop. Initially, establishing the crop, freeing it from weeds and harvesting, all demanded much hand labor; now the crop can be grown with none—establishing the crop planting to a stand or mechanical thinning, weed control with herbicides, mechanical harvesting and delivery of the roots to the factory have all been achieved.

The culture of the crop has extended throughout suitable climatic areas of Europe, North and South America, North Africa, the Middle East, India, and the Far East. In the warmer, drier climates it is grown as a winter crop, thus needing relatively little irrigation water and giving the opportunity for intensive rotation with other crop plants that can be grown in the hotter seasons.

The scarcity of sugar and its high price in the late 1960s led to greatly increased production from both beet and cane, and now we are suffering from surpluses that are offered for sale at less than production cost. This is resulting in a decline in the area of sugarbeet grown in many countries and in financial problems for sugar factories in those countries where the industry is not protected. Many countries have decided to support a degree of self-sufficiency in this basic food stuff, and find their decision justified by recent experience with oil and its continuously escalating price as scarcity develops. Increasing world population will soon need more sugar. If factory closures continue, the capacity for production will be inadequate to meet the growing demand and to construct new factories needs enormous capital investment. Will the headlines in the 1980s not be OPEC but SUPEC?

As energy problems develop we will soon doubtless be considering the efficiency of different crops as conservers of solar energy. The efficiency of conservation of incident photosynthetically active radiation into energy as sucrose recovered from plants is similar for beet and cane, averaging about one-half percent. However, cane is three times more efficient than beet in converting energy into biomass and promises to be an acceptable vehicle for converting solar energy into fuel for transport vehicles. The fibrous material from the cane will provide the heat to distill alcohol from the fermented carbohydrates. This is not so with beet, which requires more energy from fuel for distillation than is available from the alcohol produced.

World politics being what they are, sugarbeet will remain the main source of sucrose for countries in the temperate zone. Our objective, therefore, is to improve the economy of production. Growers are attempting to do so in many ways and there is great scope for the research worker to help them. As well as improving their efficiency of handling the crop at all stages, they must *regularly* achieve greater yields of sucrose per unit area. Mechanized crop establishment has been achieved by sophisticated seed drills that give regular and controlled spacing of seed. The plant breeders' contribution has been the production of monogerm seed. Herbicides applied before, during and after sowing have eliminated weed competition, but often at the expense of temporarily depressed growth of the sugar beet. We still need better and more selective herbicides whose efficacy is not critically dependent on soil and climatic conditions.

Maximum yield depends on a regular and full stand of plants. In conventional row and seed spacings in Europe, establishment of less than 70% of viable seeds sown results in yield loss from gapiness and in many crops this level of establishment is not achieved. The causes of imperfect germination are numerous—the seed not being placed in effective contact with moist soil, soil capping, seedling death from pests and diseases, toxic effects of nutrients and herbicides, and genetic inability of the seed to germinate and establish at a particular soil temperature and moisture.

One could discuss at length about how sugarbeet growing could be improved—simpler mechanization in larger fields; more effective harvesting that decreases the 10% or so loss of roots between the growing crop and factory; better yielding cultivars; more efficient water economy and irrigation; improved factory technology, and so on. Progress is being made in all these fields. But, we are a group concerned with research and development in plant protection, so what is our role?

Pests and diseases that beset the sugarbeet crop are immeasurable. The majority are local and of minor importance, and research has devised control mechanisms or ways of avoiding them. But one need not travel far to see that pest and disease incidence is a major determinant of the size and regularity of yield of sugar beet throughout the world. In spite of control measures, pests of seedlings take their toll each year above and below ground; Cercospora leaf spot defoliates crops prematurely; the yellows viruses, curly top, and yellow wilt attenuate growth in the areas where the climate favors their vectors, the aphids and leafhoppers. Beet cyst and root-knot nematodes and the ectoparasitic root feeding nematodes limit growth in many soils. *Sclerotium rolfsii* often destroys a portion of the crop in hotter climates. Leaf eating and crown-boring pests sporadiacally devastate crops. All these troubles can be controlled to some degree but better control measures are needed. We look hopefully, often with justification, to the plant breeder to develop cultivars with resistance. Every available weapon has been developed to constrain the incidence of one pathogen or another. Crop rotation is often enforced in the contract between sugar factory and grower, giving the soil several years freedom from crops susceptible to the same pathogens as sugar-beet. Much more can be done than at present about crop hygiene—for instance, avoidance of contamination of the new crop by leaves of the previous year's crop killed by *Cercospora*. This entails planting and restricting cropping density in an area, which is a significant factor in epidemiology of many pathogens. Measures are taken to avoid the overlapping of new and old crops to provide a salutory crop-free winter or summer. This is not always easy when different crops are involved; for instance, sugarbeet and spinach, seed and root crops of *Beta* species, stored roots of *Beta* species. Measures to encourage rapid growth by skillful soil cultivation, fertilizing, and watering foil the potential devstation of some pests and diseases. Choice of time of sowing may be critical in avoiding disease. Such cultural practices have been developed and their use extended in recent years to control numerous pests and diseases.

However such measures are exploited, we will always be dependent sooner or later (and usually sooner!) on pesticides as the ultimate technique of control. Fungicides and insecticides, are gradually improved, but they are expensive and their regular use generates new problems. Pesticides may eliminate predators and parasites as well as the target pest, thus resulting in pests that were of little consequence becoming troublesome. Pathogens respond to exposures to pesticides by developing resistant populations. Aphids on sugar beets have now achieved resistance—almost immunity—to organo-chlorine insecticides and resistance in *Cercospora* to benzimidazoles is common.

To avoid these complications we need to be able to assess the risk of epidemics developing, so that pesticides need not be used routinely, but be applied only when they are likely to be useful. To do this we must have a thorough knowledge of the biology of the pathogen, the factors influencing its epidemiology, the thresholds of population causing economic damage, and the reaction of the crop to changes in pest incidence. We need continuous assessments of population dynamics and population quality. Ultimately, too, we need reliable forecasts of weather or at least an assessment of changes of occurrence of a range of weather conditions, for the population dynamics of many pathogens are critically determined by weather. With such information, reliable warnings of the need or undesirability of control measures can be issued to growers. This is the trend in pest management.

New varieties with desirable characteristics of early growth, greater sucrose concentration and juice purity, resistance to specific dominant pests and diseases, are regularly produced by new breeding techniques involving polyploidy and controlled hybridization. When they prove better yielders than currently used cultivars they are rapidly adopted by growers. This widespread use sometimes reveals unsuspected weaknesses in susceptibility to pathogens that previously have occurred rarely or sporadically. For instance, the widespread use of some high yielding cultivars have led to such serious outbreaks of downy mildew (*Peronospora schachtii*) and Phoma neck rot in England that have caused their withdrawal. Powdery mildew (*Erysiphe betae*) suddenly became widespread in the United States when a cultivar that proved very susceptible was extensively grown in California. Thus the breeders' successes may create new problems for the pathologists.

Just as pest and disease incidence can be restricted by cultural practices, so changes in cultural practices can lead to new problems. An outstanding example is the problem of weed beets in Europe. The soil of many fields where sugar beet is grown is now contaminated with beet seed, so that when a beet crop is sown, it may be impossible to detect the line of the rows, because unwanted beet seedlings emerge all over. These bolt quickly so the field may soon take on the appearance of a seed crop. Some of the weed beets are true annuals, originally introduced by growing seed in areas where the weed is indigenous. Others are the product of seed from bolting beets arising in early sowings, perhaps several years previously, of cultivars with inadequate bolting resistance for that particular spring weather. The problem has developed as a consequence of mechanizing crop establishment—sowing to a stand, no thinning or hand weeding in the rows, dependence on selective herbicides for weed control, and neglecting to remove early bolters. The consequence on pest and disease incidence of these unwanted plants, not only in beet but also in other crops throughout the crop rotation, remains to be seen.

In these ways beet culture changes, pest and disease problems change, and research too must change. In these times of financial stringency, unfruitful and unexplored projects must be discontinued and new approaches to improve economy of production must be explored. As one who has spent a lifetime in sugar beet research, may I mention one problem that impresses me. Why will one field produce only two-thirds or half the yield of another nearby field when both are well farmed and have the same weather? The answer must lie below ground, and this means the root system of the plant. How little we know of what goes on throughout the depths to which beet roots can penetrate, what determines the death of these fine rootlets, their successful regeneration, their efficiency in finding moisture and nutrients? Herein lies a fertile field of exploration for pathologists as well as researchers in numerous other disciplines.

32.2   Leaf-Feeding and, Especially, Root-Feeding Pests of Sugar Beet

R. A. Dunning

*Broom's Barn Experimental Station,
Higham, Bury St. Edmunds, England.*

Sugar yield is directly related to the amount of solar radiation intercepted during the season by the foliage of the healthy sugar beet plant [1]. The total leaf area per hectare, and/or its efficiency in utilizing radiant energy, can be decreased by pest damage in three ways:

1) DECREASED PLANT POPULATION due to loss of seed (e.g. seed eaten by *Apodemus sylvaticus,* the wood mouse, before germination), or loss of seedlings before emergence or plants after emergence due to pest attack (e.g. by *Onychiurus, Scutigerella, Blaniulus, Agriotes, Limonius, Atomaria, Tetanops, Cleonus, Agrotis, Chaetocnema).*

2) DECREASED LEAF AREA due to foliage-feeding pests (e.g. *Lygus, Cassida, Chaetocnema, Bothynoderes, Pegomya, Loxostege, Scrobipalpa, Mamestra, Spodoptera),* or to root damage caused by root-feeding pests (e.g. *Pemphigus* and, when plants are not killed, those arthropod pests listed under 1) above).

3) DECREASED LEAF EFFICIENCY AND/ OR TRANSLOCATION OF ASSIMILATES due to damage by foliage-feeding pests (e.g. *Lygus, Calocoris, Aphis*) or to virus diseases transmitted by vectors (e.g. *Circulifer, Myzus, Paratanus, Piesma*).

The most important arthropod pests of beet are those that regularly attack beet in many countries, and, especially the ones whose damage is difficult to control. Despite there being very many different genera and species of arthropod pests - for example 220 recorded in southeastern Europe, i.e. Bulgaria, Hungary, Romania, Yugoslavia [2], 150 in the U.S.A. [3], but only 40 in England [4] - rather few are major pests. Their occurrence and importance differs considerably between different countries (for Europe see [5]). They are:

a) *Root feeding pests* that decrease plant establishment and/or subsequent growth (principally springtails, symphylids, millepedes, *Atomaria,* a few lepidopterous larvae, some coleopterous adults and/or larvae, root aphids and, in North America only, *Tetanops,* the sugar-beet root maggot). Because they are soil-inhabiting and are often present in the field at sowing, prophylactic treatment is necessary. Seed treatment can be effective for some but soil treatment is needed for most; this is often expensive, may be difficult to apply and is still not completely effective.

b) *Leaf-feeding pests* that attack very young seedlings (e.g. flea beetles) or that are difficult to control for various reasons (*Cassida, Scrobipalpa, Loxostege,* and birds).

c) *Virus vectors,* especially *Myzus persicae* in all countries, *Circulifer tenellus* in North America only, and *Paratanus exitiosus* in South America only. (Virus vectors are not considered in this paper).

Methods of establishing the optimum plant population (75 000/ha in 50 cm rows) have changed in recent years, are changing, or will change, in all countries, to planting-to-stand and the extensive use of herbicides. Even in North America it seems that close seed spacing, followed by thinning (usually mechanical or electronic) will be replaced by planting-to-stand.

Plant establishment is very variable in northwest Europe, where planting-to-stand is the rule rather than the exception, and averages only 50-60% (i.e. 50-60 plants per 100 seeds sown). The reasons for this are numerous; much research is in progress on all possible factors - seed and its treatment, soil physical conditions, pests, diseases, etc. - and recent results will be discussed at the Institute International de Recherches Betteravieres Congress in Brussels in February 1981.

Pests that attack the crop before the 4-6 true-leaf stage are now more important than formerly in decreasing plant establishment. The same number of pests per unit area aggregate on fewer seedlings, especially in the absence of alternative weed hosts. Furthermore, planting-to-stand tends to produce rather irregularly spaced plants, even if the population is the desired 75 000/ha; this gappiness induces greater infestations by some immigrant winged insects (e.g. *Myzus persicae* as a vector of virus yellows [6] than in a uniformly spaced crop.

Research on pests affecting plant establishment and early growth has increased in recent years - their ecology, the identification of fields at risk, methods and profitability of control, and especially the use of pesticides within integrated management programmes.

Pests that Decrease Plant Establishment

These are mainly root-feeding but severe leaf-feeding by a few species early in the season (e.g. *Chaetocnema*) can kill plants.

A recent "new" pest is the wood mouse (*Apodemus sylvaticus*), which digs out seeds that have not yet germinated and consumes the embryo. In England it is present in all fields; damage is most severe on early sown crops, especially where some seeds are not properly covered by the drill. Numbers of mice vary within and between years [7] and current surveys aim to determine populations immediately before sowing in March so as to warn growers in years of high population that control (poison baiting immediately damage is seen) may be necessary; prophylactic treatment is not recommended bewcause it may be unnecessary.

Soil-inhabiting pests - springtails (*Onychiurus*) millepedes (*Blaniulus, Brachydesmus,* etc.), symphylids (*Scutigerella*), pygmy beetle (*Atomaria*), wireworms (*Agriotes, Limonius,* etc.) - either individually or, more often, together decrease seedling establishment. They are especially important in northwest Europe and studies of some factors involved, principally seed spacing and soil compaction, have been made [8]. Pests are randomly distributed in the soil at sowing; some were shown to aggregate in the seedling root zones (e.g. for *Atomaria* 5.5 per 60 mm soil core centered over a seedling v. 1.2*** per core between the rows on 12 May, and 6.8 v. 0.3*** on 17 May).

Soil compaction was found to decrease seedling root zone aggregation of *Blaniulus* and *Scutigerella* (but not *Atomaria*); such an effect has been reported previously [9, 10].

Sugar-beet seed is treated with insecticide before sowing in most countries [11]. However, because growers fear that soil-inhabiting pests will decrease plant establishment, they have resorted to prophylactic use of soil-applied pesticides in addition. The extent of this in 1978 in some European countries is shown in Table 1.

Table 1. Prophylactic Usage of Soil-Applied Pesticides % of National Crop Treated 1978

	Ald.	Carb.	Chlor.	Terb.	Thio.	HCH	Total
Belgium	80	5			10	10	100
England	36	7			6	19	68
France	38	16	2	22	3	2	83
Germany	5	<1			<1	40	46
Greece		5	14	9		60	88
Ireland	4	4				1	11*
Italy†		38	11			6	76
Netherlands		23	6		1	3	33
Switzerland	3	22			<		27

*also Oxamyl 2%
†1977, also Phorate 21%
Pesticides are: Aldicarb, Carbofuran, Chlormephos, Terbufos, Thiofanox and HCH (Lindane)

A small proportion of this usage of carbamates is primarily for protection against damage by nematode pests and a larger proportion primarily against aphids that transmit virus yellows; however, most of the carbamate usage, and all the o.p. and HCH usage, is to protect against root-feeding insect damage [12]. It is doubtful if this extensive usage is justified in some of the countries. Recent trials on random sites in the Netherlands and England showed that seed-furrow application of aldicarb (1 kg a.i./ha) was profitable at only some of the random sites on which it was tested (Table 2), a result of the control of root-feeding and/or leaf-feeding pest damage.

Table 2. Profit from Aldicarb Seed-Furrow Treatment (1 KG A.I./HA)

	Year	Number of trials	Number profitable
Netherlands	1975	34	10
	1976	20	8
England	1976	8	3
	1977	8	2
	1978	8	4

Results such as these indicate the need for better identification of fields, and seasons, where treatment will be profitable. Sugar-beet seedlings are killed by pests in some fields in some years but not in others. The risk of damage occurring can be indicated by soil sampling to determine the resident pest populations but this is a laborious process; in any case there will rarely be a direct relationship between pest numbers and damage because of the variable effects of sowing date, pest aggregation and feeding activity, etc.

Better identification of fields at risk from soil pests is only likely to be achieved by better understanding of the pests' ecology; this is being studied currently at Broom's Barn, concentrating on springtails, symphylids and millepedes. These pests almost invariably occur together in the field and the interactions of their feeding damage are likely to be complex.

The need for more selective use of soil pesticides is further emphasized by the knowledge of their adverse effects on soil-resident beneficial insects - see [13] for recent review. For example, predators may well be important in helping to control early aphid infestations [14]; hence there is concern in Europe that overall soil treatment with gamma HCH, and even row treatment with materials such as chlormephos, might lead to increased aphid

infestation and, perhaps, virus yellows incidence. Such an effect has been recorded previously for trichlorphon foliar spray [14]. Current work at Broom's Barn seeks better integration of control measures for aphids and virus yellows, especially via more selective usage of soil pesticides against other pests.

Pests that Decrease Leaf Area

Root-feeding pest damage that does not kill the plant almost invariably decreases leaf area. Little work has been done on relationships between pest numbers, their feeding activity and the extent of root damage necessary to decrease yield, probably because attention has been concentrated on these pests' effects on plant establishment.

Virtually all foliage-feeding pests have been studied extensively. Growers readily notice the damage and are concerned that yield will be decreased; effective pesticides can easily be selected in trials and recommendations made for application when the economic damage threshold appears likely to be exceeded. Thresholds are, however, difficult to determine in field infestations and artificial defoliation trials have been made in many countries to provide basic guidelines for advice [e.g. 16].

Grazing of seedling foliage by birds, principally the skylark *(Alauda arvensis)* in northwest Europe, is an increasing problem. Recent work has provided an understanding of this pest's damage, and shown that it is significantly decreased following seed-furrow application of aldicarb against other root and foliage-feeding pests [17].

Seedling foliage damage is sometimes very severe, due to pests such as *Chaetocnema* (e.g. in Spain); foliage treatment after seedling emergence is too late and preventive treatment is necessary. Usually, however, pesticide application can be made to foliage when significant damage threatens. Correct timing of foliage treatment is important; forecasting of attacks helps but crop monitoring to follow population build up is preferable.

Conclusions

Research work in the various countries is giving a better understanding of pest damage, but pesticide usage is excessive, growers fearing that damage might occur; better advice on the need for treatment is needed. Co-operation of sugar beet zoologists seems essential and this is being attempted internationally in northwest Europe [17].

References

1. Scott, R.K. & Jaggard, K.W. (1978) Theoretical criteria for maximum yield. Proceedings 41st Winter Congress, Institute International de Recherches Betteravieres, Bruxelles, pp 179-198.
2. Camprag, D. (1973) Stetocine Secerne Repe. (Sugar beet pests in Yugoslavia and neighboring countries - Hungary, Romania and Bulgaria - with a special review to the most important species). Poljoprivredni Fakultet, Novi Sad, 363 pp.
3. Lange, W.H. (1971) Insects and mites and their control, pp 288-333, in Advances in Sugar Beet Production (edited by Johnson, R.T., Alexander, J.T., Rush, G.E. & Hawkes G.R.), Iowa State University Press.
4. Jones, F.G.W. & Dunning, R.A. (1972) Sugar beet pests. M.A.F.F. Bulletin No. 162. Her Majesty's Stationery Office, London.
5. Dunning, R.A. (1972) Sugar beet pest and disease incidence and damage, and pesticide usage; report of an I.I.R.B. enquiry. Journal Institute International de Recherches Betteravieres, 6, pp 19-34.
6. Heathcote, G.D. (1969) Cultural factors affecting colonization of sugar beet by different aphid species. Annals of Applied Biology, 63, pp 330-331.
7. Green, R.E. (1979) Wood mice taking seed. Report of Rothamsted Experimental Station for 1978, Part 1, p 62.
8. Dunning, R.A. & Baker, A.N. (1977) Some sugar beet cultural practices in relation to incidence and damage by soil-inhabiting pests. Annals of Applied Biology, 87, pp 528-532.
9. Michelbacher, A.E. (1939) Seasonal variation in the distribution of two species of symphyla found in California Journal of Economic Entomology, 32, pp 55-57.
10. Heijbroek, W. (1971) De mogelijkheden voor de bestrijding van de belanrijkste voor jaarsplagen. III. De springstaart *(Onychiurus armatus Tullb)*. Mededeelingen van het Institut voor rationelle suikerproductie, Bergen op Zoom, 38, pp 1-48.
11. Dunning, R.A. & Byford, W.J. (1978) Sugar beet seed treatments, pp 79-90 in CIPAC Monograph No. 2 : Seed treatment (edited by Jeffs, K.) Heffers, Cambridge, England.
12. Dunning, R.A. & Byford, W.J. (1979) Weed, disease and pest control: costs, profitability and possible improvements for certain programmes. Part II Disease and pest control. Proceedings 42nd Winter Congress, Institut International de Recherches Betteravieres, Bruxelles, pp 85-103.
13. Brown, A.W.A. (1978) Ecology of Pesticides, 525 pp. Wiley & Sons, New York.
14. Dunning, R.A., Baker, A.N. & Windley, R.F. (1975) Carabids in sugar beet crops and their possible role as aphid predators. Annals of Applied Biology, 80, pp 11-14.
15. Dunning, R.A. & Winder, G.H. (1972) Some effects, especially on yield, of artificially defoliating sugar beet. Annals of Applied Biology, 70, pp 89-90.
16. Green, R.E. (1979) Food selection by skylarks: the effects of a pesticide on grazing preferences. In Bird Problems in Agriculture (edited by Wright, E.N.), British Crop Protection Council Monograph No. 23. (In press).
17. Dunning, R.A. (1975) International co-operation in the development of control of pests and diseases of sugar beet. Proceedings 8th British Insecticide and Fungicide Conference, pp 1013-1018.

32.3. Sugarbeet Weed Control—Its Status and Future Direction

E. E. Schweizer

*USDA, SEA, AR, Crops Research Laboratory, Colorado State University
Fort Collins, Colorado 80523 USA*

The major annual weeds that infest sugarbeet-production areas across the world are *Amaranthus retroflexus, Avena fatua, Chenopodium album, Echinochloa crusgalli,* and *Sinapis arvensis* [5]. Other annual weeds that are troublesome in several countries are *Matricaria chamomilla, Poa annua, Polygonum aviculare, Polygonum convolvulus, Setaria viridis* and *Stellaria media.* Two perennials, *Agropyron repens* in northern Europe, Spain, and the United States and *Convolvulus arvensis* in Greece, Iran, and the United States, complete the list of the world's worst weeds in sugarbeet fields [5].

In the western United States, 51 weeds and five volunteer crops are of economic importance in sugarbeet production (personal communication). Of the 51 weeds, 45 are annuals and six are perennials. Eight of these weeds—*A. retroflexus, A. fatua, C. album, Cirsium arvensis, C. arvensis, E. crusgalli, Helianthus annuus,* and *Kochia scoparia*—infest sugarbeet fields in five to eight western states. Five volunteer crops—*Brassica rapa, Hordeum vulgare,. Solanum tuberosum, Triticum aestivum,* and *Zea mays*—are problems in five states.

In the United States, annual and perennial weeds are increasing in many sugarbeet fields under current farm management practices. The increase of weeds has resulted from shorter rotations, lack of control by certain herbicides, reduced tillage, double-cropping, monoculture, and sowing contaminated crop seed.

In the irrigated areas of the western United States, 10 to 15% of the total annual weed population is not controlled by cultivation and herbicides and competes with sugarbeets if not removed by hoeing. Low densities of weeds competing with the crop all season will reduce root yields. For example, in Wyoming, eight *S. viridis* plants or eight *A. retroflexus* plants per 30 m of row reduced yields 4 and 16%, respectively [1]. In Colorado, four *K. scoparia* plants [7] or six *H. annuus* or six *Abutilon theophrasti* plants (Schweizer, unpublished) per 30 m of row reduced yields 8, 53, and 13%, respectively.

Weeds that emerge after sugarbeets are thinned also reduce root yields. In California, *E. crusgalli* reduced yields by 4.7 t/ha compared with plots where this weed was removed once by hoeing [2]. In Washington, *E. crusgalli* and *Amaranthus* species that emerged after June were suppressed and killed by competition from sugarbeets of normal vigor in a full stand, but these weeds reduced root yields 5 to 39% when sugarbeets were spaced at ½ of a full stand compared with weed-free sugarbeets in the same stand/[4].

The cost of weeding sugarbeets in the United States varies among farms, and maximum costs are closely associated with the indigenous weed population on farms. In 1970, the federal wage scale for removing weeds from sugarbeets once with a hoe was $41/ha; in 1978, this cost had escalated to $69/ha. A second weeding cost $26/ha in 1970 and $45 ha in 1978. Since labor for weeding sugarbeets has become more expensive and scarcer each year, continued profitable production of this crop will depend on complete mechanization of all phases of production.

Complete control of all annual weed species in sugarbeets for the entire growing season by chemical and mechanical means is desirable but seldom obtainable. As weed control by chemical and mechanical means approaches 100%, weed competition is minimized and root yields often are not reduced significantly below an untreated, hand-weeded check. In field studies conducted at Fort Collins from 1971 through 1977, I found root yields in plots treated with herbicides were reduced significantly below that of the hand-weeded check only once when the final broadleaf weed density averaged five weeds or fewer per 30 m of row (14 comparisons) and twice when the final density averaged 10 weeds or fewer per 30 m of row (20 comparisons). Ideally, all weeds should be controlled. However, as the cost of hand weeding escalates and labor becomes scarcer, weed control in

sugarbeets will follow trends that have developed with other row crops, such as *Z. mays* and *Glycene max.* That is, when annual weeds cannot be controlled completely by chemical and mechanical means, they will be left to compete with the crop all season.

My research in the 1970's indicates several possible systems for controlling weeds in sugarbeets in the 1980's. A discussion of these systems requires knowledge of the terms "hand blocking" (hoeing) and "weeding". Hand blocking (first operation) is the removal of weeds and excess sugarbeet plants (thinning) with a hoe only. Weeding (second and third operations) is the removal of weeds with a hoe and by hand after hoeing (first operation), or it is the first hand-labor operation in fields that have been machine-thinned and treated with herbicides. These terms are used in the six systems presented in Figure 1.

Basically, illustrations for systems A, B, and C depict the progression of weed control and thinning from the early 1970's (system A) through 1979 (system C). System A was employed by most growers to control annual weeds and thin the crop during the early 1970's. Based on average herbicide and labor costs between 1975 and 1977, system A would cost about $175/ha. System B would cost about $155/ha, with herbicides costing $20/ha more and weeding costing $40/ha less than system A. Costs for system

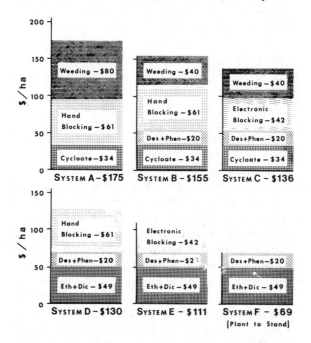

Figure 1. Comparative costs per hectare for herbicides, thinning, and weeding for six weed control systems, 1975-1977. Herbicide rates (kg/ha): Cycloate—3.4, desmedipham (Des)—0.6, diclofop (Dic)—1.7 ethofumesate (Eth)—2.2, and phenmedipham (Phen)—0.6.

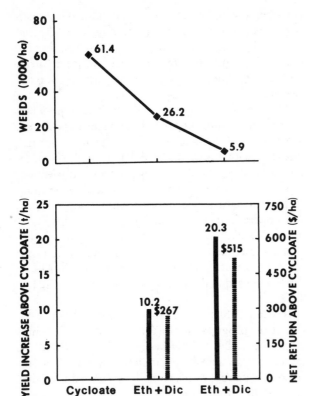

Figure 2 Relationship among weed control, root yield (solid bar) and net return above cycloate (hatched bar) for two new weed control systems. Herbicite rates (kg/ha): Cycloate—3.4, (Eth)—2.2, and phenmidpham (Phen)—0.6.

C are the lowest and can be achieved by growers who use herbicides sequentially and thin their crop electronically.

Systems D, E, and F include one of several new sequential herbicide treatments that will likely be used widely in the 1980s, and three methods of establishing a final stand of sugarbeets. The sequential herbicide treatments for systems D, E, and F—a mixture of ethofumesate plus diclofop applied preplanting, followed by a mixture of desmedipham and phenmedipham applied postemergence—effectively controls annual weeds (Figure 2). The herbicides for these systems cost $69/ha (Figure 10) or twice as much as the standard treatment of cycloate (system A). However, the cost of thinning and weeding the crop would be only $61/ha manually (system D) or $42/ha electronically (system E). System F as illustrated includes no costs for thinning or weeding, but under some conditions one weeding may be required for both systems E and F.

Weeding will be required in system E or F only when too many tall broadleaf weeds escape. For *K. scoparia* and *retroflexus*, this would be about four [7] and eight [1] plants, respectively, per 30 m of row. The competitive thresholds in terms of yield reduction are unknown for other broadleaf weeds that compete with sugarbeets.

Herbicides used in systems E and F left a total of nine weeds per 30 m of row in the first year, and 11 in the second year. This represented 98 and 99% weed control in the first and second years, respectively. The relationship among weed control, tonnage increase above that obtained with cycloate alone, and net returns above cycloate alone for the sequential treatment of ethofumesate-diclofop-desmedipham-phenmedipham is graphically presented in Figure 2. As the weed population decreases, tonnage and net returns above cycloate increase.

Several of the sequential herbicide treatments applied in the 1970's controlled weeds so effectively that root yields were not reduced significantly by the few weeds that competed all season. Opinions of weed scientists differ as to if and how a few surviving weeds should be removed. Weeds that are not removed produce seed to infest the next crop, and they cause problems with harvesting and root storage. Moreover, if they are resistant to herbicides, they can proliferate and create new weed problems.

The real issue, then is: should weeds that remain after the use of effective weed control methods be removed if their density is below the level at which the expected return from an increase in yield will not pay for the cost of labor to remove the weeds? At present, weeds that escape have to be removed by hoeing or left to compete with the crop. Although weed escapes can still be removed by hiring labor, another method as we enter the 1980s is to control weed escapes with nonselective herbicides. Nonselective herbicides have been applied to control weeds that are taller than the crop in *Gossypium hirsutum* and *G. max* by using a recirculating sprayer [6], a horizontal roller, or a rope wick applicator [3].

References

1. Brimhall, P. B., E. W. Chamberlain, and H. P. Alley, 1965. "Competition, of annual weeds and sugarbeets." *Weeds* 13:33-35.
2. Burtch, L. B. 1965. "Control of late emerging weeds can be profitable." *Spreckels Sugar Beet Bull.* 29:3-4.
3. Dale, J. E. 1979. "Application equipment for Roundup—the rope wick applicator." *Proc. Beltwide Cotton Prod. Res. Conf.* (in press).

4. Dawson, J. H. 1977. "Competition of late-emerging weeds with sugarbeets." *Weed Sci.* 25:168-170.
5. Holm, L. G., D. L. Plucknett, J. V. Pancho, and J. P. Herberger. 1977. *The World's Worst Weeds—Distribution and Biology.* 609 pp.
6. McWhorter, C. G. 1977. "Weed control in soybeans with glyphosate applied in the recirculating sprayer." *Weed Sci.* 25:135-141.
7. Weatherspoon, D. M. and E. E. Schweizer. 1971. "Competition between sugarbeets and five densities of kochia." *Weed Sci.* 19:125-128.

32.4   Controlling Preharvest Fungal Diseases of Sugar Beet

Earl G. Ruppel

AR-SEA/USDA, Crops Research Laboratory, Colorado State University
Fort Collins, Colorado 80523

Integrated pest management (IPM) entails: (i) maximum use of cultural and other natural controls to prevent pest buildup; (ii) monitoring pest populations to determine the need for further measures; and (iii), only when needed, using suppressive techniques to prevent economic damage to the crop *(Report of the President's Council on Environmental Quality* (1972). Such a program should include all the pests of a crop, interactions among other crops used in rotation, and the environmental effects of the selected control measures. Thus, IPM is based on multidisciplinary ecological principles [52]. As defined, no sophisticated IPM program is in use which includes controlling sugar beet fungal diseases; however, control measures for some individual diseases have been integrated to reduce losses in the crop.

Seedling Diseases

Pre- and postemergence damping-off is an ubiquitous problem wherever sugar beets are grown, with species of *Aphanomyces, Phoma, Pythium,* and *Rhizoctonia* being the most frequently encountered causal agents. Cultural modifications and fungicide seed treatment usually are combined to reduce seedling losses. In parts of Europe, seed lots also are routinely bioassayed for the presence of *Phoma betae* to determine the need for special seed treatment. In the USA, significant genetic resistance to *A. cochlioides* has been incorporated in some cultivars [47].

Rotation with nonsusceptible crops is a routine cultural practice in most areas. Since *A. cochlioides, Pythium aphanidermatum,* and *R. solani* cause little damage at low soil temperatures, early sowing also helps reduce stand losses [31]. Conversely, *P.*

betae is most destructive under low temperatures, and partially can be avoided by sowing at high soil temperatures [1].

The use of monogerm seed and precision planting, coupled with the universality of damping-off pathogens, has led to the almost routine use of seed protectant fungicides, which usually are applied as an aqueous mist or slurry, or as a dust in pelleting operations. Fenaminosulf, maneb, or PCNB, alone or in combination, are used in the USA. In England, where phoma black leg is important, ethylmercury phosphate (EMP) and thiram are fungicides of choice [7]; however, mercurial fungicides are not registered for use on sugar beet in the USA and some other countries. Preharvest fungicide sprays on seed crops to prevent infection by seed-borne *P. betae* have not shown much promise in England [4], but some control was obtained in West Germany with fungicides applied just before harvest [29]. Production of seed in areas of low summer rainfall (e.g. Oregon in the USA and the Mediterranean area in Europe) has greatly reduced the incidence of the fungus on seed.

Decreased use of fungicides for seedling problems will be possible only when adequate genetic resistance to the pathogens has been found and incorporated into commercial cultivars. Some resistance in seedlings to *R. solani* [12] and *Pythium* spp. [29] has been reported.

Foliar Diseases

Of about 10 foliar fungal diseases, only four require active control measures to reduce losses in sugar beet. These four include cercospora leaf spot, downy mildew, powdery mildew, and rust. Of these, only leaf spot and powdery mildew are of widespread economic importance.

Cercospora Leaf Spot

Rotations, resistant cultivars, and fungicide sprays are used to combat *Cercospora beticola* where leaf spot is a serious threat. Crop rotation of 1 to 3 years effectively reduces inoculum potential [40], and quantitative resistance protects the crop during the years of moderate epiphytotics [54].

When conditions are particularly favorable for leaf spot development, or where genetic resistance is inadequate, fungicides must be used. Applications of fixed copper or dithiocarbamate fungicides have provided moderate control [13]; however, systemic benzimidazoles [49, 55] and organic tin compounds [46, 56] largely have replaced other fungicides for control of *C. beticola*. With the development of fungus strains that are tolerant to benzimidazoles [21, 41] and triphenyltin [22], new compounds continue to be sought.

In Yugoslavia, epidemiological studies have provided data to forecast leaf spot incidence and issue spray warnings [35]. The use of such information, economic threshold data [45, 53], and the availability of genetic resistance make this disease a likely candidate for IPM system.

Powdery Mildew

Control of powdery mildew *(Erysiphe polygoni (= E. betae))* almost exclusively is effected with early sprays or dusts of sulfur. Other fungicides have controlled the disease, but none has been more effective or as economical as sulfur [8, 27]. In California, general recommendations call for aerial application of 22 to 44 kg of sulfur dust, or 11 kg wettable sulfur in 94 liters of water per hectare at the first sign of mildew; at layby, a treatment with 11 to 22 kg of wettable sulfur (in 234 liters of water) or dust, applied via ground equipment, has been effective [25]. One to three applications are used, depending on disease recurrence. In some areas of the USA, "flowable" sulfurs, with a particle size of 2.5 mu and 0.7 kg sulfur per liter, have given adequate control when applied at a rate of 4.7 liters flowable in 187 liters of water per hectare at the first sign of the fungus [16].

A recent report [18] indicated that excellent long-term control of powdery mildew was obtained with granular crown application of the systemic triadimefon. Significant yield increases relative to those obtained with sulfur treatments resulted from crown placement at 0.56 to 2.24 kg a.i./ha. Two foliar sprays at 0.28 kg a.i.ha/application, and a preplant soil application at 1.12 kg a.i./ha proved comparable to sulfur sprays. This compound is not registered for use on sugar beet in the USA.

Considerable variability in susceptibility to *Erysiphe* has been observed among sugar beet cultivars [27, 39], and resistant lines have been developed in Russia [34] and West Germany [29]. Apparently, resistance can be incorporated in commercial cultivars, but the process takes considerable time.

Substantial epidemiological, ecological, and economic threshold data have been documented for sugar beet powdery mildew [15, 16, 26, 30, 42, 43, 51], and should be invaluable in the development and implementation of an IPM program for sugar beet. Additional work is needed to clarify the role of the ascigerous stage of the fungus in the epidemiology of the disease.

Downy Mildew

Downy mildew, caused by *Peronospora farinosa* f. sp. *betae,* remains a potentially important disease in northern Europe and western coastal valleys of the USA. Control is effected primarily by cultural practices and the use of resistant cultivars. The sporadic occurrence of the disease, its systemic nature, and the need for prolonged protection are probable reasons why fungicides have proved inadequate in controlling sugar beet downy mildew [5, 9].

Sugar beet seed crops in close proximity to root crops [10] and beets left by harvest operations ("groundkeepers") [2] are considered the primary sources of downy mildew inoculum. Thus, removal of groundkeepers and separation of seed and root crops are important control measures. Decreased nitrogen application, denser stands, and early sowing are other cultural measures reported to reduce losses from downy mildew [6]. Since oospores of the fungus may persist in infected leaf debris [59], crop rotation also may be an effective way to avoid potential primary inoculum.

Use of cultivars having resistance to the pathogen may be the most practical means of avoiding losses due to downy mildew. Moderate resistance has all but eliminated the disease as a major problem in California [37], and breeding programs to select for resistance to *Peronospora* are underway in England [3] and russia [17].

Beet rust.—Uromyces betae causes a typical leaf rust of sugar beet, primarily important in Russia, but also occurring in other northern European countries and in winter-grown beets along the Pacific coast in the USA. Control measures reported by the Soviets [34] include: (i) roguing infected seed beets, followed by deep cultivation; (ii) isolation of seed beets from root crops; (iii) disinfestation of steckling roots before planting; and (iv) fungicide seed treatment. Selection for resistance has significantly reduced losses due to rust in the coastal areas of the USA [1].

Root rots: Preharvest rots of mature sugar beets are the most difficult fungal diseases to control. Fortunately, their occurrence is rather sporadic, and none is of worldwide economic importance. Nevertheless, losses in individual fields or areas can be catastrophic. Some of the diseases most frequently reported and their causal fungi include: aphanomyces tip rot *(A. cochlioides);* charcoal rot *(Macrophomina phaseoli);* fusarium rot *(Fusarium* spp.); phoma rot *(P. betae)*; phymatotrichum rot *(Phymatotrichum omnivorum);* phytophthora wet rot *(Phytophthora drechsleri & P. megasperma);* pythium rot *(Pythium* spp.); rhizoctonia root and crown root, or dry rot canker *(R. solani);* rhizopus soft rot *(Rhizopus arrhizus & R. stolonifer);* southern sclerotium rot *(Sclerotium rolfsii);* and violet root rot *(Helicobasidium purpureum).*

Since chemicals have not been effective in controlling most root rotting fungi of sugar beet, cultural practices are the chief means by which growers can hope to reduce losses due to these fungi. Generally, good farming practices that promote vigorous plant growth help reduce root rot incidence. Methods to improve soil aeration and structure should be adopted. Nitrogen fertility also is an important consideration. Land leveling is recommended to permit good drainage and uniform irrigation flow, which precludes standing water in low areas of a field [1]. All of the pathogens except *P. betae* are soil inhabitants and crop rotation does not eliminate them from the soil environment; however, long rotations with nonsusceptible crops can reduce inoculum potential to a tolerable level [11, 32, 50]. The species of crop which precedes beets also may be as important as the length of rotation, indicating some effect on other soil microflora [11, 14, 48]. Cultivation practices that cause soil deposition in beet crowns have been shown to increase losses due to rhizoctonia root and crown rot [44, 60], but the effect of soil deposits in crowns on other root diseases is yet to be determined. To reduce soil deposition in crowns, cultivations should be minimized and performed at slow tractor speed; shields can be attached to the cultivator sides, or planting can be done on preshaped raised beds.

Chemical control of *Sclerotium rolfsii,* resulting in significant incidence and increases in yield, has been reported with the systemic fungicides carboxin (2 to 2.5 kg/ha) and chloroneb (15 to 20 kg/ha) applied as a "ridge-soil drench" [38]. Bioassays to determine the level of sclerotial populations in field soil [32] may be useful for predicting the need of fungicide applications.

Genetic resistance to the root-rotting pathogens has been difficult to detect and develop. Substantial resistance to *R. solani* in two breeding lines was developed by J. O. Gaskill [19] and improved in a continuing breeding program by the US Department of Agriculture at Fort Collins, Colorado, and East Lansing, Michigan [24]. The first *Rhizoctonia*-resistant commercial hybrid incorporating resistance germplasm developed at Fort Collins was released to growers in 1979 by a private company.

Vascular Diseases

Vascular Diseases have been less important in sugar beet than in many other crops. Fusarium yellows, caused by *Fusarium oxysporum f. sp. betae,* is a problem in certain areas of Colorado and Oregon in the USA [23, 57], and in Yugoslavia [34]. Control measures have not been developed, but differences in varietal susceptibility indicate a possibility for the transfer of resistance to commercial cultivars [33, 36]. Verticillium wilt, induced by *Verticillium albo-atrum* in the USA [20] or *V. dahliae* in the Netherlands [58], has been of such limited occurrence that no control measures have been sought. In Europe, the black wood vessel disease, caused by *Pythium irregulare,* also is of limited occurrence and no control measures have been required [1].

Future Directions and Needs

The use of fungicides to supplement other control measures must continue until more effective controls are devised, or adequate genetic resistance has been incorporated into commercial cultivars. However, pesticide use must be managed more efficiently to reduce grower costs and to prevent unnecessary environmental contamination.

New, safer fungicides are needed to replace those that may be removed from our arsenal of weapons. In the USA, for example, reregistration of the ethylene bisdithiocarbamates, PCNB, and some benzimidazoles has been challenged, whereas the restricted use of mercurials probably is forthcoming in Europe. But, with new product developments, we must use chemicals only to *supplement* and not *supplant* cultural control measures. Additional research also is needed on the correct timing of applications for the most efficient, economical, and safe use of fungicides.

Breeding programs must be continued and methods updated with the goal of developing or improving tolerance or resistance to the many sugar beet pathogens. From an economical and ecological standpoint, we cannot continue to rely on costly or potentially hazardous chemicals to combat all pests. Here, pathologists can contribute their expertise in defining the epidemiological factors that can be used for the establishment of artificial epiphytotics needed to screen breeding lines for resistance.

Not all fungal diseases of sugar beet lend themselves to the ideal criteria of an IPM program but, for those that do, we need more information on the dynamics of inoculum concentration to be able to predict economic damage, i.e., the economic threshold of a given pest or pathogen. More studies are needed on the biology of the pathogens and their interrelationships with other microorganisms in their environment. Further, we must study cultural and crop management practices to determine their effect on disease incidence and severity. And, we must determine the effects of various control measure interactions on other pests and on the environment. As J. B. Kendrick, Jr. [28] so ably pointed out: "The objective of our research is to develop a pest management strategy employing all control measures in a system's context to provide the highest practical yields and maximum environmental protection." With a predicted 8 billion inhabitants crowding our earth in the next 35 years, this is our challenge . . . this must be our goal.

References

1. Bennett & Leach, *Advances in Sugarbeet Production: Principles and Practices,* Iowa State Univ. Press, Ames, p. 223, 1971;
2. Björling and Möllerström, *Socker* 19:17, 1964;
3. Brown, *Ann. Appl. Biol.* 86:261, 1977;
4. Byford, W. J., *personal communication;*
5. Byford, *Proc. Brit. Insect. Fung. Conf. Brighton,* p. 169, 1965;
6. Byford, *Plant Path.* 16:160, 1967;
7. Byford, *Plant Path.* 26:39, 1977;
8. Byford, *Ann. Appl. Biol.* 88:377, 1978;
9. Byford & Hull, *Ann. Appl. Biol.* 52:415, 1963;
10. Byford and Hull, *Ann. Appl. Biol.* 60:281, 1967;
11. Byford and Prince, *Ann. Appl. Biol.* 83:61, 1975;
12. Campbell and Altman, *Phytopathology* 66:1373, 1976;
13. Carlson, *J. Am. Soc. Sugar Beet Technol.* 14:254, 1966;
14. Deems and Young, *J. Am. Soc. Sugar Beet Technol.* 9:32, 1956:
15. Drandarevski, *Phytopathol. Z.* 65:124, 1969;
16. Forster, *Plant Dis. Rep.* 63:239, 1979;
17. Fradkina and Kazachenko, *Sov. Agr. Sci.* 4:19, 1977 (Engl. trans.);
18. Frate and Leach, *Proc. Am. Phytopathol. Soc.* 4:205, 1977;
19. Gaskill, *J. Am. Soc. Sugar Beet Technol..* 15:105, 1968;
20. Gaskill and Kreutzer, *Phytopathology* 30:769, 1940;
21. Georgopoulos and Dovas, *Plant Dis. Rep.* 57:321, 1973;
22. Giannopolitis, *Plant Dis. Rep.* 62:205, 1978;
23. Gross and Leach, *Phytopathology* 63:1216, 1973;
24. Hecker and Ruppel, *J. Am. Soc. Sugar Beet Technol.* 19:246, 1977;
25. Hills, F. J., *personal communication;*
26. Hills et al., *Plant Dis. Rep.* 59:513, 1975;
27. Hills et al., *Calif. Agr. 30* (10):16, 1976;
28. Kendrick, *Calif. Agr.* 32(2):3, 1978;
29. Koch, F., *personal communication;*
30. Krutova, *Mikolog, Fitopathol.* 3:148, 1969;
31. Leach, *J. Agr. Research* 75:161, 1947;
32. Leach & Davey, *J. Agr. Research* 56:619, 1938;
33. MacDonald et al., *Plant Dis. Rep.* 60:192, 1976;
34. Marić, *Bolesti Sécerne Repe,* Poljoprivrednog Fak. Novi Sad, 1974;
35. Marić, I.I.R.B. 7:11, 1975;
36. McFarlane, J. S., *personal communication;*
37. McFarlane, *Plant Dis. Rep.* 52:297, 1968;
38. Mukhopadhyay and Thakur, *Plant Dis. Rep.* 55:630, 1971;
39. Mumford, D. L., *personal communication;*
40. Nagel, *Phytopathology* 28:342, 1938;
41. Ruppel and Scott, *Plant Dis. Rep.* 58:434, 1974;
42. Ruppel et al., *Plant Dis. Rep.* 59:283, 1975;
43. Ruppel and Tomasovic, *Phytopathology* 67:619, 1977;
44. Ruppel and Hecker, *unpublished;*
45. Saito, *Mem. Fac. Agr. Hokkaido Univ.* 6:113, 1966 (Engl. summary);
46. Schlosser et al., *Pflanzenschutz* 9:122, 1957;
47. Schneider, *J. Am. Soc. Sugar Beet Technol.* 12:651, 1964;
48. Schneider and Robertson, *Plant Dis. Rep.* 59:194, 1975;
49. Schneider et al., *J. Am. Soc. Sugar Beet Technol.* 16:525, 1971;
50. Schuster and Harris, *J. Am. Soc. Sugar Beet Technol.* 11:128, 1960;
51. Skoyen et al., *Plant Dis. Rep.* 59:506, 1975;
52. Smith, *Calif. Agr.* 32(2):5, 1978;
53. Smith and Ruppel, *Can. J. Plant Sci.* 53:695, 1973;
54. Smith and Ruppel, *Crop Sci.* 14:113, 1974;
55. Solel, *J. Am. Soc. Sugar Beet Technol.* 16:93, 1970;
56. Stallknecht and Calpouzos, *Phytopathology* 58:788, 1968;
57. Stewart, *Phytopathology* 21:59, 1931;
58. Van den Ende, *Tijdschr. PlZiekt.* 62:21, 1956;
59. Van der Spek, *Jaarverslag 1963,* Inst. Plantzenziektenkundig Onderzoek, Wageningen;
60. Yamaguchi et al., *Proc. Sugar Beet Res. Assoc. Japan,* No. 19, p. 25, 1977 (Engl. summary).

32.5 New Approaches to Sugar Beet Virus Yellows Control

G. E. Russell

Department of Agricultural Biology, The University
Newcastle upon Tyne, NE1 7RU, England

Introduction

Virus yellows is a collective term for several different sugar beet diseases. At least three aphid-transmitted viruses are involved in different parts of the world. In addition there is an associated fungal disease, caused by *Alternaria* spp., of the leaves of virus-infected plants.

Beet yellows virus (BYV) causes a lemon-colored chlorosis on the older leaves of sugar beet and numerous very small red or brown necrotic spots usually develop on the chlorotic areas. The main properties of BYV, which is widely distributed in Europe, North America and Japan, have been described in detail elsewhere (Russell (1970) *CMI/AAB Descriptions of Plant Viruses* No. 13; Kassanis *et al., Virology 77,* 95). BYV can be transmitted by several species of aphids in which it is stylet-borne and semi-persistent. *Myzus persicae* Sulz. and *Aphis fabae* Scop. are the main vectors of BYV in the field. BYV has a rather restricted host range and infects plants mainly in the Chenopodiaceae.

The most common cause of virus yellows in Europe is beet mild yellowing virus (BMYV) which causes the older leaves of sugar beet to become orange-colored, thickened and brittle (Russell (1960) *Ann. appl. Biol. 48,* 721; Bjorling & Mollerstrom (1974) *Socker Handlingar 26,* 1; Polak & Chod (1975) *Biologia Plantarum 17,* 304). The principal vector of BMYV is *M. persicae* in which it is circulative and persistent. BMYV has a wider host range than BYV including many common weeds of arable land in Europe (Russell (1965) *Ann. appl. Biol. 55,* 245). The older leaves of BMYV-infected sugar beet plants usually become infected with weakly parasitic fungi, notably *Alternaria* spp.; this secondary disease causes premature death of these leaves which results in additional losses of sugar yield.

A third virus, beet western yellows virus (BWYV) is the most widespread sugar beet yellowing virus in the USA (1960) *Phytopathology 50,* 389) and Japan (Sugimoto & Murayama (1972) *Bulletin of Sugar Beet Research* (Japan) *13,* 1). BWYV, which is serologically related to BMYV (Duffus & Russell (1975) *Phytopathology 65,* 811), has an unusually wide host range with more than 100 host species in 21 dicotyledenous families (Duffus (1964) *Phytopathology 54,* 736). Strains of BWYV which can infect lettuce are widespread in Europe, but beet-infecting isolates of this virus do not seem to be important there at present. BWYV, like BMYV, is transmitted principally by *M. persicae* in which it is circulative and persistent.

Economic Importance of Virus Yellows

Plants infected with BYV or BMYV are reported to lose about 1.5% of their potential sugar yield for each week that they show symptoms (Russell (1964) *Ann. appl. Biol. 53,* 377; Heathcote, Russell and Van Steyvoort (1973) *FAO Crop Loss Assessment* No. 99; Heathcote (1978) *Ann. appl. Biol. 88,* 145). These effects are additive in plants which are infected with both viruses, but the extent of yield losses can vary very greatly with different isolates of BYV. Crops which become infected with yellows early in the growing season can therefore be expected to lose more than one-third of their potential sugar yield. In years when virus yellows has been common in the UK, losses caused by the disease have exceeded $20 million and between 1970-1975 the average annual loss was about 5% of the national crop (Heathcote 1978) *Plant Pathology 27,* 12). Losses at least as great as this have occurred in most of the world's sugar beet-growing areas. In California natural yellows infection has caused more than 22% loss of sugar yield (Bennett (1960) *US Dept. Agr. Tech. Bull. 1218,* 63 pp). A partial control of virus yellows increased the income of California farmers by more than $40 million a year between 1971 and 1975 (Duffus (1978) *J. Amer. Sug. Beet Tech. 21,* 1).

Present Control Measures

The main methods of controlling virus yellows have been reviewed by Hull (1974). These include the elimination of virus-infected plants, including 'groundkeeper' beets from previous crops and weed hosts, before they can act as sources of infection for nearby sugar beet crops. Most BYV infections derive from infected beet plants but weeds are the most important sources of BMYV (Heathcote & Byford (1975) *J. Agric. Sci., Cambridge 84,* 87; Jadot (1974)

Parasitica 30, 37). BWYV has a wide range of alternative host plants, including many crop plants and numerous common weeds (Duffus (1964) *Phytopathology 54,* 736). Eradication of all potential hosts of BMYV or BWYV from beet-growing areas would be impractical although their numbers can be greatly reduced by appropriate action. Other potentially important sources of aphids and virus yellows include stores of mangolds, fodder beet or red beet and seed crops.

Overwintered beet yields were common in California during the 1950s and most of the 1960s and a close correlation was found between virus yellows incidence and the proximity of overwintered beet. Beet-free periods were introduced in the late 1960s and these greatly reduced the incidence of virus yellows and significantly increased sugar yields (Duffus (1978) *J. Amer. Soc. Sug. Beet. Tech. 20,* 1). BYV and beet mosaic virus have since been almost absent from California beet fields and BWYV has moved in later than previously (J. E. Duffus, personal communication).

Virus-resistant or tolerant sugar beet cultivars have been developed both in the USA (McFarlane and Skoyen (1968) *Calif. Agric. 22,* 14) and in Europe (Russell (1972) *Proc. Roy. Soc. London Series B, 181,* 267), but most of these have not been grown on a large area because they have been of low processing quality or very susceptible to fungal or bacterial diseases (Russell (1978) *Plant Breeding for Pest and Disease Resistance.* Butterworths, London, 485 pp.).

Insecticides have been used very extensively to control the aphid vectors of beet yellowing viruses. These have generally been applied to the crop as foliar sprays, but in recent years there has been a strong trend in favor of granular formulations of insecticides applied to the soil when the seed is sown. In the UK a spray warning scheme, based on an inspection of randomly selected crops for the presence of aphids, has been used to advise growers of the best time to spray with insecticides (Hull (1974) *Brit. Sugar Beet Rev. 42,* 73). Other attempts to predict outbreaks of virus yellows in England have been based on weather data collected during the preceding winter and spring (Watson *et. al.* (1975) *Ann. appl. Biol. 81,* 181. In general, a heavy attack of virus yellows has followed a series of mild winters and springs which allowed the overwintering of numerous sources of aphids and yellows.

In spite of the present control measures and a considerable outlay on pesticides for aphid control, virus yellows remains a very damaging disease in many parts of the world. It is therefore important to seek new or improved control methods to supplement existing measures.

Future Possibilities

Improved Agronomic Measures

There is an obvious need to improve the identification and eradication of the most important sources of virus infection. It is particularly important to break the disease cycle by removing overwintering virus sources from beet-growing areas, whether these are weeds or other crop plants. There must also be a beet-free period in each main beet-growing area. These procedures could be encouraged or enforced by appropriate legislation or by stipulations in the contracts between growers and beet processors, particularly in areas where virus yellows is a serious problem in most years. However, such drastic measures are unlikely to be necessary if the farming community can be persuaded that the adoption of a stringent code of good hygiene would be of general benefit to agriculture and horticulture.

Many cultural methods have been shown to reduce the incidence of aphids and virus yellows in sugar beet (Heathcote (1972) *IIRB Journal 6,* 6). These include early sowing, because the resistance of sugar beet increases with age, and early-sown plants are therefore damaged less by virus yellows than are younger plants. It is important to grow crops with plants of uniform size and which are free from gaps, because these are less attacked by aphids and yellows than are uneven and "gappy" crops.

In countries where the incidence of virus yellows fluctuates greatly from year to year, more reliable methods of predicting virus yellows outbreaks would be particularly valuable. This would improve the effectiveness of insecticide application and reduce the number of necessary treatments.

Resistant Varieties

Following the early setbacks of using virus-tolerant cultivars, less effort has recently been devoted to the development of resistant cultivars in several countries. However, resistant cultivars can significantly reduce losses from virus yellows, and several kinds of resistance are available for the plant breeder to exploit, including resistance to the aphid vectors, a tendency to escape virus infection, and virus tolerance (Russell (1978) *Plant Breeding for Pest and Disease Resistance.* Butterworths, London, 485 pp.) Provided that sufficient resources are made available, it should be possible within a few years to develop high-yielding cultivars without the disadvantages of earlier cultivars. It should be remembered that other control methods, both chemical and biological, are usually more efficient on resistant cultivars, even where the resistance is only partly effective.

Improved Control of Virus Spread

Insecticide-resistant aphids have become a major problem in virus yellows control in several parts of the world. New, more effective, more 'durable' insecticides, which do not adversely affect natural enemies of aphids, are urgently needed. Some insecticides which were at first accepted as aphid-specific are now known to be very detrimental to insect predators of aphids. Although economic pressures on agrochemical firms seem to favour the development of broad-spectrum insecticides, aphids are such important pests that the development of truly aphid-specific compounds would be very worthwhile.

Recent work on chemical aphid repellents, including alarm pheromones, has suggested that they might be useful in virus yellows control. Hille Ris Lambers and Schepers (1978) (*Potato Research 21,* 23) applied slow-release capsules of the alarm pheromone trans-B-farnesene to the potato crop in an attempt to reduce the spread of aphid-transmitted viruses. Although they found little reduction in virus spread with the formulation used, there are strong reasons to seek more effective repellent compounds or better formulations of alarm pheromones. This is a particularly promising approach because alate aphids, which usually introduce viruses into crops, are more sensitive than apterae to these compounds (Zitter (1977) In *Aphids as Virus Vectors*. Academic Press, New York). Mixtures of alarm pheromones and contact insecticides might be very effective because the increased activity of the aphids due to the pheromone would raise their chances of coming into contact with droplets of insecticides on the foliage (Edwards *et al.* (1973) *Nature 241,* 126).

Several compounds are known to decrease aphid transmission of viruses when applied to crop plants. For example, foliar sprays of light white mineral oils partly inhibit the transmission of several stylet-borne viruses by *M. Persicae* (Peters 1977) *Proc. 1977 British Crop Protection Conf. 3,* 823). Russell (1968) (*Bull. Ent. Res. 59,* 691) showed that mineral oil sprays can significantly reduce the transmission of BYV from sugar beet to sugar beet. Field trials in Germany have recently demonstrated that combinations of mineral oil and insecticide give a better control of virus yellows than either alone (Proeseler *et al.* (1976) *Archiv Phytopath. PflSchutz 12,* 127; Kramer et al. (1978) *NachrBl. PflSchutz DDR 32,* 29).

Certain benzimidazole compounds, when applied as foliar sprays, have also reduced virus transmission in sugar beet (Russell (1977) *Proc. 1977 British Crop Protection Conf. 3,* 831). These compounds affect the settling and feeding behaviour and fecundity of aphids on treated sugar beet leaves. Combinations of benzimidazoles with certain types of insecticides might reduce spread of virus yellows more than would insecticides alone.

Certain trace elements, when applied to the sugar beet crop, can affect the settling behavior of viruliferous *M. persicae* so that virus transmission is reduced (Russell (1971) *Ann. appl. Biol. 68,* 67). Further work might show that other elements have a greater adverse effect on aphids than those which have already been examined. These could then perhaps be incorporated into formulations of insecticides.

Natural enemies of aphids can play a very important part in limiting aphid populations and spread of virus yellows. It is important, therefore, that other control measures do not harm predators such as Coccinellids, ground beetles or Syrphid larvae or parasites such as *Aphidius* spp. or *Aphelinus* spp. On the contrary, everything should be done to encourage these beneficial insects so that their contribution to aphid control can be increased.

Direct Control of Virus Yellows

Foliar sprays of mineral oil may affect both the feeding behavior of aphids and the establishment of viruses in the host plant (Peters (1977) *Proc. 1977 British Crop Protection Conf. 3,* 823). This is only one of several possible ways in which the viruses may be directly affected by chemicals. Although there was much interest during the 1950s and early 1960s in purine analog inhibitors of plant viruses, including thiouracil and 8-azaguanine, they were found to be too phytotoxic for practical exploitation.

Polyacrylic acid, when applied either to the soil as a drench or as a foliar spray, decreases the virus concentrations in, and the symptoms of, several localized and systemic virus infections (Kassanis and White (1977) *Proc. 1977 British Crop Protection Conf. 3,* 801). The effect of this compound is apparently to stimulate the natural defence mechanisms of the host plant. Unfortunately, polyacrylic acid is not very effective against beet yellowing viruses. Nevertheless, this work should encourage agrochemical companies to seek compounds which do affect virus yellows.

Another possibility is to apply compounds which suppress or mask the symptoms of virus yellows in plants that are already infected. Benzimidazole compounds can completely suppress the development of BWYV symptoms in lettuce, probably by preventing chloroplast breakdown (Tomlinson (1977) *Proc. 1977 British Crop Protection Conf. 3,* 807). Although similar effects have not been observed in virus-infected sugar beet, other compounds may already exist which could completely suppress virus yellows symptoms and this would presumably reduce losses from the disease.

Conclusions

This paper has indicated some of the many ways in which virus yellows control could be improved.

Many of the suggested new approaches are still at the experimental stage and need considerable research and development before they can be put into practice. Others could be implemented immediately. An acceptable level of control will be achieved only by using appropriate combinations of control measures, including improved hygiene and cultural methods, resistant cultivars and applications of chemicals to reduce virus spread.

32.6   Curly Top Virus Control

James E. Duffus

AR-SEA-USDA, US Agricultural Research Station
PO Box 5098, Salinas, CA 93915

Introduction

Beet curly top virus virtually destroyed the sugarbeet industry in the western USA following World War I. The disease was the principal limiting factor to sugarbeet production from the early 1900's until World War II. In recent years, curly top has been held to less than catastrophic proportions by a complex control program involving a number of facets of agriculture and which includes the use of: (1) cultivars resistant to the virus: (2) cultural practices to delay infection; (3) vector control in the crop; (4) vector control in non-crop production areas; (5) reduction of leafhopper breeding areas; (6) and reduction in virus sources. In spite of the effective reduction of economic losses induced by the curly top virus over the past several years, the virus is still present and changing along with ecology of the California production areas. These changes demand constant attention to the present control procedures as well as to future ones.

The Disease

Curly top is induced by a virus transmitted in North America only by the beet leafhopper (*Circulifer tenellus* Baker). The virus exists as a complex of strains that vary in virulence, host range, and other properties. Strains exhibit little evidence of

interference or cross-protection in hosts or the vector [2].

The disease thrives in the arid and semiarid areas of western USA and can cause serious losses of sugarbeet and other crops. Symptoms of the disease are dwarfing of plants, rolling, curling, or twisting of leaves and swelling and distortion of veins.

The beet leafhopper breeds on mustards, Russian thistle and other weeds and crop plants, producing several generations during the summer. The harvesting and drying of crop and weed hosts in the fall induce the leafhoppers to congregate on green summer annuals or perennials. As these plants dry, the leafhoppers move to foothill breeding areas and accumulate on perennials. During this fall period there is a market decrease in the incidence and virulence of the curly top virus in surviving leafhoppers.

Winter rainfall induces germination of various annuals, including filaree *(Erodium cicutarium)* and peppergrass. *(Lepidium nitidum)* in dense masses on the foothills. The leafhoppers move to these annuals and lay eggs on the plants on warm sunny slopes. Egg laying and the emergence of nymphs continue into February.

The drying of plants in the breeding areas in the spring and the maturation and drying of the annual hosts forces the migration of the new generation to the valley agricultural lands. Relatively few avirulent

508

curly top isolates are found in the migrating spring generation.

Control

Curly top virus in sugarbeet has been held to less than disastrous proportions by a complex control program involving a number of facets of agriculture. It is not consistent with modern principles of integrated pest management (IPM) and would be better termed something like SGA, or a scatter gun approach.

Resistance

Literature on the resistance of sugarbeets to curly top virus has been reviewed most recently by McFarlane in 1969 [12]. Resistance was first reported in 1908, but the first real efforts to develop this resistance was the work of Carsner in 1918 [4]. Selections of plants from commercial fields of European cultivars in the curly top areas of the west were made by government and sugarbeet company breeders. The most resistant of these selections were intercrossed in 1929 resulting in the first curly top resistant cultivar US 1 [5]. This cultivar was open-pollinated and had a low degree of resistance to curly top virus. Its introduction in 1933, however, helped to stabilize the western sugarbeet industry.

The need for seed of the new curly top resistant cultivars spawned a new industry for the USA—the sugarbeet seed industry. Prior to this time, sugarbeet seed was the product of European breeding establishments.

The cultivar US 1 was soon superseded by a series of open-pollinated cultivars more uniformly resistant to curly top and combining bolting resistance. Murphy [14] in field tests in Idaho demonstrated the dramatic improvement in curly top resistance in a few short years in comparisons of the original susceptible seed (R. & G. Old Type) to a series of resistant selections. Yields in this test ranged from 0.0 metric tons/ha for the susceptible cultivars to 37.3 metric tons/ha for Improved US 22, the most resistant cultivar at that time.

Curly top resistance has been incorporated into male sterile inbred lines used in the production of hybrid seed and in monogerm lines. It has been combined with nonbolting characteristics, *Cercospora* leaf spot resistance, downy mildew resistance, and yellows resistance.

The inheritance of curly top resistance is poorly understood. Some evidence indicates this resistance may be dominant under mild curly top conditions [1] but apparently breaks down under severe attack [16]. Disease resistance is associated with significant differences in incubation periods of infected plants rather than differences in the ability of resistant cultivars to recover from the effects of virus infection [9]. This factor of disease resistance has been incorporated into several aspects of curly top control.

Cultural Practices

Resistance of sugarbeets to infection and to injury by the curly top virus increases with plant age. Young seedlings, even those of resistant cultivars, are highly susceptible to infection and may be severely damaged by virulent strains of the virus. The avoidance of infection during the early part of the growing season has long been recognized as an important factor in preventing excessive losses [6]. With the development of resistant cultivars, the prevention of early infection was assumed to be less urgent. During the last 20 years, however, curly top isolates have increased in severity to the point that isolates considered severe in the 1950s are now considered mild [13]. With a high incidence of the more virulent strains, there is a distinct advantage to planting sugar beets in curly top areas early enough for them to attain size before they are subject to infection from viruliferous leafhoppers.

Cultural practices that promote rapid growth of the plants may reduce yield losses for several reasons. Diseased plants yield more under good cultural conditions and rapid top growth may aid in the control of spread of the virus. The beet leafhopper is a sun-loving insect and is adversely affected by shade and high humidity. Leafhopper populations will drop to very low levels under these conditions. In addition, there is a high degree of yield compensation of healthy neighbors of diseased plants and early infection rates as high as 25% may cause little or no yield reduction in properly cared for fields [17].

Leafhopper Control at Crop Site

Early attempts to control curly top with insecticides in the field were largely unsuccessful. Although there may have been high percentages of kill in these applications, they were ineffective because of the high populations of leafhoppers visiting the crop over an extended period and the inability of the insecticide to kill the leafhopper before the infection.

A high degree of control has been obtained in recent years with the use of phorate by placing the material about 12.7 cm directly below the seed at the rate of 1.12 kg active material/ha³. These applications are effective in two ways. There is an apparent repellent effect of treated areas to the leafhopper and the extended systemic action of the insecticide effectively reduces curly top incidence in the critical early period of beet growth.

Vector Control in Noncrop Areas

The concept of curly top control through the use of insecticides in uncultivated areas was tested as early as 1931 [7]. The California State Department of Agriculture adopted and expanded this program in 1943 and has continued it to the present time at a cost of up to $850,000 per year. The objective of the control program is to prevent leafhopper populations from reaching epidemic levels. Control is attempted at four critical periods in the disease cycle [10]. In the fall, large acreages (up to 81,000 ha) of beet leafhopper infested Russian thistle are sprayed to reduce the overwintering populations. If winter annuals have not emerged, the remaining leafhoppers congregate on perennial brush in the foothill areas where their numbers may be further reduced by spray applications. In January, the leafhoppers which have concentrated in the foothill areas are treated before eggs are laid. Final control attempts are made in the spring at a critical time when most of the nymphs have emerged, but before migration to the agricultural areas.

It is difficult to assess the results of the vector control program since the other control programs are in continual use and adequate checks on what would happen to curly top incidence if the spray applications were not performed have not been executed.

Reduction of Leafhopper Breeding Areas

The chief source of beet leafhoppers are weed areas that have resulted from intermittent farming, land abandonment and overgrazing [15]. Replacing these weed areas with natural vegetation (annual and perennial grasses and shrubs) would greatly reduce the leafhopper breeding areas. The replacement of weed hosts with nonhosts requires time and knowledge and the cooperation of the whole agricultural society, since the leafhopper breeding areas and crop areas are isolated geographically and economically.

Reduction of Virus Sources

The role of the reduction of virus sources in decreased incidence of virus diseases has been dramaticaly shown with the yellowing complex of sugarbeets [8]. The same principle has been suggested for the control of beet curly top [11]. The elimination of overwintering beet fields in and near the breeding areas, the destruction of ground-keepers and weed beets before the migration of leafhoppers from the breeding areas would probably greatly reduce the rapid buildup of curly top incidence in the migrating leafhopper populations.

The role of different populations of weed hosts in the breeding grounds, and/or agricultural lands or for that matter the role of beets or other crop plants in the rapid increase in virus incidence in the migrating insect population is not known. Knowledge of the principles of virus source and built up plant species, coupled with more information on the effects of host species on virus attenuation and/or increased virulence would seem to be a key for future control efforts with curly top virus.

Curly Top Control—The Future

Curly top disease has been extensively studied since it was first reported in 1888. The control and epidemiological aspects of the disease have received attention, rivaling all other virus diseases, but the constantly changing ecological conditions in areas of the west (new irrigation systems, new land development and new agricultural chemicals) where curly top is prevalent demands a new look at control measures [13].

Curly top virus is increasing in severity, but the host plant and ecological reasons for this are not understood. A continuation of the development of resistant cultivars is important to combat these severe strains. In addition, much information is needed on the life history, host plants, flight habits, population dynamics and natural enemies of the vector. New knowledge is needed on the host plants of the vector and of the virus and the interaction of these plant species on populations of the vector and incidence of severe forms of the curly top virus.

Until the ecological and climatological aspects of curly top epidemics can be determined, it will be difficult to implement the ideal criteria of an IPM program.

References

1. Abegg, F. A. and F. V. Owen. *Amer. Naturalist* 70:36.
2. Bennett, C. W. *Phytopathol. Monogr. No. 7, Am. Phytopathol. Soc.*, St. Paul, Minnesota, 81 pp. 1971.
3. Burtsch, L. M. *Spreckels Sugar Beet Bull.* 32:15-17, 24. 1968.
4. Carsner, E., *USDA Circular* 388, 7 pp 1926.
5. Carsner, E., *USDA Technical Bulletin* 360, 68 pp. 1933.
6. Carsner, E. and C. F. Stahl. *J. Agr. Res.* 28:297-320. 1924.
7. Cook, W. C. *Calif. Dep. Agr. Monthly Bull.* 22:138-141. 1933.
8. Duffus, J. E. *J. Am. Soc. Sugar Beet Technol.* 20:1-5. 1978.
9. Duffus, J. E. and I. O. Skoyen. *Phytopathology* 67:151-154. 1977.
10. Fehlman, D. R. *Spreckels Sugar Beet Bull.* 32:6-8. 1968.
11. Hills, F. J., G. L. Ritenour and L. M. Burtch. *California Sugar Beet*, pages 26-27. 1968.
12. McFarlane, J. S. *J. Int. Inst. Sugar Beet Res.* 4:73-83. 1969.
13. Magyarosy, A. C. and J. E. Duffus. *Calif. Agric.* 31(6):12-13. 1977.
14. Murphy, A. M. *Proc. Am. Soc. Sugar Beet Technol.* 3:459-462. 1942.
15. Piemeisel, R. L. *U.S. Dep. Agr. Circ.* 299, 24 pp. 1932.
16. Savitsky, V. F. and A. M. Murphy. *Proc. Am. Soc. Sugar Technol.* 8:34-44. 1954.
17. Skoyen, I. O. and J. E. Duffus, *unpublished.*

32.7 Nematodes

W. Heijbroek,

Institute of Sugar Beet Research
P.O. Box 32, 4600 AA BERGEN OF ZOOM, The Netherlands

Introduction

Integrated pest management is mainly based on manipulation of populations and this necessitates the development of techniques for the assessment of population densities. Nematodes can be extracted from soil quantitatively by different types of elutriators, but problems are caused by their uneven field distribution and their identification. This and the fact that the relation between initial densities and damage is strongly influenced by a number of external factors are the reasons why schemes for supervised control are applied only in a few countries.

The concept of integrated control of nematodes in sugar beet is rather new and mainly restricted to the beet cyst nematode *Heterodera schachtii Schm.* which is the main sugar-beet nematode pest in most European countries [1]. Predators are hardly known and only recently some effective pathogens of this nematode have been found [2]. Moreover it is very difficult to interfere in the delicate and complex ecosystems of the soil. This is the reason why most of the investigations have been restricted to possibilities for supervised control, the effect of cultural practices and resistant or tolerant varieties.

Apart from population assessments this paper deals with the different control methods, their possible application and integration, with the emphasis on *Heterodera schachtii Schm.*

Population Assessment

The distribution of nematodes in a field can be extremely uneven and therefore relative large samples from a small area are needed. This holds particularly for *H. schachtii,* where the degree of irregularity is dependent on infection time, the frequency of growing host crops and physical properties of the soil. For advisory work in the Netherlands the size of the sampled area does not exceed 2 ha and at least 50 cores to a depth of 20 cm are taken according to a standard grid pattern of diagonal lines.

The soil sample can be either processed directly in the Seinhorst elutriator or dried and subsequently extracted in, for example, the Fenwick can. For routine soil sample analyses a rotating disc automatic extraction apparatus has been developed by the Laboratory for Soil and Crop Testing at Oosterbeek (Netherlands) with a capacity of about 180 samples/hour. The collected cysts are either identified directly under the microscope or vulval preparations are made. Mixtures with *Globodera rostochiensis, Heterodera avenae, H. trifolii* and *H. cruciferae* regularly occur; to carry out this identification work properly a training period of two years is necessary. After crushing the cysts the number of eggs and larvae are estimated. The degree of infestation is determined by the numbers of viable eggs and larvae. Methods have been developed for the detection of the viability of eggs and larvae by staining or hatching tests, but they are applied in practice on a very limited scale because the reliability is not always sufficient. In some parts of the Federal Republic of Germany a biotest using oil seed rape or beet is appplied in practice [3], but quantitative interpretations of the results seems difficult.

Populations of the root knot nematodes *Meloidogyne naasi* and *M. hapla* can be assessed by the centrifugal flotation method and the ectoparasitic nematodes *Trichodorus spp* and *Longidorus spp* are extracted from soil samples by means of, for example, tenbrink elutriator. In the latter case the soil should not be disturbed. With the same type of elutriators stem nematodes (*Ditylenchus dipsaci*) can also be determined but their numbers are not relevant because several physiological races, each with its own host range exist.

Supervised Control

If we want to restrict treatments to those instances in which the expected amount of damage caused by the pest will justify the cost of control measures, tolerance limits have to be determined first. This can be done in field trials consisting of a large number of plots with different population densities. In this way the relation between the iitial densities and crop loss is established and, with the help of the Seinhorst equations [4], a tolerance limit can be deduced.

Although investigations are not completed yet we tend to believe that, for *H. schachtii*, this limit is 3-5 eggs per g of dried soil in the temperate zones [5]. There seems to be quite a variation depending on soil properties and weather conditions. In one field experiment in California a tolerance limit of about 1 egg/g of dried soil was found [6].

The determination of the economic threshold is far more complicated because at the same initial density the degree of damage is dependent on temperature, physical properties of the soil[7], weather conditions [8] and drilling date [9]. Moreover, the applied nematicides have side effects: in the case of fumigants these are on other nematodes and on weeds; systemic nematicides also control soil arthropods, aphids (and, thereby, the spread of virus yellows within the crop— and leaf-eating pests of beet. If we restrict ourselves to sugar-beet, under Californian conditions an economic threshold of 1.4 eggs/g has been established [6]. In the temperate zones this is estimated at 6–7 eggs/g but the year to year changes in weather conditions cause a wide variance. Another important factor determining the economic threshold is the degree of crop loss compensatio obtained by the application of nematicides. At economical doses and under optimal conditions this compensation is theoretically 100% only when initial densities are less than about 15 egss per g of soil.

With regard to other sugar-beet namtodes no tolerance limits have been produced, because often the degree of damage is largely determined by external factors.

Cultural Practices

Nemaotdes with a relatively narrow host crop range like *H. schachtii* can be controlled by a modified crop rotation adapted to the degree of infestation, based on pre-plant routine soil sample analyses [8]. In this way the build up of high population densities is avoided and the breakdown of present infestations can be accelarated. In modern farms the possibility for this type of control decreases because alternative non-host crops 'are few but, combined with chemical control, the prospects of this method are better. Modified crop rotation is also applied for the control of *Meloidogyne incognita,* but in the case of mixtures with other *Meloidogyne spp.* too few alternative crops remain.

Damage caused by *H. schachtii* can be reduced considerably by early drilling, because, at the time of nematode penetration, plants are larger and less usceptible [9].

Some host plants of *H. schachtii* like oil seed rape, fodder radish and mustard are not seriously damaged but the same number of larvae as in sugar beet invade the root systems. They could be used as trap crops if ploughing is done before the first generation of cysts is formed. Such a crop does not have any specific value and therefore has to be grown in the autumn as a green manure. In practice there are risks of multiplication and this is the reason why

selections are made for resistant varieites. Some of these selections show a high degree of resistance in greenhouse tests but, in field trials the decrease of the nematode population is still insufficient.

Large numbers of weed species belonging to the Cruciferae and Chenopodiaceae act as host plants [10] and can slow down the population decrease of *H. schachtii* in years when no host crops are grown. Good weed control can solve this problem.

Tolerant and Resistant Varieties

Some field selections of sugar beet and *Beta maritima* origins showed a decreased wilting, but the multiplication rates of *H. schachtii* were the same as in susceptible plants. After repeated backcrossing root yield could be improved without substantial loss of tolerance. In itself this is no solution for the nematode problem, but incorporated in resistant varieties it might be helpful. Until now no differences in susceptibility are detected between beet varieities, but individual resistant plants can be selected because of the wide variance in susceptibility. This so called partial resistance, which is also present in *Beta maritima*, seems to be polyfactorial and largely recessive, reasons why this could not be transfered to sugar beet varieites [11].

Annual wild beet of the so called "Patellaris-section" show complete nematode resistance (hypersensitivity reaction) but interspecific hybridisation is impeded by genetic barriers. Monosomic addition lines and some translocations have been produced, but transmission rates are still too low [12]. The hypersensitivity reaction is dominant but cannot be easily incorporated in the sugar-beet genome because of chromosomal aberrations. Progrss is made and it is hoped to improve transmission rates further to get homozygous lines but, in the short term, no resistant variety can be expected.

In California and the Netherlands a large number of different populations of *H. schachtii* have been tested against this type of resistance but so far no evidence of resistance breaking pathotypes has been obtained.

Biological Control

Some Collembola, such as *Onychiurus* spp., and mesostigmatid mites can feed on cyst-forming nematodes but very little is yet known about their predatory activity on *H. schachtii* in the soil. Moreover, the population density of these arthropods is rather low in light soils, where nematode damage is the most severe, and therefor they are not considered as very effective.

The migrant larvae coming from the cysts may be attacked by predatory fungi *Arthrobotrys* spp. or *Dactylella* spp. forming one or more kinds of trapping structures such as constricting rings or adhesive networks. These predacious fungi are not considered very important because of the fact that cyst nematodes are migratory only during a short period of the life cycle.

Somewhat more promising are the fungal parasites of females (*Catenaria auxiliaris*) and the egg parasites, of which the most important ones are *Verticillium chlamydosporium* and an unidentified species called the "contortion fungus" [2].

Practical Application of Integrated Control

For *H. schachtii*, systems of supervised control combined with crop rotation adapted to the degree of infestation are applied in the Federal Republic of Germany, the Netherlands and, to a lesser extent, also the United Kingdom. Soil samples are analysed by cyst extraction and identification [8] or bioassay [3]. The advice to the farmer is given according to a table of the relation between the degree of infestation and the period in which no host crop can be grown with or without the application of nematicides. A simplified version of the scheme in the Netherlands is given in Table 1.

In the Netherlands this system has been operating quite satisfactorily for more than 30 years. In the course of this period numerous modifications have

been made, but there are still some problems unsolved. Relatively new infestations of *H. schachtii* cannot be detected and more specific advice is needed for different soil types. Incorporation of resistant varieties is urgently needed as this is the only way to run this system properly in the future with the increasing tendency towards short crop rotations. In some countries strict quarantine regulations have been taken, but this seems to be a somewhat exaggerated approach because low population densities cannot be exterminated.

For the other sugar-beet nematodes no systems of supervised or integrated control are applied in practice on a wide scale at the present time.

References

1 Heijbroek, W., 1979. Results of the Sugar Beet Nematode Enquiry, organised by the "Pests and Diseases" Study Group of the I.I.R.B. Proc. 42nd Winter Congress I.I.R.B., 141-147.

2. Tribe, H. T., 1977, Pathology of Cyst-Nematodes.

3. Behringer, P., 1976, Moglichkeiten der Erfassung und Bekampfung des Rubennematoden., Zucker 29; 679-684.

4. Seinhorst, J. W., 1965, The relation between nematode density and damage to plants. *Nematologica* 11; 137-154.

5. Steudel, W.; Thielemann, R. and Huafe, W., 1978, Der Einfluss von Aldicarb auf die Vermehrung des Rubenzystenalchens (heterodera schachtii Schmidt) und den Ertrag von Zuckerruben in der Koln-Aachener Bucht.

6. Cooke, D. A. and Thomason, I. J. The relationship between Population Density of Heterodera schachtii, Soil Temperature and Sugarbeet Yields. *J. Nematology* 11; 124-128.

7. Wallace, H. R., 1956, Soil aeration and the emergence of larvae from cysts of the beet eelworm, H. schachtii Schmidt.

8. Heijbroek, W., 1973, Forecasting incidence of and issuing warnings about nematodes, especially Heterodera schachtii and Ditylenchus dipsaci. *J. Inst. Intnl.* Recherches Betteravieres 6; 76-86.

9. Weischer, B. and Steudel, W., 1972, Nematode Diseases of Sugar beet. *In Economic Nematology* ed. J. M. Webster; 60-61.

10. Steele, A. E., 1965, The host range of the sugarbeet nematode Heterodera schachtii. *J. of the Amer. Soc. Sugarbeet Techn.* 13(7); 573-603.

11. Heijbroek, W., 1977, Partial resistance of sugarbeet to beet cyst eelworm (Heterodera schachtii Schm.). *Euphytica* 26; 257-262.

12. Savitsky, H., 1978, Nematode (Heterodera schachtii) resistance and meiosis in diploid plants from interspecific Beta vulgaris x B. procumbens hybrids. *Can. J. Genet. Cytol.* 20; 177-186.

Table 1. Simplified Version of Advisory Scheme for *H. schachtii* Control used in the Netherlands

Number of Eggs and Larvae in Dried Soil		Degree of Infestation	Number of Years before Growing a Host Crop	
per 100 cc	per g		Without Nematicide	With Nematicide
<150	<1	Very light	0	0
150- 300	1 - 2	light	0	0
305- 800	2.1- 6	moderate	1-2	0
805-1750	6.1-13	rather severe	3-4	1
1755-2750	13.1-20	severe	4-5	2
>2750	>20	very severe	5-7	3-4

32.8   Storage of Sugarbeet

W. M. Bugbee

USDA, SEA, AR, Department of Plant Pathology
North Dakota State University, Fargo, ND 58105

Sugarbeets must be harvested during a short harvest period and then stored to await processing when grown in climates where winter temperatures fall below freezing. Most of the sugarbeet acreage in the world is harvested then stored for several to 150 days. In the USA up to 75% of the crop may be stored outdoors in piles 5 to 7 m high × 55 to 67 m wide. Respiration causes about 70% (Wyse, Proc. Beet Sugar Dev. Foundation Conf. p 47, 1973) and decay about 10% of the sugar loss (Bugbee, et al. J. Amer. Soc. Sugar Beet Techol. 19:19-24, 1976). Sucrose also is lost through fermentation under conditions of low oxygen content in poorly ventilated piles.

The amount of sugar lost from stored sugarbeets as a result of pathogen activity has been assessed in the USA (Bugbee, et al. ibid.). The surveys were made at a Moorhead, Minnesota factory in the processing seasons of 1974-1975 and 1975-1976. During a 128-day survey of the first processing season there were 414,427 tons of sugarbeets processed of which 1.2% was decayed tissue. Thus nearly 5000 tons of rotted tissue went into the factory. This rotted tissue had a potential sugar yield of over 500 tons. Impurities such as reducing sugars are present in rotted tissue which interfere with crystalization of sucrose. It was estimated that these impurities from rotted tissue caused an additional 809 tons of sucrose to go into molasses; therefore, the total sugar loss due to rotted beets at this one factory was over 1300 tons. Similar losses probably occured at five other factories in the Minnesota-North Dakota growing region. This was a potential loss of 7800 tons of sugar during the processing season of 1974-1975 in the Red River Valley of Minnesota and North Dakota. Results from a 117-day survey of the 1975-1976 processing season indicated that the losses were only 27% of that in 1974-1975. The reason for less rot during the second season probably was due to environmental factors during the growth period which are known to affect resistance to storage pathogens.

A 24% reduction in the amount of sugar extracted from stored sugarbeets after January 1 was reported for the processing seasons of 1957-1958 and 1961-1962 in USSR. (Zhigaylo, Sak. Prom. 43:47-50, 1969)

Most of the decay at the Moorhead factory was caused by the fungal pathogens *Phoma betae* and *Penicillium claviforme*. Observations in the Red River Valley since 1969 have not supported the observation of others with respect to the importance of *Botrytis cinerea* as an important storage rot pathogen. The low frequency in the Moorhead survey probably was due, in part to antagonism of *P. claviforme* toward *B. cinerea* (Bugbee, Can. J. Plt. Sci. 56:647-649, 1976). Sugarbeet isolates of *P. claviforme* inhibited the growth of *B. cinerea* in vitro and in vivo. Others also have shown that *P. betae* and not *B. cinerea* predominated among isolates taken from diseased sugarbeet roots in Alberta, Canada (Cormack et al., Phytopathology 51:3-5, 1961). Possibly the low level of *B. cinerea* reported in Canada also could be due to this antagonism. Other storage rot fungi of lesser importance include species of *Penicillium, Aspergillus, Fusarium, Rhizopus,* and *Pythium ultimum*. Bacterial pathogens are important only when the beets are stored under warm conditions. Once the pile temperatures have cooled down, bacteria become important only as fermenters under low oxygen levels.

Phoma betae is potentially the most damaging storage rot pathogen because its disease cycle is closely associated with the life cycle of the sugarbeet. This pathogen infects the seed, especially in humid seed production areas or when windrows of seed stalks are rained upon. Seedlots with 40 to 50% infection are not uncommon under these conditions (Byford, Plt. Path. 21:16-19, 1972). Seedlots with over 95% infected seeds have been observed in the US. When the infected seeds are planted, some of the seedlings may damp-off, but others may continue to grow and produce healthy-looking roots. The fungus, still inside the crown tissue of the sugarbeet after the beet is harvested and placed in storage, will begin to decay the roots usually about 80 days later. The sugarbeet also requires 70 to 90 days of exposure at low temperatures to induce flower and seed development upon regrowth of the root. The similar time required for both of these events suggests the need for research to determine if this is a coincidence, or if increased susceptibility to rot and induction of flowering are related. The fungus is able to survive in soil for 26 months after the infected sugarbeet seed have been planted. It survives continuously in storage yard soils (Bugbee, Phytopathology 64:1258-

1260). It was detected in sugarbeet debris that fell from piler booms as sugarbeet piles were being constructed (Bugbee, Plt. Dis. Reptr. 59:396-397). Therefore, *Phoma*-laden debris being brought in from an infested field could be deposited on beets brought in from noninfested fields during the piling operation.

Many species of *Penicillium* cause rot generally referred to as mold, but *P. claviforme* is the most prevalent fungus isolated from stored sugarbeets in the North Central Region of the USA. Rot is usually associated with wounds, and the fungus can occur in tissue rotted by *P. betae*. Rot caused by *P. claviforme* can be identified by coremia that are produced on brown, rotted tissue.

Botrytis cinerea, the most aggressive of these fungal pathogens, is capable of rotting tissue quickly over a wide temperature range. Rot caused by this fungus can be identified by the occurrence of sclerotia and gray masses of spores. Rotted tissue is dark brown or black. *P. betae* also may be present in *B. cinerea*-infected beets.

Decay is greater in roots that have been wilted, frost damaged, wounded during harvest procedures, or diseased during the growing period. Wounds predispose roots to initial infection. Before modern-day mechanical harvesting procedures, sugarbeets were lifted from the soil with a hooked knife. Decay originated often in the damaged areas caused by this hook.

A study in 1930 reported that most of the rot in stored roots was in the crown, and the amount of this rot, caused mostly by *P. betae* was proportional to the amount of crown that was removed. Traditionally, crown tissue has been removed because it has less sucrose than in the main taproot and it contains a high concentration of impurities, which contribute to the formation of molasses and interferes with the crystalization of sucrose during processing. But the rot that developed during storage in wounded crown tissue lowered the value of what was considered a properly crowned root to less than that of what was considered an improperly or high crowned root. The suggestion was made that roots destined for storage would maintain a higher quality if topped high rather than low (Tompkins, Phytopathology 20:621-635, 1930).

Artschwager and Starrett (Agri. Res. 47:669-674, 1933) reported that little decay resulted from roots cut through the crown because of specialized anatomical structure of this region. Recently, extensive data have shown that removeal of the crown tissue exposes pith tissue within the crown and that the pith is the most susceptible tissue in the root. Two to three times more decay may be expected to occur in storage in roots that have been partially crowned to expose susceptible tissue (Bugbee, Can. J. Bot. 53:1347-1351, 1975). Russian work more than 35 years ago recognized increased storage rot in roots that were topped by the conventional method. They recommended crown removal in such a way as to not expose pith tissue. The amount of decay originating at wounds at the tip of the taproot and body of the root does not increase during the storage period, but decay in the crown area continues to advance and eventually accounts for a major portion of the total decayed area (Bugbee et al., J. Amer. Sugar Beet Technol. 19:19-24, 1976). Moreover, removing crown tissue also causes a sharp rise in respiration which further contributes to sucrose loss. Recent research has shown that with proper management of available nitrogen in the field, sufficient sucrose can be extracted from intact crowns to warrant discontinuance of crown removal from roots intended for storage, especially those roots to be stored for 60 or more days (Akeson, et al., J. Amer. Soc. Sugar Beet Technol. 18:125-135, 1974; Cole, et al., J. Amer. Soc. Sugar Beet Technol. 19:130-137, 1976).

Soil fertility affects storability of the sugarbeet. Roots grown in soil low in phosphate were more susceptible to *P. betae* than roots grown in soil with adequate amounts of phosphate. Phosphate fertilization also seemed to reduce the loss of sucrose due to respiration (Larmer, J. Agr. Res. 54:185-198, 1937). Sugarbeets grown under low nitrogen fertility were more susceptible to *P. betae* than those grown under adequate nitrogen fertility (Gaskill, Amer. Soc. Sugar Beet Technol., Proc. 6:680-685, 1950). Results from Russia have shown that roots grown in adequately fertilized soil are more resistant to *B. cinerea* than roots grown under low fertility (Khovanskaya, Sakh. Svekly 1:35, 1962).

A standard practice, where feasible in the United States, is to make available only enough nitrogen to provide adequate top growth during most of the growing season. The intent is to have the nitrogen supply depleted from the soil a few weeks before harvest. This reduces top growth and causes more sucrose to accumulate in the root instead of being used as an energy source. The production of roots with a low nitrogenous content is highly desirable because these compounds interfere with the extraction of sucrose during processing. But the above research indicates that high quality roots desired by the processor are more susceptible to storage rot than roots of lower quality. This interaction is recognized in Russia where they suggested that selection for resistance to storage rot pathogens proceed simultaneously with selection of sugarbeet lines that have low levels of impurities.

Storage losses have been reduced by utilizing standard methods of genetic resistance and fungicides. In the USSR, with their vast acreage of sugarbeets, of which millions of tons must be stored each year, storage rot destroys roots intended for sugar production and mother roots intended for seed production. Thus, Soviet scientists began in the 1930s developing cultivars with resistance to storage rot pathogens. Commercial cultivars are in use which sustain 1.5 to 2 times less damage from storage pathogens than nonselected cultivars (Kornienko, Zash. Rast., Moscow 6:21, 1975). They screen for resistance only to *B. cinerea* because they claim that resistance to this fungus also conditions adequate levels of resistance to other storage pathogens (Shevchenko, Sel'khoziz:Moscow, pp. 375-379, 1961). However, in the USA it was shown that individual roots resistant to *B. cinerea* may be susceptible to *P. betae* or *P. claviforme* (Bugbee, unpublished data).

The development of breeding lines resistant to storage rot was begun in the 1950s in the USA but was not continued (Gaskill, Amer. Soc. Sugar Beet Tech. Proc. 6:664-669, 1950). The need for resistance took on new importance in 1970 and breeding lines with resistance to *P. betae, B. cinerea* and *P. claviforme* have been developed (Bugbee, Crop Sci. 18:358, 1978). A program is underway to combine storage rot resistance and low respiration rate into a storage-type cultivar.

Attempts to control storage rot with chemicals have had moderate success. Milk of lime or calcium hydroxide, a by-product of the factory process, has been applied to sugarbeet piles to increase the pH of the surface of the sugarbeets and render this area unsuitable for growth or establishment of a pathogen. The white surface also reflects sunlight and helps to keep the roots cool. This practice is routinely used in the USSR early in the storage season. The usefulness of thiosulfonic acid esters against storage rot has been shown (Khelemskii et al., Fiziol. Akt. Vesh. 4:110-113, 1972). In the USA thiabendazole controls rot caused by *B. cinerea* and *Penicillium* spp. and is gaining wide use.

Storing sugarbeets in a desirable environment obviously requires extensive expenditure of labor and capital. The first, and most important step is to lower the temperature of the stored root to reduce respiration and retard microbial activity. A root temperature of 10 C with relative humidity not below 95% and air velocity of the ventilation system of 0.08 to 0.25 m/sec for 10 to 14 days will enhance wound periderm development on the injured roots (Khelemskii, Sak. Prom. 45:46-48, 1971). The

temperature for extended storage should then be adjusted to 3-5 C with over 95% relative humidity. The least sucrose loss, sprouting, and fungal growth occured when roots were stored in 6% CO_2 and 5% O_2 (Karnik et al., J. Amer. Soc. Sugar Beet Technol. 16:156-167, 1970).

Sugarbeets have been stored under various types of protective covers and ventilated to protect them from freezing and dehydration (Wyse, ed., Proc. Beet Sugar Dev. Foundation Conf., pp. 5-87, 1975). These include rigid, insulated buildings, plastic film, and inflated, plastic film structures supported by air pressure. In the 1974-1975 processing season, initial investment construction costs for these structures ranged from $1.18 to $18.00/metric ton of sugarbeet. The annual cost of operation was $1.12 to $2.03/ton. The labor required to store the roots under these structures was 0.011 to 0.048 man hr/ton. The loss of recoverable sugar/ton/day was 86 to 91 g under the structures. Expected losses from conventional storage is approximately 206 g/ton/day. Obviously, sugar losses can be reduced by storing the roots under protective cover but economics dictate which, if any, of these storage alternatives will be implemented.

Trash is second in importance to temperature as a factor affecting the storability of sugarbeets. Trash might consist of petioles, leaves, weeds, and soil. This material can restrict air movement within the pile resulting in low oxygen concentrations thus favoring fermentation and the generation of heat. Trash is difficult to remove but an attempt at this has been used with moderate success by Michigan sugar companies. Barrels welded end-to-end with a fan at one end are mounted beneath the piler boom. This giant air gun blows trash from the beets as they fall off the end of the piler. The trash comes to rest on top of the pile, not within it. In Europe, delivered beet loads are classified as to trash content then assigned to one of three storage piles: heavy trash loads placed in small piles to be processed as soon as possible; moderate trash loads placed in moderate sized piles to be processed within a few weeks; and clean loads placed in the largest piles to be processed last.

The storage of sugarbeets generates problems common to the storage of many other crops. Knowledge is available to store these roots in a proper environment. Sugarbeet cultivars with properties that would classify them as "storage types" are not yet available, but when they are, the combination of a good storage environment with a storage genotype has the potential of saving a large part of the sugar the farmer has spent considerable time and money to produce. Sugarbeet cultivars that must be stored should possess two qualities in addition to desirable agronomic characters: a low respiration rate and resistance to storage rot. Some concern exists that combining these two characters into a commercial cultivar may lower the root or sucrose yield of the resultant hybrid. This phenomenon has occurred where resistance has been bred into commercial cultivars. However, the value of a storage-type cultivar cannot be assessed until it has been in storage for at least 80 to 100 days and compared with a standard commercial cultivar. A cultivar with desirable storage characteristics may have a root and sucrose yield at harvest lower than that in a standard cultivar. But the value of a storage-type cultivar over a standard cultivar will not be apparent until both have been in storage. Obviously the storage-type cultivar must yield more extractable sucrose than the standard cultivar after storage.

If one were to employ all of the methods known to reduce storage rots and thus affect a maximum savings of sucrose in the storage piles, he would: (1) not increase harvest injuries by removing crown tissue; (2) ventilate piles to lower root temperatures thereby lower respiration rates, lower microbial activity, provide adequate oxygen concentrations, and enhance wound repair; (3) reduce trash content of long-term storage piles by storing only the cleanest loads; (4) store roots under protective cover and ventilate them to reduce damage from freeze-thaw cycles and dehydration; (5) grow roots in soil with adequate fertility; (6) use *Phoma*-free seed and practice a 4-year rotation so as to not plant in *Phoma* infested soil; (7) treat storage-yard soils with fungicides to eliminate *P. betae,* which survives continuously, and to reduce or eliminate other pathogens suspected to survive on sugarbeet debris; (8) and finally, grow storage-type cultivars (when they become available) on a percentage of the land proportionate to the tonnage intended for long-term storage and store the roots from these hectares in long-term storage piles. This set of controls is extensive and expensive and may partially explain the reluctance of the sugarbeet industry in the USA to fully practice these measures. The economic status of our industry governs whether investments will be made to alleviate this complex problem.

33.1   Current Status in Pest Management of Soybean Pathogens

D. P. Schmitt and M. A. Ellis

Department of Plant Pathology, North Carolina State University
Raleigh, NC 27650
and
The Ohio Agricultural Research and Development Center, Ohio State University
Wooster, OH 44691

Soybean diseases cause an estimated 5-10% annual loss worldwide. Losses in various countries around the world have ranged from 0-20%. [5, 9]. However, accurate estimates are difficult to obtain because of the complexity of disease loss assessment.

The goal of plant pathologists is to minimize these losses without adversely affecting other management units in the agroecosystem. Pathogen populations are managed primarily through preventive tactics such as lowering the mean population density of pathogens, raising the "threshold" by using a pathogen-tolerant or resistant cultivar, or protecting the crop from infection.

An optimal soybean pathogen management system must be based on an understanding of the biology, population dynamics, and economic importance of the various pathogens as well as their interaction with the various management techniques used. We should avoid using control tactics where they are not justified economically and where they may have adverse side effects. Within this framework, one of the major limiting factors in soybean disease management is our inability to accurately characterize yield and/or quality losses due to specific pathogens and/or pathogen complexes. A group effort is being made by the Southern Soybean Disease Workers and some individuals in the North Central USA to establish accurate crop loss assessments utilizing seed treatment, foliar fungicide, nematicide tests, and disease surveys ([9]; R. E. Ford, *personal communication*). Currently, specific economic thresholds are used for some nematodes, whereas more general economic thresholds are used for most of the other soybean pathogens whereby disease management tactics are recommended on the basis of disease incidence of the previous crop. For example, the extension plant pathologists in Illinois have developed the "Disease Risk Area" concept for each soybean pathogen based on disease history (R. E. Ford, *personal communication*). Whether based on general or specific thresholds, effective management strategies have been developed for many soy-

bean pathogens [8]. Some of the tactics employed include the use of geographically adapted disease-resistant cultivars, crop rotation, planting of high-quality seed, various tillage practices, altering fertility, time of planting, time of harvest, and other various cultural practices, quarantines and pesticides.

Soybean Disease Management Tactics

1.   Selection and Breeding for Disease Resistance

The most economical and effective control of specific soybean disease pathogens is achieved by using disease resistant cultivars. However, the cost of developing resistant cultivars may be high. Disease resistance is available for many viruses, bacteria, fungi and nematodes. It is essential to maintain active disease resistance breeding programs because pathogens often overcome the resistance. The development of a resistant cultivar with a yield potential equal to or greater than that of high yielding susceptible cultivars grown in the absence of the disease has been and is still the goal of many breeding programs. Under certain situations, resistant genotypes might be practical to grow in spite of a somewhat lower yield potential than that of nonresistant cultivars.

2.   Cultural Practices

a.   Crop Rotation.   Crop rotation is an effective management tactic for most soybean pathogens, with the exception of many nematodes, because they cannot infect other crops, especially nonlegumes. Most important soybean pathogens can survive between growing seasons in crop debris and/or soil. Thus, continuous culture of soybeans may result in severe disease, especially under tropical conditions.

b.   Tillage Practices.   Methods of tillage that result in considerable soil disturbance and burial of crop debris can effectively reduce populations of bacterial, fungal, and nematode pathogens. However,

519

no till and minimum till practices now becoming popular for energy and soil conservation reasons may result in an increase of disease incidence.

c. Altered Fertility. Sufficient and balanced fertility can be an important factor in reducing soybean disease losses. Inadequate levels of phosporus or potassium have resulted in increased losses from several root infecting pathogens, as well as pod and stem blight of soybean.

d. High-Quality, Disease-Free Seed. Many important fungal, bacterial and viral pathogens can infect soybean seed resulting in reduced germination as well as distorted and discolored seed. In addition, the seed provides an excellent mechanism for transmitting pathogens from one growing season to the next and from one region or country to another. No seed certification program currently insures freedom of seedborne plant pathogens of soybean. Establishment of such programs could make a significant contribution to IPM programs for diseases of soybean.

e. Growing Location, Date of Planting and Harvest. Many soybean diseases can be controlled by growing the crop under environmental conditions in which the pathogens cannot grow, reproduce or be disseminated. The selection and use of arid regions with supplemental irrigation for seed production has great potential for minimizing seed production problems associated with bacteria and fungi. Adjusting planting date so that crop maturity coincides with dry seasons could also by beneficial in reducing seed quality losses due to pathogens, particularly under tropical conditions where soybeans can be produced all year and seasonal variations in rainfall are aften quite distinct.

Seed infection by fungi often increases with time as harvesting is delayed past maturity [3, 10]. Harvesting seed promptly at maturity is beneficial in limiting seed infection by fungi and is an essential practice in the production of high quality seed.

3. Quarantine

The soybean cyst nematode *(Heterodera glycines)* is the only soybean pathogen for which a quarantine had been established. The Federal Quarantine was terminated because it was ineffective in preventing the nematodes spread. Certain states still have quarantines, but they are relatively ineffective because intrastate spread of the nematode continues.

4. Chemical Control

Chemical control is effective only for certain fungal and nematode pathogens of soybeans. Fungicides are used as a seed treatment for control of seed rotting fungi or as a foliar application for control of diseases such as *Septoria* brown spot, pod and stem blight, stem canker, *Cercospora* (frogeye) leaf spot, and anthracnose. Fungicide seed treatments have increased soybean stands when fungus infected poor quality seeds were used for planting. Foliar application of fungicides has resulted in yield increases of 10 to 15% under certain conditions, and have been beneficial in increasing seed quality under severe disease development conditions [3]. Yields in fields treated with foliar fungicides have been variable, but in years when environmental conditions are optimal for disease development, yield increases may be significant.

Nematicides effectively control many soybean pathogenic nematodes. However, they are generally quite costly and may be economical only where serious nematode pathogens are present that cannot be controlled through alternate, more economical tactics.

Much more information is needed in the soybean agroecosystem on the impact of pesticides regarding beneficial and non-target organisms as well as effects on the crop. Foliar applications of fungicides reduced populations of the entomophagous fungus, *Nomuraea rileyi*, resulting in reduced yields in fungicide treated plots due to increased populations of the velvet-bean caterpillar [6]. Organic phosphate nematicides, when used in combination with the herbicide metribuzin, often cause adverse affects to the soybean plant (Corbin & Schmitt, *unpublished data).*

Application of Management Tactics through Predictive Systems

The general threshold concept that has been used in the past has allowed for considerable progress in pathogen control, but predictive systems can be more precise and result in the wiser choice of control tactics [4].

A rating system (point system), developed by some soybean pathologists in the USA, is used to determine when to apply foliar fungicides (W. E. Moore, M. C. McDaniel, *personal communication).* This effort is a first and an important step in moving away from the so-called calendar sprays. The rating system is based on environmental monitoring, cultural conditions of the field, and field location. Monitoring of the pathogen has not been done, however, relating population densities of many soybean pathogens to the decision of implementing management tactics may be impractical because of their rapid rate of reproduction and wide area of dispersal. Soil-borne pathogens are spatially limited and relatively immobile. Consequently, assays of soil to determine the numbers of species present are practical. Assay techniques are being developed for a few soilborne fungal pathogens that will aid in the study of their population dynamics. Much has been learned about the population dynamics of nematodes infecting various crops, including soybeans, and the relationship of nematode populations to crop damage [2]. Reliable assay procedures have been developed and are dependent on sampling time, sample size and laboratory assay [1]. Threshold numbers have been established for several soybean-parasitic nematodes [2, 7]. These threshold levels must be refined for soil types, geographic location, climate, etc.

Systems science has not been utilized to any great extent in soybean disease management. We have considerable data and certain mental concepts of systems that could be translated into mathematical models. Systems science may be a great asset in studying the complex problems with pathogen population dynamics and environmental systems in the soybean ecosystem, allowing greater and more rapid progress in the future.

Needs and Future Challenges

Some of the most obvious needs for developing IPM programs for soybean pathogens are:

1. Obtaining reliable disease loss estimates for establishing priorities among the more important soybean pathogens;
2. establishing damage and economic thresholds for priority pathogens;
3. studies on the epidemiology and spread of priority pathogens under field conditions;
4. develop practical methods for reliable and timely detection of pest populations under field conditions for use in monitoring or scouting programs; and
5. the incorporation of this information (perhaps with the aid of empirical or mathematical models) into IPM programs which use all available information and managment tactics in an integrated approach directed toward minimizing losses caused by plant pathogens, without negatively interfering with other elements in the crop management system.

Plant Pathologists in cooperation with other plant scientists have done and are currently doing much toward limiting disease losses of soybean. Through developing a better understanding of the soybean ecosystem, the interacting forces of soybean pests, the impact of production systems on the pest complex, and how each control discipline interacts with each other, soybean disease losses in the future should be reduced substantially.

Literature Cited

1. Barker, K. R., and C. J. Nusbaum. 1971. "Diagnostic and advisory programs," pp. 281-301, In *Plant parasitic nematodes*, B. M. Zuckerman, W. F. Mai and R. A. Rohde, eds., Vol. I. Academic Press.
2. Barker, K. R., and T. H. A. Olthof. 1976. "Relationships between nematode population densities and crop response." *Annu. Rev. Phytopathol.* 14:327-353.
3. Ellis, M. A., and J. B. Sinclair. 1976. "Effect of benomyl field sprays on internally-borne fungi, germination, and emergence of late-harvested soybean seeds." *Phytopathology* 66:680-682.
4. Ferris, H. 1978. "Nematode economic thresholds: derivations, requirements, and theoretical considerations." *J. Nematol.* 10:341-350.
5. Ford, R. E. 1978. "World soybean losses caused by disease." *Proc. 5th Annual Southern Soybean Disease Workers Conf.*, Fort Walton, Beach, FL. March 23-25, 1978.
6. Johnson, D. W., L. P. Kish, and G. E. Allen. 1976. "Field evaluation of selected pesticides on the natural development of the entomopathogen, *Nomuraea rileyi*, on the velvetbean caterpillar in soybean. *Environmental Entomol.* 5:964-966.
7. Schmitt, D. P., and K. R. Barker. 1978. Soybean diseases caused by plant-parasitic nematodes. North Carolina State University Plant Pathol. Information Note 209.
8. Sinclair, J. B., and M. C. Shurtleff. 1975. "Compendium of soybean diseases." The American Phytopathological Society, Inc. St. Paul 69 p.
9. Southern Soybean Disease Workers. 1978. "Southern states soybean disease loss estimate—1976." *Plant Dis. Reptr.* 62:539-541.
10. Wilcox, J. R., F. A. Laviolette, and K. L. Athow. 1974. Detoriation of soybean seed quality associated with delayed harvest. *Plant Dis. Reptr.* 58:130-133.

33.2   Pest Management for Soybeans—Weed Control

Ellery L. Knake

University of Illinois
Urbana, IL

There is a certain amount of energy available on each hectare of land for production of plants. This energy is in the form of nutrients, moisture and light. The amount of energy used by weeds means that much less energy available to the crop and a proportional decrease in crop yield. Therefore, controlling weeds is a primary concern for soybean growers.

Pest management for a farm should include not only the various pest control disciplines, but it should include all crops and also non-cropland such as fencerows and ditchbanks. For example, annual morningglory is one of the most difficult weeds to control in soybeans, but can be easily controlled when the field is in corn. Rotating soybeans with corn can help to control corn rootworm, but success is altered if volunteer corn is allowed to grow in soybeans. Weeds in non-crop areas can harbor insects and diseases as well as producing weed seed that can move to adjacent fields.

Although "pest managment" is a popular term today, the concept will likely be more readily accepted if consultant practitioners provide more breadth by including such aspects as soil fertility, irrigation scheduling, and variety selection.

For the weed control phase, the first consideration is the species of weeds present or likely to be present. The texture and organic matter content of the soil should be considered simultaneously. Observing or scouting fields can provide information for making weed maps to show what weeds are located where. Soils maps may already be available or can be developed. With this information, coupled with the use of weed susceptibility charts, the most appropriate herbicides can be selected, as well as the rates necessary for optimum weed control with minimal risk of adverse effect on the crop.

Although herbicides are used on the majority of soybean fields today, consideration should also be given to cultural practices such as adequate tillage, rotary hoeing and row cultivation. These practices are not only good for controlling weeds, but can also influence the performance of herbicides. For example, tillage and proper placement of incorporated herbicides should be considered concomitantly. Not only can the rotary hoe destroy many weed seedlings very quickly and effectively, but it can help to give slight incorporation of the herbicide, which may be beneficial, especially during relatively dry periods. And in some areas, the tolerance of specific soybean varieties to herbicides should be considered.

Herbicides

Chemicals can now be selected to do almost any weed control job required in soybeans. Consideration should be given to annual broadleaf and grass weeds and perennial broadleaf and grass weeds. Proper application methods and time of application should be considered. Will the herbicide be incorporated or surface applied? Will it be applied preplant, preemergence, postemergence, banded, broadcast or directed?

For control of annual grasses, the dinitroaniline and acetanilide herbicides are quite effective. Chloramben gives good control of annual grass weeds as well as controlling many broadleaf weeds.

Control of annual broadleaf weeds has been more of a challenge. Pigweed and lambsquarters can be rather easily controlled with most soil-applied herbicides. Smartweed can be controlled with linuron, metribuzin or chloramben. For velvetleaf and jimsonweed, linuron, metribuzin or chloramben can be helpful but are sometimes a little variable. Common ragweed can usually be controlled with linuron, metribuzin or chloramben but giant ragweed is a little more difficult to control. Fortunately, giant ragweed is usually not serious in soybeans and there are some postemergence options.

As we have increased soybean hectareage and decreased use of 2,4-D in corn, annual morningglories and cocklebur have become more serious. For annual morningglories, the dinitroanilines or vernolate can often give some suppression. Bifenox, while not outstanding, is better than most other preemergence materials. Postemergence application of dinoseb, naptalam + dinoseb, or 2,4-DB can offer some help. Acifluorfen (Blazer) also appears promising. For cocklebur, metribuzin can help if rates are adequate; and, postemergence applications of bentazon, naptalam + dinoseb or 2,4-DB are effective.

For many perennial grasses and broadleaf weeds in soybeans, glyphosate now offers considerable help. Use of the recirculating sprayer will likely increase dramatically since it provides a very timely, effective, and economical control method. Glyphosate with the recirculating sprayer can also provide a good answer for annuals like wild cane and volunteer corn as well as some annual broadleaf weeds growing taller than soybeans.

Special Practices

Double cropping with soybeans immediately after wheat has proven quite successful. Paraquat is usually used to kill existing vegetation, although dinoseb or glyphosate might sometimes be considered. Linuron or metribuzin for broadleaf control and an acetanilide or oryzalin for grass control can be helpful in approved combinations.

"Solid drilling" or very narrow rows are now more feasible than formerly with new help for broadleaf weeds from metribuzin and bentazon. And diclofop may offer postemergence help for grass control.

We much prefer controlling weeds early, but when early control measures are not adequate and a desiccant is needed to dry green weeds, paraquat and be used as a harvest aid.

We have considered above, some of the major weed problems and practices of the midwest United States. Additional weeds and additional practices should be considered in other soybean production areas of the world.

33.3 A General View of Plant Protection in the People's Republic of China with Special Reference to Soybean Pests and Diseases

Shen Chi-yi

Beijing (Peking) University of Agriculture
Beijing, China

Mr. Chairman:
Ladies and Gentlemen:

I am very pleased to have an opportunity to talk on "A General View of Plant Protection in the People's Republic of China with Special Reference to Soybean Pests and Diseases" in the soybean section, arranged by the Program Committee of the Congress.

Since the founding of the People's Republic of China in 1949, achievements have been made in both research and the practical control of plant pests and diseases. The age-old calamities of the oriental migratory locust have been practically eliminated. Cereal smuts, wheat rusts, rice stem borers, soil insects etc., have been greatly reduced. However, the effective and economical control of insect pests, diseases and weeds is still our long-term problem.

"Prevention first and then integrated control" is the policy we adopted on plant protection. By "prevention" we mean the strengthening of plant quarantine, reform of ecological conditions to suppress the development of diseases and pests, and the organization of a nation-wide systems of predicting and forecasting for timely control. By "integrated control" we mean the use of cultural control measures and resistant varieties as a basis, and the rational utilization of chemical, biological and physical measures to make the control economical, safe and efficient. The main measures adopted are as follows:

1. To carry out strictly both external and internal plant quarantine regulations both at ports and in crop production areas in order to exclude the invasion and spread of dangerous diseases and pests.

2. To reform ecological conditions in order to eradicate or diminish the sources of diseases and pests. For instance, we have laid equal stress on both "reform" and "control" to transform the former locust-breeding grounds into flourishing agricultural areas. It is estimated that about 75% of the locust-breeding areas, or 300 million ha. have already been reformed up to 1978. Consequently, the age-old calamities of the locust have been practically eliminated in all the reformed areas.

3. To organize a nation-wide prognosis system. That is about two thirds of the counties in our country have established forecasting stations and forecasts are now being carried out all together on about sixty diseases and pests of national or local importance. Researches on the occurrence, spread, migration and control of major diseases and pests have been conducted to make the prognosis more effective. For instance, in addition to the utilization of resistant varieties, the over-summering and over-wintering bases of yellow and stem rusts of wheat, the routes of their spread, as well as the division of epiphytotic regions have been determined, which makes it possible to issue forecasts both in winter and in early spring every year. Thus, the epidemics of wheat yellow and stem rusts have been controlled for more than one decade, however, it was estimated that in 1950 about 6 million tons of wheat were damaged by wheat yellow rust alone. We have also conducted the research on species, distribution and ecology of the army worm, and the brown plant hopper as well as the trends and time of their migration. On this basis, long-term and medium-term forecasts could be issued for effective control.

4. To breed resistant varieties. Emphasis is laid on the research on the occurrence and variation of biotypes of the pathogens. We have revealed 21 biotypes of the yellow rust of wheat, 13 biotypes and sub-biotypes of the stem rust of wheat, three of the Fusarium wilt of cotton and twenty-four of the rice blast. Effective control has been attained by the use of disease-resistant varieties. Insect-resistant varieties such as 6208 provide successful control of the wheat midge (*Sitodiplosis mosellana*).

5. To use rational cultural practices. For instance, before liberation the paddy stem borer (*Tryporyza incertulas*) was very destructive, generally reduced the yield of rice about 10-20%, and now it can be controlled firstly by plowing and irrigation in spring in order to diminish the source of the pest and reduce its population density or by adjusting the seeding date and using early-maturing varieties so that the most vulnerable booting stage may escape the attack of the pest. The rate of the loss at present is less than 1% of the total crop.

6. Rational application of pesticides. About a

hundred pesticides are now produced and used in China. Restrictions have been placed on the use of organochlorine compounds. Efforts have been made to develop high-effective-less-toxic-lower-residual chemicals such as dipterex, malathion, rogor, DDVP, kitazine, carbendazol and nitrofen, etc. Efforts are also being made to investigate and produce pyrethroids, juvenile hormone, sex-pheromone and insect sterilants. Techniques of ground and aerial application of ultra low volume and low volume sprays have been extended on a large scale. Studies on both the pest population density and its developmental stage and determination of both the target and the proper date of control are of importance to obtain economical and effective results in chemical control.

7. Biological control. The utilization of natural enemies in China was recorded as early as more than 1,600 years ago. Biological control including mainly the use of insects, fungi and bacteria has been applied to about 7 million ha. in recent years. *Trichogramma* wasps are used widely in combating the corn borer (*Ostrinia nubilalis*), the grey borer of sugarcane (*Eucosma schistaceana*) and the rice leaf folder (*Cnaphalocrosis medinalis*). With the utilization of the large eggs of the cynthia moth (*Samis cynthia*) and the perny silkworm (*Antheraea pernyi*) to rear wasps, and with the help of simple machines, the extension of biological control has been greatly promoted. The "seven spotted ladybird" (*Coccinella septempunctata*) provides successful control of cotton aphids at their early stage. Lacewings (*Chrysopa* spp.) have good effects on the control of aphids, the greenhouse whitefly (*Trialeurodes vaporariorum*) and the cotton bollworm. Control of various insect pests of food crops, cotton and vegetables by *Bacillus thuringiensis* has been widely popularized.

The antibiotics "Tinggangmycin" is extensively used to control the rice sheath blight (*Pellicularia sasakii*) and also "Chunleimycin" is used to control the rice blast (*Pyricularia oryzae*).

Here I should like to touch on one more problem related to the control of several important soybean diseases and insect pests in China.

1. The soybean mosaic virus (SMV-Y) is prevalent in North-east China. The symptoms usually expressed are mosaic, yellowing, top necrosis, crinkles and brown spots of diseased seeds according to various enviromental conditions and the time of infection. The virus is transmitted through seeds, the sap of diseased plants and by aphids. High temperature, drought and abundance of aphids are responsible for serious damages. The disease is effectively controlled by the use of virus-free seeds

and resistant varieties such as Tiefang 18 and 19.

2. The soybean cyst nematode (*Heterodera glycines*) is prevalent in arid areas with sandy alkaline soils in Northeast China. Reduction in yield is up to 30-50% in serious cases. The disease can be effectively controlled by resistant varieties, Beijing Black and other five varieties being used as the source of resistance for breeding. A 3-5 years' rotation, manuring and soil treatment with nematocides also provide good effects.

3. The soybean pod borer (*Leguminivora glycinivorella*) is one of the major pests on soybean in Northeast China and Shantong Province. Serious attack causes seed damage up to 30-40%. Resistant varieties with tight cell tissues on the epidermal layer of the pod, such as "Four-grained Iron-pod Yellow," cause a higher mortality of the entering larvae. The damage may be more serious if successive cropping of soybean is adopted and the peak of pod formation coincides with that of oviposition. The use of resistant varieties, rotation, cultivation and application of insecticides, such as DDVP and IPSP used as fumigants, constitute the control measures.

4. The soybean aphid (*Aphis glycines*) causes serious damage in Northeast China and Inner Mongolia. The aphid population reaches its peak from seedling stage to the blooming stage of soybeans. The population declines quickly once the growing point of the plant ceases to grow. From late autumn the aphids migrate back to *Rhamnus davuricus* to deposit eggs for overwintering. A total of 15 generations develops on the soybean annually. The large number of overwintering eggs and the high aphid population density in the soybean seedling stage cause serious damage. During the period from June to July, the average temperature of 22-25°C and relative humidity of below 78% are found to favor the aphid development, on the basis of which forecasts may be made. Application of insecticides is the main control measure at the present time.

Although great progress has been made in the field of plant protection since the establishment of the PRC, we now still lag behind as compared to those of the developed countries. It is our aim to learn from the scientists present here at this Congress. By the use of modern science and technology, by the fully utilization of natural resources, and by the extension of international cooperation, we can certainly speed up the development of the science on plant protection and reduce the damages of pests and diseases to meet the demand of food supplies of the increased world's population. The Chinese scientists are willing to cooperate with scientists of the world in all these fields to fulfill this aim.

34.1 Early-Maturing, Wilt-Resistant Cotton for Narrow-Row Cultivation in California

Stephen Wilhelm, J. E. Sagen, and Helga Tietz

Department of Plant Pathology University of California, Berkeley, CA 94720

Recent results of our research on Verticillium wilt-resistant cotton cultivars suitable for California agriculture were presented during the Evening Discussion session. The problem of developing such cultivars is essentially determined by three parameters. First, available sources of wilt resistance with useful agronomic characters, such as 'Seabrook Sea Island', 'Waukena White', and a few other long-staple forms of *Gossypium barbadense,* as well as some wild races of *G. hirsutum,* are merely field-resistant to Verticillium wilt. That is, these cottons possess a physiological mechanism that restricts internal colonization of vascular tissues of the stem by *Verticillium* and prevents infection of the leaves, and thus, though infected, they usually grow in the field without showing wilt symptoms once they have passed the seedling and juvenile stages. However, they lack resistance to root infection and sustain a measure of internal colonization of stem-vascular tissues, and this susceptibility is transmitted to their hybrid offspring. Consequently, they cannot be solely relied upon as a means of achieving wilt-resistant cultivars.

Second, while *G. hirsutum* is a poor source from which to select for high wilt resistance, the tolerance of some Upland cultivars to infection by *Verticillium* as well as to wilt symptom expression may be increased by in-row crowding of plants or by close spacing of rows. For maximal utilization of this phenomenon, such cultivars must be bred and selected to grow and fruit optimally under the conditions of this type of cultivation.

In our studies, increased wilt resistance in response to high-density planting occurred in certain Upland cultivars that have substantial wilt tolerance and are distinguished by a short-branching, arching phenotype. The characters of short branching and arching facilitate one-time harvesting by stripper-picker. The arching, which is due to the weight of green bolls, causes a partial girdling of the stem at the ground line and positions leaves favorably to the sunlight, thus improving uniform boll maturation up to the shoot tips. As the maturing bolls dry, the stalks return to an upright position. When stripper-harvested, the bolls readily snap from the short lateral branches.

With two experimental lines of such wilt resistant stripper cottons, designated 'UC-12' and 'UC-13', we have over the past five years obtained yields averaging nearly 1200 kg/ha (Fig. 1). They were densely planted on wilt-infested land, two rows per bed instead of the conventional single row, with resulting plant populations of 115,000 to 180,000 per hectar. The achieved yields are more than twice those obtained in previous years on the same land with wilt-susceptible 'Acala' cottons grown and harvested by traditional methods. In addition, the production costs were reduced by about 40 percent.

A single-harvest, early-maturing stripper cotten also has clear advantages in the management of wilt inoculum. If, as in California, the shredding of stalks immediately following the harvest can be accomplished before microsclerotia of *Verticillium* have formed in the infected tissues, and if the

Figure 1. Experimental Cotton 'UC-13' at Full Maturity. Two rows per bed and close spacing of plants in the row optimize resistance to Verticillium wilt. The stripper picks 95% of the bolls in a single harvest.

527

shredded stalks are allowed to remain on the soil surface as a mulch, and dry quickly, the formation of microsclerotia is checked, and the carry-over infestation by *Verticillium* is thus reduced. The use of a mechanical procedure whereby the underground plant parts could be removed and shredded as well would further slow the buildup of inoculum.

Third and finally, sources of wilt resistance, for the most part, are late-maturing. In our researches, this difficulty has been overcome in some experimental lines that were derived by crossing otherwise desirable, wilt-resistant lines into early-maturing Soviet cultivars such as 'S-4727', a major parent of the 'Tashkent' family of cottons.

34.2   Cotton Pest and Disease Management in Black Africa

J. Cauquil

Institut de Recherches du Coton et des Textiles Exotiques
Mission de Recherches Cotonnieres
B. P. 997 - BANGUI - République Centrafricaine

Introduction

The overall cotton production of the African continent constitutes less than 10% of the world's production. Black Africa (not including Egypt and Maghreb) comes up to only 5.5% of the overall total (table 1). Yet, economically speaking, cotton is of utmost importance for Black Africa. It is indeed the number one export of Chad, Sudan and Mali which in 1973 amounted to 66, 54 and 49% of those three countries' overall exports. It is also the second most important export of four other countries: Uganda, Tanzania, Upper Volta and the Central Africa. In most producing countries, seed cotton often bought by State Organizations is the only cash crop for cotton growers.

Thirty years ago cotton farming in Black Africa essentially consisted of small fields that were tilled and then abandoned to fallow with several different crops grown on the same plot of land. Cotton was grown without using fertilizers or pesticides.

Since that time things have changed considerably. When referring to the data of the 11 French-speaking countries (Table 2), nearly half the cultivated areas are tilled with plows (animal traction or tractor) and

Table 1. Cotton fiber production of African countries (1000 metric tons) (origine CCIC)

	Year	
Country	1967 / 68	1977 / 78
Algeria, Morocco	6	9
Egypt	437	401
Sudan	195	163
Ethiopia	10	24
Angola, Mozambic	54	34
South Africa, Rhodesia	32	76
Zaire, Burundi	11	16
Benin, Cameroon, Central Africa, Chad, Ivory Coast, Malagasy Republic, Mali, Niger, Senegal, Tobo, Upper Volta	122	208
Kenya, Malawi, Nigeria, Tanzania, Uganda, Zambia	169	158
Others	0	4
Total Africa	1 036 (% 9,6)	1 093 (% 8.8)
Total Black Africa	593 (% 5,5)	693 (% 5,5)
Total World	10 823	12 486

Table 2. Data concerning the cotton farming in 11 countries of Black Africa : Benin, Cameroon, Central Africa, Chad, Ivory Coast, Malagasy Republic, Mali, Niger, Senegal, Togo, Upper Volta (origine CFDT)

Years	1971 / 62	1968 / 69	1977 / 78	
Total surface (1000 ha)	565	751	795	
Production of seed cotton (1000 t)	128	427	531	
Yield of seed cotton (kg/ha)	226	568	668	
Plowed surface (animal traction or tractor) (1000 ha)	4,0 (%0,7)	7,6 (%10)	381	(% 48)
With mineral fertilizers (1000 ha)	5,4 (%1)	150 (%20)	420	(%53)
With insecticides (at least 3 sprayings) (1000 ha)	6,8 (%6,8)	163,6 (%22)	437,5	(%5,5)

528

more than half profit from mineral fertilizers and are protected by insecticides.

Yields have increased and average nearly 700 kg of seed cotton per hectare. Countries like Ivory Coast and Senegal have an average production of more than 1 metric ton per hectare.

Cultivated varieties (*Gossypium hirsutum,* except the Gezira in the Sudan where it is *G. barbadense*) adapt well to the various ecosystems. The fiber produced answers the requirement of the international market as most of the production is exported. Pests and diseases often constitute the main factors limiting cotton growing. Let us now give a short description of the way these two problems have been dealt with.

The Cotton Pests

There is a great variety of cotton pests in Black Africa, due to the large number of existing ecosystems. However, the species which have been listed do not differ fundamentally from those of other producing countries.

There is however a typically African Lepidoptera whose larva is a formidable parasite of buds and bolls: *Diparopsis watersi* (*D. castanea*). But it must be noted that the boll weevil *Anthonomus grandis* does not occur on the continent.

Leaf parasites are numerous: *Acarina, Hemitarsonemus latus, Empoasca fascialis, Lygus vosseleri, Aphis gossypii, Bemisia tabaci, Spodoptera litura, Cosmophila flava, Sylepta derogata.*

The direct damages they cause justify a special intervention only in a few cases. On the other hand, their indirect damages are more serious (for example honey-dew caused by whiteflies in dry areas). They also cause great concern for the role they play as a disease vector (leaf curl, mosaic, blue disease. . .).

However, the pests of the cotton fruiting period (buds, flowers and bolls) need an organized fight for they cause great economic damage. It concerns above all the Lepidoptera larvae: *Diparopsis watersi* (*D. castanea*), *Pectinophora gossypiella, Heliothis armigera, Cryptophlebia leucotreta, Earias insulana* (*E. biplaga*). An Hemiptera *Dysdercus volkeri* and others species play an important part in tropical and equatorial zones and are involved in boll rotting.

Pest control essentially requires the use of insecticides. In African Research Stations numerous programs have been carried out on the possibilities of biological control. Although major results have been obtained in the laboratories, it has not yet been possible to put them into practise in farmer's fields.

In host plant resistance to insects, the leaf pilosity is widely used. This genetically transmitted charac-teristic enables reduction in damage caused by Jasside. Today the vast majority of varieties cultivated in Africa possess hairy leaves due to adequate crossings. Cultural control always plays a major role in fighting cotton pests and some practices have been recommended for many years. For instance, most of the time, early sowing reduces the incidence of boll-worms in the areas infested by *Diparopsis*. The uprooting of stalks after picking is an important prophylactic measure that permits stopping the proliferation of certain insects in a continent that does not have a cold winter. This process is made compulsory by law in many countries.

1) - *The Pesticides used.* Their main target is to protect cotton plant during the fruiting stage. Their specifications are as follow: (i) A small number of types of active ingredients. As in many countries, they are bought under bid contract by State Organizations and Collectives. It implies a yearly checking of the quality and quantity of imported chemicals; (ii) A moderate utilization in quantities, since they are often made in foreign countries and expensive with regard to the peasant income. In 1979, one litter of imported insecticide cost $4-5. That is to say $10-15 for a single treatment. In some countries the purchase price of seed-cotton includes protection by pesticides. The active ingredients constituting the commercial insecticides are mixed in a binary or ternary way with DDT to eliminate *Heliothis* (800 to 1,000 g/ha) sometimes with parathion methyl (150 to 300 g/ha). The main ingredient may be endrin (200 to 600 g/ha), endosulfan (400 to 800 g/ha), carbaryl (400 to 1,000 g/ha). Less frequently: monocrotophos, toxaphene, polychlorocamphene may be used. The synthetic pyrethroids have been used for a few years: decamethrine (12 to 18 g/ha), cypermethrine (30 to 50 g/ha), fenvalerate (50 to 100 g/ha) and (27% of the market in 1979 for the French-speaking Africa). These new compounds are particularly efficient in the control of boll-worms but must often be mixed with an organic phosphate in order to make up for their lack of efficiency against *Acarina* and *Hemiptera.*

2) *The way the insecticides are applied.* Aerial spraying is done mainly in areas where farms are mechanized (Angola, Rhodesia, South Africa, Malagasy Rep.) or in the large irrigated scheme of the Gezira in the Sudan. Such treatments are nearly impossible in traditional farming areas due to the concentration of trees and scattered fields.

Most applications are with knapsack sprayers equipped with a lance or a boom. The boom fitted with four nozzles can either be horizontal (West

Africa) or vertical (East Africa). The quantity of liquid sprayed varies from 75 to 120 liters per ha, which makes work hard and tiring for the cotton grower. For about 10 years hand held ultra low volume atomizers, powered by eight torch batteries, have been used with great success. In quantity 66% of insecticides bought in 1979 are ULV. Thanks to a small motor that runs at 8,000 rpm it is possible to obtain a micronization of the active ingredients mixed with a special solvent. The quantity of commercial insecticides atomized amounts to 2 or 3 liters per ha. This technique which is both quick and efficient requires less work than the conventional one. It is in great demand among cotton farmers: one or two batteries are necessary to treat 1 ha. It takes about 3 hours of work to spray 1 ha with a knapsack sprayer equipped with a boom able to treat two cotton rows at a time. The ULV atomizer using the drift can treat 5 or 6 cotton rows at a time in a single operation and it takes 1 or 1 and a half hour to treat 1 ha. For these two different techniques one apparatus is needed to protect 10 to 15 hectares of cotton. Generally, the number of treatments done by cotton growers varies from 3 to 6. They take place during the fruiting stage of the cotton plant to protect floral organs. Flowering occurs regularly every 12 or 15 days and the date of the first application is timed without regard to parasitism.

As in some countries only a part of the cultivated areas is protected, the quantity of pesticide used remans low.

In 1976, it amounted to 10.6 liters per ha in West Africa and 3.9 liters in Central Africa.

Countries, such as Senegal and Ivory Coast, which treat all the cotton surfaces reach an average of 17.1 liters and 16.3 liters per ha. A scouting and warning system has been set up in Ivory Coast to determine the timing of the treatments; but it has not worked more than a few years. It is indeed difficult to determine the economic threshold of each pest because of the scattering of fields, poor communication means, and a lack of scouting experts.

The Cotton Diseases

In Africa, cotton diseases are not economically speaking as serious as cotton pests. But to control them important breeding work has been carried out in two different Research Stations during the last 30 years.

1) *Bacterial blight.* The bacterial blight (*Xanthomonas malvacearum*) is found in Africa in all cotton-producing countries. It is responsible for seedling blight, angular leaf spot, blackarm, and boil rot. This disease is transmitted by seeds, fuzz, and crop debris and often transported by rain splashes or insects such as *Lygus, Dysdercus, Xanthomonas malvacearum;* it can be controlled by means of seed treatment which eliminate the seedborne infection. If seed delinting with acid is difficult to realize in Africa, seed disinfection is often used with: organic mercurial compounds, copper formulations, bronopol.

For 25 years, IRCT breeders have worked to obtain resistance to bacterial blight. According to Roux, the source of resistance is obtained either from the Allen genetic background, major gene B9 L with one (several) minor genes, either from the N'Kourala genetic background (genes B2 + B3).

The origin of these genes for resistance is not well known. It seems that *Gossypium hirsutum race punctatum* with an American ascendance introduced in Africa and grown for hundreds of years has a natural resistance to bacterial blight. A part of these genes for resistance has been transmitted by natural crossings to more recent varieties. We can also think that the good resistance of Allen grown in Africa since the beginning of the century can be attributed to a favorable mutation.

Varieties cultivated in French-speaking Africa are all resistant to bacterial blight. The most famous REBA B50 which was bred in Central Africa (gene B9 L) seems to possess the best resistance to different races of *Xanthomonas malvacearum.* Its efficiency has been proved in the field in Thailand, India, Paraguay, and Argentina. The variety BJA 592 (genes B2 + B3) is essentially grown in Africa. Both these cultivars have been used as a basis for numerous crosses in various producer countries.

In English-speaking African countries, research of the same kind has led to the series of Albar whose resistance comes from Allen.

The results thus obtained for many years prove that resistance to bacterial blight can be combined with good yields and top quality fibers without any problem.

2) *Fungus diseases.* They are numerous and varied. We shall examine here only those which are of importance.

Seedling diseases cause damage in the humid zones and forest areas of tropical and equatorial climates. Among most common pathogens we found: *Colletotrichum gossypii, Rhizoctonia solani, Macrophomina phaseoli, Fusarium moniliforme,* and *Pythium spp.* Chemical control by seed disinfection is the most interesting method. Mercury and copper compounds which at the same time have a bactericide activity are currently used. Systemic fungicides have been also introduced. In French-speaking

African countries seeds are treated with the mixture of a mercury compound and of a chlorinated hydrocarbon insecticide. However, the use of these products has been prohibited in some countries.

Fusarium and Verticillium wilts occur. *Fusarium oxysporum f. vasinfectum* occurs in the Sudan, Tanzania, Angola, Zaire, and Central Africa. In Tanzania there is the American race of Fusarium. The Sudan and Egypt each have particular races. The races of the other countries have not been indentified. IRCT breeders have obtained two varieties resistant to *Fusarium* wilt: Reba B 50 which is now being replaced by 761. Their character of resistance comes from Coker 100 wilt. In Tanzania UK 71 is the most widely used. *Verticillium albo-atrum (= V. dahliae)* has been reported in the Eastern and Southern countries of Africa and the Malagasy Republic. Breeders have obtained tolerance to this wilt from the Acala lines bred in the USA.

Boll rots caused by fungi and bacteria often exists in the most humid zones which have annual rainfall exceeding 1,000 mm and where rains occur and overlap with the fructification of cotton. The fungi are most numerous and the best know. The influence of bacteria is often underestimated because of difficulties in isolation and identification of specific organisms. Both of them differ only by a few species of those found in the USA. The most important are: *Colletotrichum gossypii, C. indicum, Diplodia gossypina, Fusarium moniliforme, Rhizopus stolonifer,* and *Xanthomonas malvacearum.*

All of these organisms cause damage but of great interest is the means by which they gain entry into the boll because few are capable of penetrating the pericarp by their own means. The role of rain is preponderant as it permits their entrance through the intercarpellary sutures. A constant infestation by cotton stainers (*Dysdercus sp.* Hemiptera) favors boll rotting.

Preventive measures consist of planting cultivars resistant to bacterial blight. Chemical means in the form of fungicides is not foreseen but the insecticide treatment will be timed to eliminate the *Dysdercus* during the critical period of fructification.

3) *Disease attributed to viruses or mycoplasms* has for some years developed greatly in Africa. Only the most current diseases will be examined here.

Leaf curl, which also attacks *G. hirsutum,* is mainly dangerous to *G. barbadense.* Thorough research on this virus has been carried out at the Gezira Research Station in the Sudan. The whitefly *Bemisia tabaci* is the vector. The causal agent of leaf curl has not been purified or identified. Two types of symptoms can be distinguished: small vein thicken-

ing (SVT) and main vein thickening (MVT). In the Sudan there are cultivars resistant to leaf curl. Applications with a systemic organophosphorate (dimethoate, monocrotophos) at the beginning of cotton growth limit rapid multiplication of whiteflies.

Mosaic occurs in West and Central Africa. In 1969 and 1970 in Chad, it caused great devastation on BJA 592 and the cultivation of this cotton cultivar had to be abandoned in the south of the country. This disorder transmitted by *Bemisia tabaci* is due to a presumed virus which has not yet been characterized. Mosaic symptoms in Africa seem to be similar to those of "Mosaico" in Central America.

The lines bred from Allen and N'Kourala have good tolerance. The genetic process of this tolerance is not known: anyway it is easily transmissible by crossing. According to Roux, a high susceptibility to this disease is found on the African varieties with a Triumph background and many other American varieties: Coker 417, Deltapine 16, Del Cerro, Acala 1517 Br., Acala SJ 1. Besides, tolerant varieties are all of those issued from Allen: HG-9, Y 1422, SRF-4 (Chad), Reba BTK 12 (C.A.E.), Stoneville 213, Coker 310, Hopi.. *G. hirsutum var. punctatum* is considered as resistant.

"The blue disease" is caused by a virus that has neither been isolated nor identified. It is transmitted by *Aphis gossypii.* Its African epicentrum is the Central Africa. But similar symptoms occur in many other countries: Thailand, the Phillipines, Paraguay. The symptoms are typically a leaf roll with dwarfing (internode shortening and zigzag growth). The spreading of this disease which happened about 12 years ago seems to be due in Central Africa to the repeated applications of endosulfan-DDT in cotton fields. As a matter of fact this insecticide which is inefficient against aphids has promoted their rapid multiplication by destroying their natural predators.

The resistance to this disorder was found in 1973 in lines bred from the triple hybrid HAR (*G. hirsutum, G. arboreum, G. raimondii*). This resistance is certainly coming from *G. arboreum.*

The line HAR 1186 is practically immune. The SRF-4 cultivar obtained in Chad is tolerant to blue disease; that is why it is grown more and more in Central Africa.

Phyllody occurs in Upper Volata, Mali, and Ivory Coast. It is a mycoplasma disease whose vector is a Cicadellidae (genus *Orosius*). The symptoms include floral virescence with sterilization of cotton plants.

Up to now, no cultivar resistant or tolerant to this disease has been found. Chemical control against its vector is the only mean to reduce damage. However, for economical reasons it has not been used in traditional farming systems.

Discussion

Although in Black Africa there are no integrated methods of control other than experimental, positive elements in chemical control against cotton pests must be underlined: (i) protection of cotton restricted to the phase of fructification except in special occasions, (ii) moderate doses of applicated insecticides, and (iii) limited number of treatments.

The entomofauna is also far less perturbed than in other producer countries. Instances of acquistion of resistance to certain active ingredients has been observed with *Diparopsis, Crytophlebia, Earias*. But these observations have often been carried out in the vicinity of Research Stations. Instances of pollution with an incidence on the wild fauna especially birds and fishes has been reported; but these rare incidents cannot compare by far with what has been observed elsewhere. It is to be hoped that the wise use, which up to now has been made of insecticides will not change in the coming years, especially with the intensification of cultures and the wide use of ULV atomizera which could favor in the cotton fields an increase in the number of treatments. The great success pyrethroids have met with, should not hide the fact that they may be dangerous if used alone. These insecticides promote the multiplication of certain pests: Acarina, Aphids, and Whiteflies.

In some cases its is necessary to add organic phosphate with pyrethroids to protect the natural balance of cotton entomofauna.

As far as diseases are concerned many results are satisfactory. The economical incidence of bacterial blight has been reduced considerably thanks to the breeding of resistant cultivars and seed disinfection.

Interesting solutions have been found to *Fusarium* and *Verticillium* wilt on the varietal level.

Boll rotting is still important in the zones with a high ambient humidity but the character of resistance to *Xanthomonas malvacearum* has reduced its incidence. Besides, programs of insecticides treatment to control *Dysdercus* help to get better results.

The recent spreading of diseases attributed to viruses or mycoplasmas still causes concern in Africa. In the case of blue disease it comes from the destruction of the natural balance by the rapid multiplication of the vector *Aphis gossypii*. Things have been set right in C.A. by using insecticides efficient against aphids to protect the cotton fields. In Chad, the cultivation of a too susceptible cultivar BJA 592 had increased the incidence of mosaic. By switching to new tolerant cultivars: HG-9, Y 1422, this disease has been economically controlled.

References

BINK, F. A. 1975. Leaf curl and mosaic disease of cotton in Central Africa Cott. Grow. Res. 52, 233-241.

CAUQUIL, J. 1975. Cotton boll rot: Laying out of a trial of a method of control. ARS-USDA ed. 143 p.

CAUQUIL, J. 1977. Etudes sur une maladie d'origine virale du Cotonnier: la maladie bleue. Cot. Fib. Trop. 22, 259-278.

C.F.D.T. 1978. Rapport d'Activités 1978 - Compagnie Francaise pour le Développement des Fibres Textiles PARIS 123 p.

EBBELS, D. L. 1975. Diseases of upland cotton in Africa. Rev. Plant. Path. 55. 742-763.

DELATTRE, R. 1972. Elément de base pour une lutte intégrée dans les cultures cotonnieres d'Afrique I.R.C.T. non publié 24 p.

DELATTRE, R. 1972. Evolution des traitements phytosanitaires sur cotonnier dans la zone Franc. CR. Semaine d'Etude Problémes Intertripicaux. Fac. Sc. Agr. Gembloux, Belgique 503, 521.

DELATTRE, R. 1973. Parasites et maladies en Culture Cottonniere. I.R.C.T. PARIS.

PEARSON, E. O. 1958. The insects pests of Cotton in Tropical Africa. Emp. Cot. Corp. LONDON, 355 p.

ROUX, J. B. 1977. Recherche de caracteres de résistance aux maladies du cotonnier. I.R.C.T. 6 p.

34.3 Verticillium Wilt and Seedling Diseases of Cotton in Greece and Neighboring Countries

E. C. Tjamos, H. Kouyeas, A. Chitzanidis

Benaki Phytopathological Institute,
Kiphissia, Athens, Greece.

N. Galanopoulos, E. Kornaros, and K. Elena

Hellenic Cotton Board,
Athens, Greece

Abstract

This presentation gives a short account of the current status of cotton cultivation in Europe and deals with recent experimental work on Verticillium wilt and seedling diseases of cotton in Greece. Research work on Verticillium wilt in Greece is aimed at evaluting the tolerance to Verticillium wilt of new cotton cultivars to select the most appropriate for Greek conditions. Tests involved parallel studies of the virulence of Greek *V. dahliae* isolates on tolerant and susceptible cotton cultivars. The role of growth retardants in mitigating Verticillium wilt symptoms is also an aspect of the field studies. Experiments in controlling seedling diseases of cotton centered on the development of a laboratory technique for rapid and accurate testing of fungicides against the soil borne pathogens.

General Information

Cotton constitutes one of the most important fiber crops in Europe. Greece, Trukey and, to a lesser extent, Bulgaria are the main cotton-producing countries of Europe. Cotton is also cultivated in a restricted area of South Yugoslavia. Information concerning acreage, cultivated varieties and the main cotton growing regions in Europe is given in Table 1. It can be seen that the plains of the south, central and north part of Greece are extensively cultivated, while in Turkey cotton is cultivated along the Mediterranean coast. In Bulgaria cotton is cultivated around the town of Chirpan where the Cotton Institute of the country is located; in Yugoslavia cotton is cultivated only around Strumitsa near the Greek border. Information referring to the most harmful pests and most serious diseases is given in Table 2. *Thrips tabaci,* spider mites, White fly and Pink boll worm are the most dangerous and destructive pests in Greece, while Verticillium wilt and seedling diseases

the most important diseases. Other pests and diseases of minor importance are also listed.

Table 1. Approximate figures and general information concerning acreage, cultivated varieties and main cotton growing areas in Europe

Country	Average acreage	Main cultivars	Areas
Bulgaria[a]		Thrakia	Chirpan
Greece	500,000	4-S, Coker 210 (Delcot 288, R-253F)[b] SJ-1	Boeotia, Thessaly, & Central Macedonia
Turkey	1,300,000	Coker 100/2 Deltapine 15/21 Caroline Queen	Adana, Ege region & Antalya
Yugoslavia	5,000	Local	Strumitsa

[a]acreage not available
[b]under experimental scale

Table 2. Serious pests and diseases of cotton in Greece

Pests

Very serious	Occasionally serious	Not serious
Thrips tabaci	Wire worms	Lepidoptera (*Agrotis* sp.)
Spider mites (in South Greece)	*Hylemyia* sp. (during humid weather)	*Aphis gossypii*
Pink boll worms (where it persists)	European boll worm	Jassidae

Diseases

Verticillium wilt	Fusarium wilt	
Seedling diseases (*Pythium ultimum Rhizoctonia solani Thielaviopsis basicola*)	Angular leafspot (*Xanthomonas malvacearum*)	

Research Work

Verticillium Wilt of Cotton

The harmfulness of Verticillium wilt of cotton in Bulgaria and the dissemination of the fungus by vario of cotton plants has been reported by Savov [6, 7, 8] working in the Cotton Institute of Bulgaria at Chirpan. *Verticillium dahliae* is the dominant pathogen of cotton wilt in the Ege region around Smyrna as reported by Esentepe [2] and Saydan et al. [9]. They showed that the disease causes serious losses particularly in Antalya than in Adana regions and give information on various aspects of the problem.

Verticillium wilt has also become a limiting factor for cotton cultivation in many traditional cotton growing areas of Greece [10]. The problem is rather acute in the south belt of cotton cultvation although fields up to the north of the country could be highly affected.

Losses of cotton production caused by *V. dahliae* is mainly attributed to the extensive cultivation of highly susceptible cultivars. One of these, called 4-S, is very productive, early maturing, producing cotton fiber of excellent quality and well adapted to Greek weather conditions. Although the Cotton Institue at Sindos near Thessaloniki had tried and selected some wilt tolerant cotton cultivars years ago, the actual study of Verticillium wilt of cotton has started in an expanded scale from 1976 onwards. The Hellenic Cotton Board in collaboration with the Cotton Institute and the Benaki Phytopathological Institute studied the tolerance and productivity of new cotton cultivars selected by Dr. Chlichlias of the Cotton Institute at Sindos [3]. Among the tested cultivars were: R-153F, two selections of SJ-1 (Acala 71039 & Acala 71042-460 and Delcot-288. These cultivars were compared with the susceptible 4-S and Coker 210 in 8 main cotton growing areas. The experiments were done in naturally infested fields. The fields had been selected during previous summer field surveys and had severe symptoms of Verticillium wilt.

Experimental plots consisted of 4, 10-meter rows, 1 meter apart. Five replications were included in a randomized complete-block design. Disease severity was based on a scale from 0 to 4 including wilting, leaf chlorosis, necrosis and defoliation. Twenty plants with symptoms rated from 0 to 4 were marked twice in each plot. All plots were harvested by hand picking to obtain data of total yield, boll size and weight and fiber properties.

Three year field trials (1976-1978) showed that the cultivars Acala 71042-46, R-153F and Delcot-288 were tolerant to *V. dahliae* in Greece compared with the susceptible cultivars 4-S and Coker 210.

Differences in yield between *Verticillium* tolerant cultivars and the susceptible 4-S expressed as percentage over the control 4-S are presented in Table 3.

The figures presented in Table 3 show that tolerant cultivars proved to be productive and worthy of cultivation on an expanded scale. Yield differences were higher by 5-37% in tolerant compared with the susceptible 4-S. Zero differences could be attributed to the unfavorable environmental conditions for disease development in some of the field trials. Table 4 shows the *Verticillium* tolerant cotton cultivars proposed for further field evaluation in an expanded scale.

Having had the first evidence concerning the tolerance of the tested cotton cultivars to *V. dahliae* in Greece, a study of the virulence of cotton *V. dahliae* isolates on the tolerant R-153F and susceptible 4-S was attempted. The aim of this research was to find out their ranges of pathogenic variation and to thus facilitate selction of the most suitable cultivars for Greek conditions. The pathogenicity of a representative sample from a large collection of cotton *V. dahliae* isolates obtained from all over the country by using a water culture technique was tested. Tjamos & Kornaros [11] presented evidence indicating that cotton *V. dahliae* isolates in Greece are of lower pathogenicity

Table 3. Differences in yield between the *Verticillium*-tolerant cotton cultivars (SJ-1, R-153F and Delcot-288) and the susceptible 4-S expressed as percentage over the control (4-S)

Counties[a]	SJ-1 (Acala 71042-46)	R-153F	Delcot-288
Boeotia	34	23	34
Fthiotida	0	14	0
Karditsa	7	5	5
Larissa	15	17	24
Serres	0	12	11
Thessaloniki	0	0	13
Trikala	12	10	12
Veroia	32	37	14

[a] The data were obtained from 8 main cotton growing areas in Greece during 1976-1978.

Table 4. Verticillium wilt tolerant cotton cultivars proposed for further field evaluation in an expanded scale in Greece

Cotton growing areas	Latitude	Cultivars
Boeotia (South Greece)	38°	Delcot-288 and Acala 71042-46
Fthiotida and Thessaly (Central Greece)	39°-40°	R-153F, Acala 71042-46 and to less extend Delcot-288
Macedonia (North Greece)	40°-41°	R-153F and Acala 71042-46

compared to those found in USA, Russia, or elsewhere. It was shown that none of the isolates studied belonged to the severe defoliating strain known as T-1 [10]. This is possibly due to the fact that cultivars highly tolerant to strains similar to SS-4 have not been cultivated on a commercial scale.

Greek strains severe to both cultivars (R-153F & 4-S) but not causing defoliation could be considered as the Intermediate 2 described by Schnathorst [10]. Mild strains that cause slight or moderate symptoms resemble SS-4. Most of these isolates were non-pathogenic to tomatoes except for a group of isolates originating from areas where cotton was cultivated in close vicinity with solanaceous plants. An isolate belonging to T-1 strain kindly provided by Prof. J. DeVay, Davis, California, caused defoliation on the susceptible cultivar 4-S under the same conditions used for all experiments reported herein. A rapid experimental test was also used to check the virulence of 400 cotton V. dahliae isolates by growing the fungus in plain-agar medium following a technique applied by Schnathorst [personal communication]. Although some 30 isolates formed linear, rather than globular microsclerotia, none of them showed the ability to defoliate the plant providing an evidence that the defoliating T-1 strain either does not exist or is very rare in Greece.

The effect of the growth regulators Cycocel and Pix on the incidence of the disease was also investigated in the field on the susceptible 4-S cultivar. Two kinds of experiments were carried out in naturally infested fields by foliar application of the chemicals at a dose of 25 g a.i. per ha at the beginning of flowering. In one of these experiments plots covering an area of half an acre each, were treated with Pix or Cycocel but no differences in early maturity or cotton seed yield were found.

In a second experiment, consisting of 15 experimental plots of 40 sq. m. each with three treatments and five replicates, again no difference was detected in seed yield. When, however, we attempted to elucidate the effect of the chemicals on cotton seed yield of 100 cotton plants (20 from each plot) randomly collected and showing various degrees of wilt symptoms rated from zero to 4 (Table 5), we found that there was a difference in yield fluctuating from 10-26% over their respective unsprayed controls.

The contradictory results could be explained by the fact that the inoculum was not evenly distributed in the experimental plots. Thus, the expected difference already reported by Erwin et al. [1] was not observed in our tests.

The early maturing cultivar used and the earliness of the Boeotia county, could also mask the effect of

Table 5. Effect of Pix (DPC-Dimethyl piperidinium chloride) and Cycocel (CCC-Chlormequate) applied by foliar Spray (3 July 1978, dosage 25 g a.i./ha) on cotton seed yield of 100 randomly collected cotton plants (cultivar 4-S, 20 plants per plot) showing various degrees of symptoms rated from 0 to 4.

Disease[a] Severity	Total Yield in g		
	Control	Pix	Cycocel
0	2,454.5	2,850	2,829
1	2,287	2,520.5	2,689.5
2	2,548	2,656.5	2,746.5
3	2,100	2,662	2,459.5
4	1,282	1,539	1,518

[a] 0 Healthy plant
 1 Wilting in lower leaves
 2 Chlorosis in lower leaves of the plant
 3 Chlorosis and necrosis of the plant
 4 Necrosis, defoliation or dead plant

the chemicals on the disease. It can be concluded that field trials of growth regulators must be done in late maturing cotton growing areas in fields with evenly distributed microsclerotia population for more reliable results.

Screening technique for the evaluation of cotton seed protectants against soil borne fungi

The ban of organomercurials, used as seed protectants, created the problem of selecting effective substitutes safe for the environment. Extensive field experimentation which started in 1973 throughout Greece, gave inconsistent results. This, however, was inevitable due to variations in the environmental and local conditions, to the variety of pathogens involved and to differences in inoculum potential. These parameters greatly influenced the behavior of the fungicides.

Thus, the problem led the group studying the seedling diseases of cotton to proceed with a new method for screening chemicals under glass conditions before finally testing them in the field.

The routinely used methods for evaluating chemicals are mainly based on the comparison of their effectiveness at a single dose or a series of doses against a certain inoculum concentration of the pathogen but they present a weak point: they do not take into consideration the "inoculum potential." The role of this factor was recently stressed by Garber et al. [4]. As it has been pointed out by Garrett [5] the inoculum potential does not depend only on the number of the propagules of the pathogen but on its energy of growth, which is directly proportional to the density of the inoculum, the vigor of the inoculum and the collective effect of the environmental conditions.

The developed screening technique assesses the ability of a certain dose of the chemical to protect the

Table 6. *Pythium ultimum:* Percent survival of cotton seedlings from differentially treated seed grown for 30 days in various levels of inoculum density (Test 1)

Fungicides[a]	Dose (g or cc/100 kg of seed)	Inoculum levels percent (in soil of 13,5 % moisture content)[bc]				
		0	0.08	0.75	6.25	12.5
Control		100 ab	52 bc	2 d	0 d	0 d
Busan 30 E.C.	400	107 ab	102 a	57 abc	0 d	7 d
+ Demosan 65 % W.P.	360+360	85 bc	90 a	82 ab	2 cd	7 d
Demosan –C W.P.	624	112 a	102 a	87 ab	15 cd	7 d
Dexon 50 % W.P.	220	105 ab	107 a	87 ab	85 a	85 a
+ PCNB 75 % W.P.	262,5+250	92 abc	82 ab	90 a	80 a	70 ab
+ Demosan 65 % W.P.	262,5+360	90 abc	105 a	70 ab	95 a	72 a
+ Vitavax 75 % W.P.	262,5+250	97 abc	97 a	85 ab	82 a	95 a
+ Daconil 75 % W.P.	87 + 100	102 ab	97 a	95 a	92 a	40 bc
Vitavax–C W.P.	380	97 abc	107 a	57 abc	0 d	5 d
Kathon 70 % S.P.	240	75 c	35 c	30 c	45 b	15 cd
+ Dithane M45 80% W.P.	240+300	115 a	77 ab	62 abc	30 bc	7 d
Terracoat L-21 L.	1000	105 ab	80 ab	50 bc	7 cd	10 d

[a]The number of plants in the control (zero inoculum level) that is 40, was taken as 100 % survival

[b]Statistic evaluation was done according to Duncan's Multiple Range Test at the 5 % level, among various treatments in the same inoculum level

[c]Survival over 50 % is underlined

Table 7. *Rhizoctonia solani:* Percent survival of cotton seedlings from differentially treated seed grown for 30 days in various levels of inoculum density (Test 1)

Fungicides[a]	Dose (g or cc/100kg of seed)	Inoculum levels percent (in soil of 12 % moisture content)[bc]				
		0	0.08	0.75	6.25	12.5
Control		100 ab	27 d	23 e	0 b	0 d
Busan 30 E.C.	400	89 b	49 cd	18 e	7 b	7 cd
+ PCNB 75% W.P.	360+250	98 ab	68 abc	34 cde	64 a	34 b
+ Demosan 65% W.P.	360+360	86 b	84 ab	61 abc	16 b	27 bc
Demosan –C W.P.	624	104 ab	102 a	79 a	77 a	102 a
Vitavax –C W.P.	380	111 a	98 a	89 a	70 a	84 a
Kathon 70 % S.P.	240	68 c	50 cd	7 e	11 b	20 bcd
+Dithane M45 80% W.P.	240+300	100 ab	45 cd	11 e	4 b	2 d
Terracoat L-21 L.	1000	104 ab	77 abc	59 abcd	70 a	84 a
Benlate 50 % W.P.	452	102 ab	59 bcd	54 bcd	77 a	86 a
Daconil 75% W.P. +Dexon 50% W.P.	100+87	102 ab	57 bcd	32 de	14 b	9 cd

[a]The number of plants in the control (zero inoculum level) that is 44, was taken as 100% survival

[b]Statistic evaluation was done according to Duncan's Multiple Range Test at the 5 % level, among treatments in the same inoculum level

[c]Survival over 50 % is underlined

Table 8. *Pythium ultimum* and *Rhizoctonia Solani* (1:1) : Percent survival of cotton seedlings from differentially treated seed grown for 30 days in various levels of inoculum density (Test 1)

Fungicide[a]	Dose (gr or cc/100 kg of seed)	Inoculum levels percent (in soil of 13 % moisture content)[bc]				
		0	0.08	0.75	6.25	12.5
Control		100 abc	54 b	0 e	0 d	0 d
Terracoat L-21 L	1000	119 a	110 a	103 a	103 a	59 a
Demosan –C W.P.	624	108 ab	102 a	92 ab	61 b	25 bc
Kathon 70 % S.P.	180	88 bc	56 b	10 de	3 d	0 d
+Dithane S 60% W.P.	180+400	97 abc	58 b	7 de	14 cd	2 d
Dexon 50% W.P. +Demosan 65% W.P.	262,5+360	80 c	83 ab	96 bc	78 b	53 a
+Vitavax 75% W.P.	262,5+250	98 abc	93 a	75 bc	31 c	10 cd
+ Daconil 75% W.P.	122+100	117 a	81 ab	25 d	20 cd	15 cd
+ PCNB 75 % W.P.	262,5+250	108 ab	61 b	64 c	61 b	41 ab

[a]The number of plants in the control (zero inoculum level) that is 59, was taken as 100% survival

[b]Statistic evaluation was done according to Duncan's Multiple Range Test at the 5 % level, among treatments in the same inoculum level

[c]Survival over 50 % is underlined

seed against the series of increasing inoculum densities of the pathogens *Rhizoctonia solani, Pythium ultimum,* and a mixed inoculum of both fungi. The cotton seed was treated by the slurry method 3 months before sowing. Inoculum of the pathogenic fungi was gown for about 28 days in a mixture of sand and corn meal (sand + 30% corn meal + 60% water of the waterholding capacity of the sand) in Erlenmeyer flasks at 22°C.

The levels of the inoculum density were usually 0.0, 0.08, 0.75, 6.25 and 12.5% w/w in a soil with 7-13% moisture content. The same quantities of equal parts of the pathogens were also used in the case of mixed inoculum. The soil-sand-inoculum mixture content was 250 g per plastic cup. The plants were kept in an air-conditioned room in which the temperature fluctuated between 16° and 19°C. The results were taken one month after sowing by counting the surviving plants in each treatment and replication. The number of plants in the control was taken as 100% survival.

All data obtained from the application of the described method are fully presented in Tables 6, 7 and 8. Examination of these tables reveals that high protection against *P. ultimum* was given only by Dexon and its combinations. Two combinations Demosan-C, Vitavax-C, Terracoat L-21 and Benlate give the best protection against *R. solani* while Terracoat L-21, Dexon + Demosan and Dexon + PCNB were highly efficient against mixtures of *P. ultimum* and *R. solani.*

References

1. ERWIN, D.C., TSAI, S.D. & KHAN, R.A. 1979. Growth retardants mitigate Verticillium wilt and influence yield of cotton. Phytopathology 69:283-287.
2. ESENTEPE, M. 1974. Investigations on determination of the cotton wilt disease agent and its distribution, severity, loss degree and the ecology in Adana and Antalya provinces. J. Turkish Phytopathol. 3:29-36.
3. GALANOPOULOU, S. N., LEUKOPOULOU, S. S. & CHLICHLIAS, A. G. 1978. Verticillium wilt of cotton (in Greek) p. 12, Sindos, Thessaloniki (mimeo).
4. GARBER, H. R., DeVAY, F. E., WEINHOLD, R. A. & MATHERON, D. 1979. Relationship of pathogen inoculum to cotton seedling disease control with fungicides. Plant Dis. Rep. 63:246-250.
5. GARRETT, S. D. 1960. Pathogenic root-infecting fungi, Cambridge University Press, 294 p.
6. SAVOV, S. G. 1977. Harmfulness of Verticillium wilt to cotton in Bulgaria. Rastenievadni Nauki (Plant Science) Vol. XIV, 6, 95-103.
7. SAVOV, S. G. 1978. Influence of some agrotechnical measures on Verticillium wilt on cotton. Rastenievadni Nauki (Plant Science) Vol. 15, 4, 122-127.
8. SAVOV, S. G. 1978. Dissemination of *Verticillium dahliae* Kleb in different organs of cotton plants. Rastenievadni Nauki (Plant Science) Vol. 20, 7, 121-125.
9. SAYDAM, C., COPCU, M. and SEZGIN, E. 1973. Studies on the inoculation techniques of cotton wilt caused by *Verticillium dahliae* Kleb. J. Turkish Phytopathol. 2:69-75.
10. SCHNATHORST, W.C., and D. FOGLE. 1976. World distribution and differentiation of *Verticillium dahliae* strains pathogenic in *Gossypium hirsutum.* (Abstr.) 2nd Int. Verticillium Symp., Univ. Calif., Berkeley. P. 39.
11. TJAMOS, E. C., and KORNAROS, E. 1978. Virulence of Greek *Verticillium dahliae* isolates on susceptible and tolerant cotton cultivars. Plant Dis. Rep. 62:456-458.

34.4   Integrated Pest Management of Cotton in China

H. F. Chu

Institute of Zoology,
Academia Sinica, Peking, China

Cotton is an exotic crop in China and has been cultivated more than seven hundred years since its introduction from foreign countries in the thirteenth century. At present, there are five cotton areas scattering about latitude 19-44° North and longitude 76-124° East. They are the Yellow River basin cotton area, the Yangtze River basin cotton area, the South China cotton area, the Northwest inland cotton area, and the Liao River basin cotton area. Each area possesses its own climatic, soil, vegetative and other ecosystem characteristics. For a dozen years, the author and his colleagues have worked in several major cotton producing regions on the subject of cotton insects.

In China, the high priority placed on agriculture has brought progress both in production and in protection. Integrated management is a traditional tactic in combating injurious insects in China. For the cotton crop, the practices generally center around the cultural manipulation including field sanitation practices, chemical, physical, biological and some other suppressing measures. This paper deals with the integrated management of cotton insects in the major cotton areas, especially the Yellow River basin cotton area.

Ecosystem Diversity and Insect Complex

Since five large cotton areas are distributed widely in China and with its own natural conditions, the knowledge of agroecosystem is the foundation of integrated pest management of cotton. A lot of information has been accumulated over the years. In a previous paper [Chu, 1978], the climatic, soil, vegetative and ecosystem characteristics of five cotton areas are given. It suffices to say that there are 300 and more species belonging to 8 orders and 53 families recorded in China as cotton pests. Most of them, however, are secondary or minor pest species and a few are playing more important roles on cotton production. The major or key pests are the cotton aphid, *Aphis gossypii*, bollworm, *Heliothis armigera*, spotted red spider mite, *Tetranychus cinnabarinus*, Spiny bollworms, *Earias* spp. and pink bollworm, *Platyedra gossypiella*, and some others. Some of them are widely distributed in various cotton areas, such as the cotton aphid, spotted red spider mite and cotton bollworm. Of the three species of spiny cotton bollworms, *Earias cupreoviridis* occurs in all areas except the N.W. inland cotton area; *E. fabia* extends from South to North China; while *E. insulana* limited to South China only. Since cotton was introduced from abroad, the question of how did the insect pests immigrate into cotton fields is of utmost concern in studying Chinese cotton insects. The reasonable answers are: 1. The insect members of a primary community shifted into the secondary community—These include many omnivorous species and some oligophagous species originally inhabiting host plants of the mallow family which are indigenous in China. 2. A few exotic cotton pests immigrated from foreign countries.

Following the reformation of cultural system, some major insect pests have changed their status. The cotton inchworm, *Ascotis selenaria* and the cotton leaf roller, *Syllepta derogata* which were major pests in the Yangtze River basin area are scarce in recent years. On the other hand, the secondary pest in the Yellow River basin area, the cotton bollworm becoming more serious. In Yunnan cotton growing region, the winter is not cold and the cotton stubs grow new sprouts for producing the second crop. This ratooning or stub cotton provides a continuous source of food for the spiny bollworms, aphids, mites and some other cotton pests. When no stub cotton is allowed in culture, the food sources of these pests will be interrupted, and the pest population suddenly reduces.

Cultural Tactics

The cultural control provides the first line of defense against pest species. In the labor available agricultural system, cultural control is particularly useful. Field sanitation and weeding greatly reduced the populations of red spider mites, cutworms, thrips and cotton aphids as proved in some large scale field experiments in Hopei and Shansi provinces. Management of alternate host plants of cotton aphids, cotton bollworms and spiny cotton bollworms reduced the migrants into cotton fields. Planting of pest attracting plants, such as sesame, clover or corn in neighboring fields or in alternating strip with cotton reduced damage by cutworms, cotton bollworms and plant bugs on cotton. Roguing infested seedlings resulted in suppressing aphid spreading. Pruning vegetative branches and topping cotton plants in fruiting stage suppressed aphid propagation and bollworm oviposition. The pink bollworm problem is eliminated in Liao River basin cotton area and northern Hopei province, since seed cotton is stored in large warehouses or in open fields, the overwintering larvae are killed by the cold temperature.

Successful utilization of cultural controls requires a detailed knowledge of the pest's biology and ecology in order to couple the techniques with agronomic practices.

Biological Tactics

Use of beneficial insects to control crop pests is a traditional measure in China. However, it has received effective emphasis since the collapse of broad spectrum persistent organic insecticides. Biological control is becoming an important part of integrated control along with cultural control. Using lady beetles, mostly *Coccinella septempunctata*, and *Leis axyridis* in controlling cotton aphids is popular in North China. About mid-May, lady beetles were collected in wheat fields and released into cotton fields. They may be bred in the laboratory and the egg masses released on cotton plants. It was reported that each bettle larva can consume about 200 aphids within three days. *Trichogramma* spp., mostly *T. dendrolimus*, are used in controlling lepidopterous pests, especially the cotton bollworm, with parasitism reaching 70-100% as reported in Shansi. The pink bollw a parasite, *Dibrachys cavus*, was used in cotton storehouses both in Shanghai and Hupei for more than ten years with consistent success. There are two advantages of this approach: 1. the work is done indoors with conditions easily controlled; 2. breeding material is the pink bollworm larvae which are plentiful in the storehouses. Entomologists have released 5-14 million wasps annually in Hupei from 1957-1973, and around 3 million wasps each year in Shanghai from 1960-1972. Both places had fruitful results. *Chrysopa* spp. are also studied for controlling aphids, mites, bollworms, but not very popular yet. The hunting wasps, *Polistes* spp. were tried to control *Heliothis armigera* and *Anomis flava* in experimental fields. It was reported that when 100 wasps per *mu* were introduced, 70-80% of pest larvae could be killed. Protecting spiders in the cotton field seems very effective in suppressing pest's population. Studies on the nuclear polyhedrosis virus were conducted with high mortalities on cotton leafworms.

Pesticide Treatment

It is recognized that once pest population reach economic thresholds demanding treatment there is little recourse except the use of pesticides or insect pathogens. This situation will happen when cotton farmers placed unilateral dependence for insect control on routine pesticidal treatments. The problems associated with overdependence on chemicals are becoming apparent in causing insect resistance and pollution. Besides, insecticides kill beneficial insects and spiders, and relieve major-pests from their natural enemies. The pests then surge to damaging levels and sometimes cause outbreak of the classical patterns. Ill-times, too frequent, or unnecessary chemical treatments have created many pest problems. In the fifties and sixties we used large amounts of organochlorines and organophosphorus. Because of the factors of environmental persistence and insect resistance, some of them are being phased out of use (e.g., DDT, BHC). However, it is still impossible to avoid using DDT completely, which is used in a quite limited way for bollworm control in

certain places. The use of BHC is being much reduced and lindane is available. Some organophosphorus insecticides are also being gradually phased out of use (e.g., systox, methyl parathion, etc.), but dichlovos (DDVP), malathion, trichlorfon (Dipterex), dimethoate (Rogor), and some others are used widely in cotton fields. There are a few varieties of carbamates produced in China, and carbaryl (Sevin) is used occasionally for controlling the cotton bollworm. The development of new and safe insecticides is much desired with emphasis on the high effect and low toxicity (e.g., phoxim, permethrin). In the integrated control, sometimes, infested areas are spot treated with spraying or dusting. Microbial insecticide, such as *Bacillus thuringiensis* is sparingly used to meet the need of integrated control.

Miscellaneous Measures

Attracting insects with black light is very common in China. The effectiveness of this method was tested in several localities [Hopei, 1977; Hupei, 1975; Nantung, 1975] with positive results in reducing pest populations, especially those of the bollworms and cutworms. Use of tree stick bundles to attract the bollworms and cutworms was found effective.

Experiments on the insect pheromones, such as gossyplure for the pink bollworm are carrying on with good progress. Preliminary tests on the application of juvenoids against cotton aphids were reported by Kwantung College of Agriculture and Forestry and revealed that the aphid population showed no increase in a period of seven days.

At last, but not the least, the monitoring and forecasting system is well organized at various levels of the agricultural organizations. Based upon the biological and ecological knowledge of the major pests in coordination with climatic and agronomic conditions, the fluctuations of pest populations are forecasted. With the knowledge when the population will reach the threshold level, control measures, especially the insecticidal treatment could then be taken at the proper time. Forecasting therefore is the commander of the whole integrated control system.

34.5   Achievement in Hybrid Cotton in India and Related Pest Management Problems

R. R. Mishra
D. B. Desai

Maharashtra Hybrid Seeds Company Ltd.
19 Raj Mahal, 84 Veer Nariman Road, Bombay 400 020 (India)

Introduction

Cotton is one of the major crops of India with the largest coverage of area under cultivation in the world among cotton growing countries. Presently it is cultivated in about 8 million hectares in 20 states.

India is perhaps the only country where all the four cultivated species of cotton are under commercial cultivation. Presently *hirsutums* occupy approximately 60%; *arboremus* 22%; *herbaceums* 18%; and *barbadense* types occupy a few thousand hectares. In the 1950's the area under *hirsutum* cottons was only about 20% of the total, whereas the position now has reversed and presently about 60% of the total area belongs to American (*G. hirsutum*) cottons.

India's cotton production has made giant strides since Independence. The total output of cotton has registered a three-fold increase from 2.29 million bales (170 kg each) in 1947-48 to 7.1 million bales in 1977-78. There is an increase of 72% in the per hectare cotton yield. We are also producing 2.7 million bales of long and extra long staple cottons whereas not a bale of these cottons was produced until the 1950's.

Review of Cotton Research in India

These achievements have been brought about mainly by: (i) development and extension of new high yielding hybrid cotton cultivars, (ii) expansion of irrigated cotton area, and (iii) spread of improved agronomy. Of the above three factors, we shall discuss in this paper only the development of hybrids, their spread and related pest management problems.

It will be observed from Table 1 the area under hybrid cottons increased from a mere 23,000 has in 1970-71 to a record level of 755,000 ha in 1975-76. Subsequently there has been a decline, the main reason being the low price realization for seed cotton by the farmers. It is to be particularly noted that the performance of the hybrids was excellent and farmers have accepted them. In 1974-75 when the country produced a record crop of 7.16 million bales, the prices crashed down causing severe losses to the producers, who had to invest much more and wait for

longer periods for returns from growing hybrid cotton. For the major part of 1975-76 also, the prices were not favorable. Disappointed, the growers had to cut down the area under hybrid cottons which are capital intensive. Thus, the decline in the coverage under hybrids in 1976-77 to 565,000 ha. However, the momentum picked up and scaled to a new peak of 758,000 ha in 1978-79. Presently India has potential to bring about 1.2 million hectares immediately under hybrids with a requirement of 3 million kilograms of hybrid seed.

Research on cotton breeding was started in India in 1920's. It received further fillip since 1967 when the Indian Council of Agricultural Research sponsored an All India Co-ordinated Cotton Improvement Project.

The quest for high yielding hybrid cottons commenced in the 1940's. The pioneering research efforts initiated in Surat in 1948 led to the identification of a few cotton hybrids in the early 1950's. However, none of them survived the harsh realities of large scale cultivation on the farmers' fields to emerge as commercially acceptable. Their main defects were their unstable field performance, high neppiness of yarn and the difficulties in producing F_1 seed on a large scale, as one of the parents of the most promising hybrid happened to be a perennial. Undaunted by these initial problems the cotton scientists persisted in their efforts and the crowning success to these efforts came in the late 1960's. It came in the shape of

Table 1. Commercial hybrid cotton and hybrid seed cotton hectarage and production per year in India

| Year | Area under cultivation | | F_1 hybrid seed (quintals) |
	Commercial (10^3 ha)	Hybrid seed (ha)	
1970-71	23	243	680
1971-72	53	1016	6957
1972-73	206	1754	13910
1973-74	523	2183	23060
1974-75	749	3616	20026
1975-76	755	976	7744
1976-77	565	1027	12588
1977-78	624	2344	20996
1978-79[a]	758	2504	19249

[a]Estimated

Table 2. Pedigree and important fiber properties of some released hybrids

Name of Hybrid	Pedigree (parentage) Female	Male	Staple length mm	Ginning %	Micronaire value mc	Pressly strength index	Average spinning capacity counts	States in which grown
Hybrid -4	Gujarat 67	American Nectariless	27.9	34	4.6	6.5	50s	Gujarat, Andhr, Pradesh, Madhya Pradesh, Mahararashtra & Karnataka
Varalaxmi*	Laxmi	SB 289-E	29.6	31	2.8	8.1	60s to 80s	Gujarat, Andhra Pradesh, Madhya Pradesh, Marharashtra, Karnataka & Tamil Nadu
C.B.S. 156*	Acala glandless	SB-1085-6 E	33.0	32	3.0	8.8	100s	Tamil Nadu
Hybrid 5*	Gujarat 67	SB 289E	33.5	28	2.9	8.6	60 to 80s	Gujarat
R.H.R. 253 (Savitri)*	Kopergaon 203	SB 1085-6 E	32.5	32	2.7	8	60 to 80s	Maharashtra
J.K. H-1	Khandwa-2	Reba B-50	25.4	33	–	–	40s	MP, Maharashtra
J.K. H-11*	Khandwa-2	ERB 4492	29.5	29.5	–	–	65s	Madhya Pradesh

*Interspecies hybrids; female parent is *hirsutum* and male parent is *barbadense*.

Hybrid-4 an intraspecies hybrid between two *hirsutums*, viz. G-67, an elite variety of Gujarat and Nectariless, an exotic cultigen from the USA. Its forte was its prolific bearing habit. It made its debut on farmer's fields in 1967-68 and was an instant success. Indeed, it was the harbinger of the hybrid cotton era in India and perhaps in the world.

In the wake of the success of H-4, another equally outstanding hybrid Varalaxmi was released from Dharwar, Karnataka. This is an interspecies between widely adaptable *hirsutum* variety Laxmi and SB 289-E a *barbadense* developed from Russian material. Later on various other hybrids have been released from Tamil Nadu, Madhya Pradesh and Maharashtra (Table 2).

Cotton Research in Mahyco

At the dawn of Hybrid era in India in 1964, Mahyco (Maharashtra Hybrid Seeds Company Limited) was established to breed, produce, process and market hybrid seeds of corn, sorghum, cotton and pearl millet. Mahyco established its research program in the year 1966 to evolve better hybrids and to develop better technology for hybrid seed production. The cotton research nursery was started in the year 1971.

The source of genetic male sterility was received from Dr. J. B. Weaver. The company has successfully transferred genetic male sterility to the potential lines adaptable to Indian conditions. These genetic male steril lines were utilized for making several hybrid combinations to evolve long staple and extra long staple cotton hybrids. Mahyco is

proud to be perhaps the first to use male sterile line on a commercial scale to evolve an interspecies (*hirsutum*×*barbadense*) hybrid MCH-1 in the year 1974 for commercial cultivation. After testing the hybrid at our research station it was tested on farmers' fields for 2 years. Another landmark was achieved by the release of Intraspecies (*hirsutum* ×*hirsutum*) hybrid MCH-11 in the year 1976 again based on genetic male sterility. These hybrids are very promising and combine good fiber qualities and high yield potential. Characters of Mahyco hybrids MCH-1 and MCH-11 are compared with two existing popular hybrids H-4 and Varalaxmi (Table 3).

The basic advantage of using male sterile seed parent in hybrid seed cotton have been experienced by us as (i) higher per acre seed production, (ii) higher

Table 3. Characteris of Mahyco hybrids MCH-1 and MCH-11 in comparison with hybrids H-4 and Varalaxmi

Name of Hybrid	Avg yield qtls per ha	Staple length (2.5% span length mm)	Ginning %	Micronaire value MC	Pressly strength index	Avg spg counts
MCH-1	34-45	40.0	30.0	2.9	9.2	80 to 90
Varalaxmi	30-40	37.0	31.0	2.8	8.1	60 to 80
MCH-11	35-45	29.5	38 to 39	4.1	8.0	50s
Hybrid 4 (H-4)	30.40	30.0	34 to 35	4.6	6.5	50s

Table 4. The average seed yield (delinted) and the minimum percent hybridity in practice of four cotton hybrids in 1978-1979

Hybrid	Yield (kg/ha)	Hybridity (%)
H-4	275	90
Varalaxmi	378	90
MCH-1	740	98
MCH-11	735	98

percentage of hybridity (Table 4), and (iii) comparatively low cost of production per unit of seed.

Hybrid Seed Production Technology

Production of quality hybrid seed of cotton is a highly skilled and labor intensive job. India with its huge manpower resource could economically produce hybrid cotton by hand emasculation and pollination. Table 1 gives the area and production of hybrid seeds in the country during 1970.

The seed thus produced in 1977-78 i.e. 20996 qtls is sufficient to plant about 0.84 million hectares at a seed rate of 2.5 kg/ha (1 kg/acre).

Methodology

The seed rate for femal plants is 1.125 kg per hectare and for male plants 0.625 kg per hectare. The female parent is planted in the main crossing field. On the same day 10% of the female seed is planted in the polethylene bags containing a fine mixture of oil and FYM (the seedlings to be used for gap fillings). The male parent seed is planted in an adjacent plot with time staggering two/three times, to have a continuous supply of pollen. Female plants are stecked by sticks to give the plants sufficient support. This stecking reduces losses due to breakage of branches, lodging of plant and by preventing lower bolls from soil contact.

Hybridization

It is necessary to inspect and rogue all offtypes and volunteers from both male and female plants. The buds which are likely to open next day are selected. The operation is done in the afternoon. Emasculation is done by two methods: the blade method and the nail method. In the blade method a half portion of blade is taken and a circular cut is given to the calyx at half stage so that a complete circle of corolla also gets the cut. After giving a circular cut the calyx is gently twisted to remove the corolla. Anthers are removed by using thumb and fore finger and cleaned perfectly.

In the nail method the worker has to grow the nail of thumbs. Thumb nails are gently pierced in the bottom portion of calyx without injuring the ovary and the bud is opened completely. By this process the staminal tube gets removed from ovary and style. There is no need to remove individual anthers in this process. In this method the ovary is exposed.

Experienced workers can emasculate the bud effectively. After completing the emasculation the stigma is protected by fixing a piece of straw tube (5-6 cm) and this method is followed commonly. Red colored butter paper bags are also used to protect the stigma from contamination. Covering with butter paper bags is time consuming as well as comparatively more expensive.

Early in the morning flowers of male plants which are likely to open in the next two to three hours are plucked and distributed to workers. Corollas are removed and the flower is used for pollination. The colored straw tube or red butter paper bag is removed and all the lobes of the stigma are pollinated. The stigma is covered with a white-colored straw tube or white butter paper to distinguish between emasculated and pollinated buds. A string is tied to the stalk of the boll after pollination as a mark of identification. Generally one male flower could be used for pollinating four buds. Application of proper pollen on all sides of stigma is important, otherwise the boll will not have a proper development and only fertilized sectors will develop.

Cotton Picking

In a hybrid cotton seed production program after hybridization is completed, flowers and buds from female parent are removed regularly so that there is no selfing of bolls. At the time of picking of cotton an individual boll is checked for the presence of thread and scar of circular cut on calyx. Cotton is picked frequently 4 to 6 times to avoid losses due to rains. After picking the cotton it is dried and packed in jute bags. Bags are labelled and sent to ginning centers.

Ginning, Delinting and Cleaning

Seed is ginned on double roller gins specially installed for this purpose. Seed is hand cleaned and picked to remove broken, infested or diseased seed. The ginned seed is acid delinted, cleaned and graded on a gravity table. After treating the seed with a recommended fungicide and insecticide, it is packed in 750 g (non-delinted 1.0 kg) packings sufficient for one acre.

Thus packed and sealed hybrid seed is released to the market only after getting a satisfactory report of hybridity (by grow-out test), physical purity and germination.

Use of Male Sterility in Hybrid Seed Production

Hand-emasculated and pollinated hybrid seed is expensive and becomes a limiting factor in more effective coverage under hybrid seed. Use of male sterile female cotton lines could help in reducing the cost of hybrid seed and boost the production of the cotton crop if planned properly.

At MAHYCO, we started the transference program of genetic male sterility as well as cytoplasmic male sterility since 1971/72. These male sterile lines were utilized to bring about some promising hybrid combinations. Presently two hybrids, viz MCH-1 and MCH-11, are released for commercial cultivation.

Seed production technology has been developed for producing hybrid seed on genetic male sterile lines and successful seed production is being undertaken. While utilizing the male sterile female parent for hybrid seed production, the seed rate differs (female seed=2.5 kg/hac, male seed=0.5 kg/hac).

Fertile plants are removed after testing and the male sterile plants are maintained in desired proportion for hybridization in the main field. There is no need of emasculation but at present seed is produced by hand pollination. Female flowers are covered by a straw tube and pollinated as described in production technology. Cost of cultivation even by adapting the same process is comparatively low and average high yields of hybrid cotton seed per hectare are obtained when compared to other hybrid seed production. Due to lack of desired isolation in Kharif season we are required to undertake hand pollination. It would be interesting to study the yield potential difference between hand-pollinated and insect/natural pollinated plots. We laid out a small scale trial in a non-cotton growing area and were able to produce a small quantity of seed by insect pollination but economics of this production need to be further worked out to make it a practical/commercial practice.

Presently we have transferred cytoplasmic male sterility into several *hirsutum* lines. The transference of male sterility into *barbadense* lines also is in advance generations. We have identified lines that completely restore fertility. Transference of restorer genes is in progress. However, the results of transference vary from line to line. We have not yet been able to identify the exact factors for varying percentage of fertility. We have identified hybrids which, we hope, shall be adaptable to large growing areas using cytoplasmic male sterility and restorer. Next year, we hope to take extensive trials with these hybrids in the main cotton growing areas of India.

Utilization of male sterility in cotton for production of hybrid seed shall revolutionize the cotton production not only in India but all over the world. We see a time, not far away, when cotton hybrids will be produced in the fields just like any other hybrids produced by insect pollination.

Pests and Diseases of Cotton

In the quest for increased production, by bringing more area under irrigation and high yielding varieties/hybrids, pest management has become an integral part of cotton cultivation. Although American cottons have inherent potential for higher yields they are relatively more susceptible to a variety of pests and diseases. While the total cotton area treated against pests was negligible about 15 years ago, it is estimated that in 1977-78 as much as 8.32 million hectares (gross) were covered by plant protection measures. The extent of losses due to damage by pests is estimated to be 1 to 1.5 million bales valued at about Rs. 15 to 25 million annually.

The experiments on chemical control of the pest complex on cotton are being carried out at various centers include: (a) testing of insecticides as foliar spray in control of sucking pests like aphids, jassids, thrips and mites, (b) testing of systemic insecticides as in-furrow application in control of sucking pests, (c) study of efficacy of several insecticides in the control of boll worms, and (d) evolving an effective and economic schedule of control for the cotton pest complex.

Cotton Pests in India
Cotton Jassid (*Armasca biguttula biguttula*)
Aphid (*Aphis gossypii*)
Tailed mealy Bug (*Ferrisia virgate*)
Cotton mite (*Tetranychus* sp.)
Cotton Semi-looper (*Anomis flara*)
White weevil (*Myllocerus* Sp.)
Leaf roller (*Sylepta derogate*)
Bihar or Black Headed Hairy Caterpillar (*Diaerisia oblique*)
Red Cotton Bug (*Oxycarenus laetus*)
Pink bollworm (*Pectinophora gossypiella*)
Spotted bollworm (*Earias insulana* and *E. vittella*)
American Cotton bollworm (*Heliothis armigera*)

Cotton Diseases in India
Fusarium wilt (*Fusarium oxysporum* and *F. vasinfectum*)
Verticillium wilt (*Verticillium dahliae*)

Black arm or Bacterial blight (*Xanthomonas malvacearum*)

Root rot (*Rhizoctonia bataticola, Macrophomina phaseoli*)

Anthracnose (*Colletotrichum capsici, Glomerella gossypii*)

Greymildew (*Ramularia areola*)

Stenosis or small leaf (mycoplasmal disease)

Root knot nematode (*Meloidogyne incognita* var. *acrita*)

The cotton disease investigations have been intensified to cover genetic and chemical methods for control of diseases specially the wilts and bacterial blight. The National Centre at Poona for screening of cotton cultivars for resistance to the wilt disease caused by *Fusarium* has made significant contribution to the identification and building up of resistant cultivars. In the breeding program more emphasis has been given to evolve resistant cultivars for multiple diseases. The cultivar Sujata (*G. barbaense*) is resistant to bacterial blight and Verticillium wilt.

In the chemical control program, seed treatment including plantvax, vitavax, and other common organomercurials (Agrosan, Ceresan) have been found useful in controlling seed-borne infections.

Number of economic and effective plant protection measures has been suggested to control the pest and diseases for commercial cotton cultivation and hybrid seed production program. A general plant protection schedule is shown in Table 5.

The usual precautions should be observed in using pesticides, such as following the recommendations on the labels. If the weather is cloudy, fungicide sprays can be used but if rains are frequent, dusting may be better. During the boll bursting period, spraying should be done after cotton picking. Depending on the incidence of pests and diseases, after 120 days suitable plant protection measures should be extended.

Acknowledgements

The authors thank officials of the Directorate of Cotton Development (Ministry of Food and Agriculture), Bombay, for statistics on cotton cited in this paper.

Table 5. Plant protection schedule for pests and diseases of cotton in commercial fields in India

Age of crop, days after sowing	Pest/Disease	Pesticide and/or Fungicide
Before sowing	White grub	5% Heptachlore 35 to 50 kg per hac. should be mixed thoroughly in soil
8	Aphids Jassids Mites	Soil application of Thimet/ Solvirex/Systemic granules by ring method
16	Aphid Jassids Leaf roller	Spray Endrin/Thiodan/Ekalux
30	Aphid Jassids Bollworm Leaf roller	Soil application of Thimet/ Solvirex, by rind method and spray Rogor/Metasystox/ Anthio
45	Bollworm Anthracnose Bacterial blight	Carbaryl/Sevin + Bavistin/ Copper Oxychloride/Blue Copper + Agrimycin + Edalux/ Labsid/Thiodon
60	Bollworm Mites Anthracnose Bacterial blight	Spray Rogor/Metasystox + Sulphur + Copper Oxychloride /Diathane M-45 Bavistin
75	Bollworm Caterpillars	Folidol dust/Eklux Dust/ Carbaryl dust or misture of B.H.C. 10% + D.D.T. 10%
90	Bollworm Anthracnose Bacterial blight	Repeat the schedule followed at 45 days' stage
105	Bollworm Anthracnose Bacterial blight	Repeat the schedule followed at 60 days' stage.
120	Bollworm Anthracnose Bacterial blight	Repeat the schedule followed at 75 days' stage

34.6   Cotton Pest Problems in Egypt

Abdel Latif Isa

Ministry of Agriculture, Egypt

Introduction

Cotton is the most important agricultural crop in Egypt. The cotton area has varied, during the last few years, between 1.25 and 1.5 million acres per year. It is still considered the backbone of the national economy in spite of the recent attempts at industrialization and crop diversification. This crop, however, is highly susceptible to arthropod infestations which attack all parts of the plant at all times of the growing season. Cotton diseases are of relatively minor importance.

The program for cotton pest control starts yearly before cotton is planted. It has become well-organized and one can say that the cotton yield is no longer subjected to heavy losses regardless of the size of infestation by different pests.

Cotton Insects and Mites

The important arthropod pests in Egypt are classified as follows: 1) Early season pests (cutworms, cotton thrips, cotton aphids, spider mites); 2) Mid-season pests (cotton leafworm, lesser cotton leafworm, American bollworm); 3) Late season pests (pink bollworm, spiny bollworm). Practically, the cotton leafworm and the pink bollworm are considered as key pests of cotton in Egypt.

Cutworms

Several species of cutworms occur in Egypt. The greasy cutworm *Agrotis ipsilon* is the most common. This insect is known to cut off seedlings of cotton and many other plants at or near the soil surface. Infestation with this insect increases when preparation of cotton land is carried out in a hurry without allowing the soil and the remnants of the previous crop (mostly clover) to dry up. It is recommended, therefore, the plowing of cotton fields should be done early enough to give the soil ample time to dry before cotton is sown.

However, a chemical treatment is applied when necessary either as a poison bait or by spraying of cotton seedlings. Of the 1.5 million acres or so, the area sprayed annually against cutworms ranges between 30 and 80 thousand acres. This area was 40,000 acres in 1979.

The Cotton Thrips: Thrips tabaci

Thrips attack the leaves and terminal buds of cotton seedlings. Infestation may be slight, or sporadic, or it may spread over the whole field. Severely infested plants may be stunted and the stand of cotton can be reduced. Early sown cotton is less subject to infestation than late sown cotton. Proper irrigation and fertilization allow the plant to tolerate thrips infestation.

Spraying of cotton with insecticides to control cotton thrips is common. However, an economic threshhold for the chemical control of this pest has been determined (8-12 individuals per seedling according to plant age). The area treated chemically against thrips ranged between 8 and 25% of the total acreage in the last 5 years.

The Cotton Aphid: Aphid gossypii

Cotton seedlings are subjected to infestation with the cotton aphid during April and May. However infestation on developed cotton plants with this pest is not uncommon. From 20 to 30% of the area is treated annually against cotton aphids.

Cotton Spider Mite: Tetranychus spp.

The cotton spider mite better known as the red spider mite is a common pest of cotton and many other field and truck crops. Before 1950, mites were not known as economic pests on cotton. After the introduction of the chlorinated hydrocarbons in the early 1950's, mites became a major pest. However, its importance declined after shifting to organophosphorus compounds in the 1960's. Infested fields are treated and the extent ranged between 15 to 60 thousand acres annually during the last 5 years.

One chemical treatment is usually sufficient against all seedling pests. A ready formulated mixture of helthane and dimethoate (referred to as helthane S) is commonly used in the fields infested with mites, aphids and thrips.

The Cotton Leafworm: Spodoptera littoralis

The cotton leafworm is the most serious pest in Egypt. It is extremely phytophagous having, beside cotton, a wide host range of field and truck crops.

This insect has three generations on cotton. The first is the most important. Its peak usually occurs within the last 10 days of June. The second and third occur in July and August.

Clover is the main host of the insect during winter. Clover fields harbor the insect population and is regarded as the main source of the first generation on cotton. With this fact in mind, two measures are taken to manage the insect population in clover fields. First, law stipulates that the last irrigation of clover fields should not take place later than 10 May. In this way, clover as a crop will terminate early and it loses its attraction to the egg laying moth; hence no infestation is found in clover beyond that date.

In addition, the soil of clover fields become so dry that the insect's full grown larvae have difficulty seeking the soil for pupation. Second, kerosene is added to the water of the last irrigation of clover at the rate of 30 lt. per acre.

Eggs of the insect's first generation on cotton is laid in masses of 200 to 300 eggs each on the lower leaf surface. Egg laying starts by late May and increases to reach a peak by late June as mentioned earlier.

The population density of the insect differs enormously from one year to another. A routine count of immature stages in representative samples of clover soils is made in April and May to predict the size of the infestation during the first generation on cotton. However, in most cases, no sound relationship is found between the figures obtained from clover fields and those for the size of infestation on cotton. This suggests that the ecology of this insect is not well known.

Egg masses of this insect are collected by boys working in groups in a way that one third of the whole cotton area should be cleaned daily of insect's eggs. This process, although laborious, reduces the need of periodic chemical treatments for control. Hand-picking of egg masses is done with reasonable efficiency during June and early July covering the first generation of the insect on cotton. It becomes much less efficient afterwards due to the advanced growth and branching of cotton plants.

However, when the insect population is high in certain parts of the country, the number of egg masses per acre may reach as much as 20 thousand or more. In this case a good number of egg masses is left behind by the pickers. Besides, it would be impossible in such a case to cover the usual one-third of the area daily. As the eggs of the insect hatch after 3 days, it is imperative that the egg masses be removed within this period, otherwise hatched larvae could spread on the plants and start to feed. Should the egg masses hatch, chemical treatment should be applied immediately to kill the young larvae and thus save the plants. Advanced grown larvae are less affected by most insecticides.

The area of cotton treated with insecticides differs greatly from one year to another according to the intensity of infestation. It ranged between 13% of the cotton area in 1976 and 65% in 1977.

Our policy is to minimize the use of insecticides in cotton fields during June and early July so as to preserve the populations of predators and parasites that reach their peaks during that period.

The Lesser Cotton Leafworm: S. exigua

This insect is similar in its life history and feeding habits to the cotton leafworm. Nevertheless, it is still of secondary economic importance as a pest of cotton. It appears in cotton fields early in the season. Its egg masses are collected with those of the cotton leafworm.

The American bollworm: Heliothis armigera

As a result of the extensive use of insecticides for a long period, *Heliothis* started to show up as a pest in cotton fields in a limited area in the 1972 season. It spread during the following seasons to cover most of the cotton-growing area in the country causing considerable damage in certain areas. At that time it became a threat to cotton growers and caused tremendous concern among the pest control policy-makers in the country as it interfered with the organized program of cotton pest control.

A tentative economic threshold level of 5 to 10 eggs and larvae per 100 plants (imported from Sudan) was applied though not evaluated locally. According to this level the area of cotton treated with insecticides against this pest reached 25% of the whole area in 1975, i.e. over 300,000 acres.

Intensive research on this pest led to the following three facts: 1) the insect is very susceptible to predators in cotton fields, 2) the population of the insect builds up in the field following the use of certain chemicals during June and July, and 3) in all experiments to test certain insecticides against this pest, the untreated plots were less damaged after a period of time than were the treated plots.

Accordingly the tentative economic threshhold was raised to 20 small larvae per 100 plants. As a result the area treated against this pest was brought down to between one thousand and 20 thousand acres for the years 1976 to 1979.

At present this insect is no longer considered a serious pest of cotton in Egypt.

The Pink Bollworm: Pectinophora gossypiella

Infestation of cotton with this pest is common and regular all over the country. It is actually the main cotton insect in Egypt because of its direct effect on yield. Furthermore, about 84% of the insecticides used for cotton pest control is mainly directed against this pest.

Starting in July, and at 1 week intervals, groups of technicians collect samples of bolls (100 bolls each) from representative areas concentrating on fields around the villages where the main source of infestation lies in the leftover stalks of previous crop.

When the infestation with bollworms in the samples approaches 10% in any area, spraying starts. Usually three sprays at 15 to 21-day intervals are applied. At the end of the season the same group of technicians assess losses in cotton yields due to infestation with bollworms. The estimated yield loss ranges usually between 6 and 9% in the different years. Experimental data show that the loss would be around 30% in the untreated plots.

By decree all seed cotton must be ginned before the end of March every year, and all seeds are treated by heating between 55 and 58°C for 5 minutes. In this way the diapausing larvae are killed within the seeds without affecting the vitality of the embryo.

Early cultivation of cotton helps in evading the high infestation by bollworms which usually prevails at the end of the season (towards September).

The Spiny Bollworm: Earias insulana

In spite of the fact that the individual spiny worm is more destructive than the individual pink worm, the density of the spiny worm is very much less. However, the spiny worm is more common in Upper Egypt and the ratio of both insects might be equal in certain areas in the far South.

Measures already discussed to control the pink bollworm also affect the population of the spiny bollworm.

From the foregoing outline, it would appear that the annual program of pest control which starts before cotton is sown is based on the integration of cultural, mechanical, physical, legislative, natural and chemical control measures, in a more or less well-balanced pest management system. All chemical pesticides used in Egypt undergo a thorough scientific program of screening, testing and evaluation.

34.7   Cotton Diseases in Egypt

M. Y. El-Sawah

Deputy Director, Institute of Plant Pathology
Agricultural Research Center, Ministry of Agriculture, A. R. E.

Cotton diseases in Egypt are one of the limiting production factors; however, they usually come in the second order of importance relative to other cotton pests. The major diseases of cotton in Egypt are those caused by *Fusarium oxysporum* f. sp. *vasinfectum, Rhizoctonia solani* and other soil-borne pathogens attacking cotton seedlings, root knot nematodes, boll rots and Alternaria leaf spot, respectively.

Fusarium Wilt

This disease is more damaging than all other cotton diseases in most of cotton growing countries. Many of the famous long staple Egyptian cotton cultivars namely; Sakalaridis, Karnak, Amoun, and Menoufi have deteriorated due to this disease. Its vascular nature, makes chemical treatment even of systemic action of no practical value. Breeding for resistance is the ideal way of control. Intensive efforts in this area have led to the development of highly resistant varieties to replace the ancient ones. Presently, long staple resistant cotton cultivars such as Giza 45, 67, 68, and 69 are planted for commercial production. Other medium and short staple resistant cotton cultivars, such as Giza 70, 72, and 75 are grown as well. These different cultivars are distributed geographically all-over the country as each is adapted to different climatic conditions. However, breeding for resistance is a continuous practice where other inbreds and lines are being propagated for replacement in case of deterioration of today planted cultivars.

Seed-Rot and Seeding Damping-Off of Cotton Seedlings

This is a disease complex caused by a group of pathogenic soil fungi which develops seriously under certain climatic conditions. In Egypt, *Rhizoctonia solani* is found to be the most frequently isolated fungi from damped-off seedlings associated with one or more of *Fusarium solani*, *Fusarium spp.*, *Sclerotium rolfsii*, *Macrophomina phaseolina*, and *Pythium ultimum*.

The disease is mostly restricted to the Northern Governorates of the Nile Delta and certain provinces of Middle-Egypt, where the soil and environmental climatic condition are favorable for disease prevalence. It was reported that reduction in seed germination may reach 40% in certain localities, while reduction in yield may be from 5 to 10% in some areas. The disease hazards surpass the use of high seeding rates and costs of resowing. It was observed that resowing results in late maturing plants which are consequently more exposed to the attack of Cotton Leaf Worm (*Spodoptera littoralis*), the Egyptian Spiny Bollworm (*Erias insulana*) and the Pink Bollworm (*Pectinophora gossyppiella*); the major cotton pests. Beside the geographical location and climatic conditions that affect the disease prevalence, crop rotation was found to have an apparent effect on the disease development by its direct influence on the inoculum potential and activity of the causal fungi or indirectly by affecting the antagonistic reactions of the soil fungi. Certain agricultural practices, e.g., sand sowing, seed-treatment, are common and are followed by many cotton growers for disease control. Organo-mercurials, captan and Quintozene formulations were previously used for cotton seed treatment. Carboxin; a systemic chemical, is now-a-days used with captan (Vitavax 300), at a rate of 300 g per 100 kg of cotton seed. This treatment has so far been reported as satisfactorily efficient by virtue of its prolonged systemic action. An approximate of 250,000 Feddans (one feddan = 4200 m square) of chosen localities mostly at the northern parts of the country and where the disease develops more seriously are treated annually.

Root Knot Nematode of Cotton

Meloidogyne incognita is not a common pest in all cotton-growing fields. Through surveying, fields with a high population density of nematodes are localized and treated at sowing time with a granular formulation of Aldicarb 10% (Temik) at a rate of 9 kg/Feddan. The treatment was reported to have a side controlling effect on *Gryllotalpa vulgaris*, cut worms and other cotton seedling pests; aphids and thrips. An approximate of 10,000-12,000 Feddans are treated annually.

Boll Rots

The disease is caused by a group of mucor fungi that develop in certain areas and under certain bad drainage conditions, excessive irrigation in dense cultivation and boll worm injuries at the bolling stage specially on bolls carried on the lower branches. Controlled irrigation and opening of cotton lines to evacuate the cotton plants are common agricultural practices to minimize losses. On the other hand chemical control measures by the application of systemic fungicides are being evaluated; (Benomyl, Carbendazims, Thiophanates and Carboxins).

Alternaria Leaf Spot

The causal agent of this disease is *Alternaria tenuis*. The disease is reported to be of minor importance because of the arid conditions that coincide with cotton growing and that the attacked old big leaves fall naturally; consequently the pathogens are rarely transferred to the bolls.

Various chemical treatments were evaluated. Spraying with copper oxychloride formulations at a rate of 500 g/100 litres or maneb formulations at a rate of 250 mg/100 litres applied twice at 15-day intervals starting the middle of June were reported as effective. However, such treatment has never been used as an expanded control measure.

With respect to Verticillium wilt, no apparent damage has been noticed. It could be that the prevailing temperature is the limiting factor. However, some experimentation has been initiated to uncover the reaction of the local cotton cultivars in view of the fact that this fungus has been isolated from hosts other than cotton.

35.1   Integrated Plant Protection in Oil Palm

C. W. S. Hartley

Three Gables, Amberley, Stroud,
Glos., England

The oil palm is a native of West Africa and it was not until shortly before the first World War that a plantation industry began to develop. Until that time palm oil and kernels were extracted from the semi-wild palm groves which maintained themselves in symbiosis with the human populations. Palms were rarely planted or felled, pests were uncommon and the life of the palm was only terminated by a disease of senescence, *Ganoderma* Trunk Rot. This idyllic situation could hardly be expected to prolong itself into the plantation age, particularly when the palm was transported to the other two continents lying athwart the tropics; for although Africa is peculiarly poor in palm species, both Asia and America have over a thousand species and in the latter continent reside the majority of Cocoids of which *Elaeis quineensis* is one of only two African species. Thus it is largely in Asia and, more latterly, in America that the most serious diseases and pests have appeared.

The concept of integrated control has been introduced into oil palm cultivation largely by Wood [Pests of Oil Palms in Malaysia and their Control, 1968] who defines such control as "measures which attempt to supplement natural control factors rather than ignore them" and "choosing chemicals which have less effect against natural enemies than against pests" or "applying chemicals using specialized techniques such as restricted distribution or careful timing." These definitions are not as wide as many workers might demand for "integrated plant protection," but disease and pest control with the oil palm has often been piecemeal, consisting mainly of combatting dangers as they arise, and it is only recently that sophisticated measures of protection have been considered. Moreover, the nature of the crop presents special difficulties and limitations in the control of both diseases and pests. For instance, the fact that the palm is growing and producing continuously means that the timing factor mentioned by Wood can only be employed in relation to the pest or to the season but not to the palm itself, since the latter is always presenting all its organs for attack. Nevertheless there have been a number of very important diseases, pests and disease-pest combina-

tions to which the integrated concept has been applied and to which I will now refer.

Africa

Africa, the home of the oil palm and with its paucity of palm species, has remained remarkably free from pests and diseases. There have been two major disease scares. Firstly, *Fusarium oxysporum*, causing Vascular Wilt disease, devastated plantations in Zaire and its presence in certain parts of West Africa led to efforts being made to breed tolerant progenies; but in the last twenty years no really serious outbreaks have occurred. Similarly Dry Basal Rot (*Ceratocystis paradoxa*) appeared suddenly in Western Nigeria in the late 1950s, devastating certain fields but later subsiding. As with Vascular Wilt, certain progenies have shown marked tolerance.

Serious pest damage in Africa has been largely confined to the attack of one species, *Coelaenomenodera elaeidis*. This Hispid leaf miner is found throughout West Africa in the endemic state; all stages of development can be found simultaneously, but populations are so well controlled by natural enemies that little damage is done. This comfortable situation has, in several countries, passed to a state of swarming in which a numerically increasing cyclical development is accompanied by correspondingly increasing foliar devastation. These sudden upsurges are thought to be due to abnormal falls in temperature, deficiency of egg parasites, or perhaps to increases of reproductive potential. Eventually, through competition for food and increasing parasitism, the population falls and the endemic state is reestablished; but the damage done, with yield reductions of over 50%, is unacceptable and for this reason extensive studies of the pest and its parasites have been made by Mariau and Morin [*Oléagineux, 25,* 11; *26,* 83 & 373; *27,* 496; *29,* 233 & 549; *33,* 153, 1971-78].

An indispensable prelude to integrated control is the monitoring of the size of the egg, larval, pupal and adult populations. This is done on a sample of one palm per ha with rounds which cover the plantation every two months. If, however, any abnormal population increase is noted, special checks are instituted in the area concerned. These are done monthly when the larval index (larvae per sample leaf per palm) is between 10 and 40, and fortnightly with larval indices above 40, a level at which insecticidal intervention is considered necessary [*Oléagineux, 33,* 277; *33,* 429, 1978].

The eggs of this leaf miner are parasitised by the wasp *Achrysocharis leptocerus,* and there are three Eulophid larval parasites among which *Pediobius setigerus* dominates. These parasites tend to kill only larvae in the late stage of development or pupae, and, during swarming, when there are several weeks with an absence of larvae or with only larvae of a size unsuitable for parasitism, the parasites become less able to control the population. Two courses are then open. Firstly, insecticides may be used. To minimize the fall in the number of parasites, application is localized in areas of high infestation and concentrated on the period after hatching and before the last larval instar is reached. Lindane has been most extensively used and Parathion has also been successful [*Oléagineux 28,* 167, 1973]. However, chemical treatment is often insufficient to wipe out a focus of infestation; biological methods are therefore being sought and attempts made to introduce hispid parasites which will attack a broader spectrum of larval life.

Asia

Although there are many recorded diseases in Asia only *Ganoderma* Trunk Rot (Basal Stem Rot) has caused serious havoc. In fields replanted from coconuts or oil palms on coastal soils there have been devastating outbreaks emanating from the massive sources of inoculum built up in old palm stumps. Oil palms planted after rubber or forest only contract the disease at a much later stage [Turner, P. D. *Ann. Appl. Biol., 55,* 417, 1965]. The primary means of protecting the crop has therefore been the expensive process of ridding the fields of infective foci, whether stems, boles or roots. Poisoning the palms to hasten rotting is followed by root raking, cutting up and windrowing and then burning [Simpson, K. M. S. & Rasmussen, A. N., *in* Advances in Oil Palm Cultivation, 1973]. However, apart from the considerable expense of these methods, there has been doubt whether they will prove sufficiently effective. Therefore, attention has been turned to other measures. Firstly, substantial progeny differences in disease incidence have been recorded experimentally in Sumatra [Umar Akbar *et al, Oléagineux, 26,* 527, 1971]. Secondly, a start has been made on trials of

systemic fungicides [Loh, C. F. *in* Int. Developments in Oil Palm, 1976].

Pests of the oil palm in Asia have been more numerous than in Africa, but two groups only, bagworms and nettle caterpillars, have caused serious leaf damage. Bagworms are an example of a group whose spread has been encouraged by contact insecticides. In the interwar period *Mahasena corbetti* was occasionally troublesome and extensively studied [Corbett, J. H., *Malay. Agric. J., 15,* 338, 1927]. Until 1956 *Cremastopsyche pendula* and *Metisa plana* were usually seen in small numbers only [Wood, J. B., *op. cit.*], though *C. pendula* would sometimes attack the palm severely but in restricted areas. Following the use of contact insecticides, resurgences occurred and *M. plana* became the dominant species. Wood [*Planter, 49,* 367, 1973] demonstrated this effect experimentally; successive sprayings with small amounts of dieldrin resulted in the pest increasing in and spreading from a 2-acre plot in which a low, parasite-controlled population had hitherto subsisted. In Sumatra and Sabah larger attacks of both *M. plana* and *C. corbetti* also followed the use of contact insecticides [Hutauruk, Ch. & Situmorang, H. S. *in* Crop Protection in Malaysia, 1971; Mackenzie, R. *in* Int. Developments in Oil Palm, 1976]. In Sabah, it was seen that not only was there a general spread of the pest, but with the migration of the young larvae to the younger leaves spraying became more difficult.

These attacks, together with those of nettle caterpillars to be mentioned later, made it essential to introduce the systematic monitoring of pest spread through a regular census. For Malaysian conditions a well-tried system consists of monthly *detection* rounds in which either all the palms or alternate rows are covered and any pest or disease noted as being of high, medium or low intensity; detection is then followed by *enumeration,* in the area of detection, by sampling about 12 palms per ha and 6 leaves per palm [Syed, R. A. & Speldewinde, H. V. *Planter, 50,* 230, 1974].

Bagworms are parasitized in Asia by a large number of wasp species some of which attack the larval and others the pupal stage. The assassin bug, *Sycanus dichotomus,* and the beetle, *Callimerus arcufer,* are important predators. Broad spectrum contact insecticides must therefore be avoided. When the census shows a larval number greater than 10 per leaf, spraying with trichlorfon is timed to coincide with the emergence of young larvae; spraying when the proportion of pupae is high is of little use.

The method of interrow cultivation has an important effect on maintaining the population of parasites and predators. In Sabah, extensive weeding by accidental burning and by herbicides has been followed by severe *M. corbetti* attacks [Syed, R. A. & Shah, S. *in* Int. Developments in Oil Palm, 1976]. Mackenzie [*op.cit.*] cites a number of species in the natural flora which should be maintained, and he claims that pest populations have been much smaller since this advice has been followed.

The trunk injection of systemic insecticides against *M. plana* has recently been successfully tried [Wood, B. J. *et al, Oléagineux, 29,* 299, 1974]. Monocrotophos at a dosage of 6g per palm was used. Injection early in a generation kills that generation while injection after cocoon formation is sufficiently residual to repress the explosion of the next generation.

The spread of nettle caterpillars has often been encouraged by the use of contact insecticides [Wood, B. J. *op.cit.*]. These caterpillars eat the whole leaf blade, leaving a skeleton of midribs. The most serious is *Setora nitens,* but infestations of *Darna trima* and *Thosea asigna* sometimes occur, and other species are often found in mixed colonies. The eggs, larvae and pupae of *S. nitens* are all heavily parasitized by wasps and flies and their common predators are *Sycanus dichotomus* and other bugs and beetles. When census counts show more than 5 caterpillars per leaf, either hand picking or localized spraying may be resorted to. The use of either lead arsenate or trichlorfon in integrated programmes was at first recommended, but these insecticides have often failed. As a result there have been widespread efforts to discover alternative methods. These have included the rearing of predators for release, the dissemination of disease and the discovery of more suitable insecticides [Wood, B. J. *et al., in* Int. Developments in Oil Palm, 1976]. Attempts have been made to increase the population of *Sycanus dichotomus.* Both *S. nitens* and *D. trima* have been reported as subject to virus infection, and the spraying of the crushed bodies of diseased individuals was early successful [Syed, R. A. *in* Crop Protection in Malaysia, 1971]. In Sarawak this was taken a step further by spraying a diluted suspension of virus inoculum prepared from dead larvae [Tiong, R. H. C. & Monroe, D. D. *in* Int. Developments in Oil Palm, 1976]. This quickly decimated the population which never returned to its original high level. A wide range of insecticides have been tried, bearing in mind the importance of a good kill and reasonable selectivity, so as to avoid recurrence. Of all formulations tried, *Bacillus thuringiensis* insecticides (Dipel 100), dicrotophos, quinalphos and endosulphan have been most effective as sprays while, as with bagworms, monocrotophos has shown promise in trunk injection [Wood, B. J. *et al., op. cit.*].

America

The cultivation of the oil palm in tropical America started in 1943 and only began to expand in the 1960s; since then an increasing number of pest and disease problems have arisen. The palm has, with some exceptions, been planted in relatively small blocks in widely scattered locations from southern Mexico to Brazil on both sides of the continent; and it is often planted in close proximity to other Cocoid palms with their large indigenous insect fauna. But apart from the more visible insect pests which have attacked the palm, two lethal and undiagnosed diseases have arisen which, in one country at least, have discouraged planting. The first of these is Lethal Bud Rot (Pudricion de Cogollo) which has decimated large plantings in Colombia and elsewhere but has only been sporadic in many other regions. Investigations have not so far revealed the cause, and the only method of protection where severe attacks have been felt is the planting of the hybrid *Elaeis quineensis* × *E. oleifera* which has so far proved immune to the disease [*Oil Palm News, 18,* 1, 1974].

The second important disease is Sudden Wither or Marchitez Sorpresiva characterized by browning of the petioles, drying out of the leaves and fruit bunches, and rotting of the roots. A plantation in Colombia was destroyed by this disease and it has appeared sporadically in Ecuador, Peru and Surinam. The etiology of the disease is obscure but there seems to be an association both with flagellate protozoa, which have been isolated from sieve tubes of roots and peduncles of diseased plants [Dollet, M. & Lopez, G., *Oléagineux 33,* 209, 1978], and with the root-mining caterpillar of *Sagalassa valida* which has done considerable damage in areas, particularly near forest and rivers, where Marchitez Sorpresiva has occurred. There is evidence from Peru and Ecuador that, where treatment against *Sagalassa* has been successful, Marchitez incidence has diminished, and it is claimed that in Ecuador it has been rapidly decreasing since *Sagalassa* control, through applications of endrin around the base of the palm, became general [Dzido, J. L. *et al., Oléagineux, 33,* 55, 1978]. The life history of *S. valida* has been studied by Genty [*Oléagineux, 28,* 59, 1973] who considers that, apart from the danger of Marchitez, the damage done to the roots is a direct cause of debility. Nevertheless, no study of the natural enemies of *S. valida* has been reported and the long-term effect of endrin applications on the natural control of this pest, and perhaps other potential pests, in not known.

Another association between insects and pathogens has been found in the widespread increase of

Pestalotiopsis Leaf Wither in areas of Colombia, Ecuador and Honduras. In Colombia the suppression of a Tingid, *Leptopharsa gibbicarina*, by applications of chlordimeform or endosulphan has caused a rapid recuperation of the palms [Genty, Ph, et al., *Oléagineuax, 30,* 199, 1975, & *33,* 407, 1978]. Other insects may be implicated elsewhere, especially the caterpillar of *Peleopoda arcanella* in Ecuador.

As expected, the foliage has been widely attacked by nettle caterpillars of which species of *Sibine* are most important. Genty [Oléigineux, 33, No. 7, 1978] has given the critical threshold for most species in caterpillars per leaf; above these thresholds control with trichlorfon or carbaryl, or in some cases toxaphane, is recommended. With *Sibline fusca* it has been possible to effect control by the dispersal of ground-down larvae infected by a virus [Genty, Ph. *Oléagineux, 28,* 255; *30,* 349; *32,* 357, 1974-77].

The interesting associations in Latin America between pests and pathogens, together with the recent discovery in Africa of the role of the Jassid, *Deltocephalus (Recilla)* sp., in the transmission of Blast disease of nursery seedlings, excentuate the importance of team work in plant protection. So far in Latin America there have been too few pathologists and entomologists working together for sufficient time to solve the more serious problems.

Abstract

The control of important diseases and pests in three continents is reviewed with special reference to integrated methods. The palm is native to West Africa where it was growing in symbiosis with human populations until plantations were started at the beginning of the century. Plantings in Africa have suffered from Vascular Wilt and Dry Basal Rot, for both of which breeding for tolerance is possible. Special monitoring of the crop is required for the one serious pest, *Coelaenomenodera elaeidis,* and insecticides are used at the early larval stage only. Asia and America, with their larger populations of Palmae, present graver problems. In Asia, integrated methods are successfully used against bagworms and nettle caterpillars, while protection is obtained against *Ganoderma* Trunk Rot by stringent cultural methods at replanting. In America, protection of the palm is difficult because of the widely separated planting localities and the large palm insect fauna. There have been large losses from Lethal Bud Rot and Marchitez Sorpresiva. The former is as yet undiagnosed; the latter is associated with the presence of flegellate protozoa and with *Sagalassa valida,* control of which has proved an important measure. Severe outbreaks of Leaf Wither have followed attack by the Tingid *Leptopharsa gibbicarina,* protection against which has reduced incidence. It is increasingly apparent that the introduction of integrated methods with the oil palm requires a greater deployment of pathologists and entomologists working together.

35.2   Integrated Plant Protection in Cocoa

G. A. R. Wood

Cadbury Limited, Bournville, Birmingham, UK.

Background

The cocoa bean of commerce is the fermented and dried seed of the tropical tree, *Theobroma cacao.* This tree had its origin in the head waters of the Amazon basin from which area it spread naturally and with the help of man, to surrounding areas of Central and South America. Following the discovery of the Americas, a market for cocoa beans developed and ultimately this lead to the spread of the tree in West Africa and south-east Asia.

In 1900, when production was about 100,000 tonnes, the major producers were Ecuador, Brazil, San Thome and Trinidad. Subsequently West Africa took over as the major producing area, Ghana becoming the prime producer in 1910, a position she held until a year or two ago. Nigeria, Ivory Coast and Cameroon are the other major producing countries and these four countries together produce 60-65% of total world production, which in the current crop year amounts to 1.4 m tonnes.

During the 1970s cocoa production has been more

or less static at 1.4 to 1.5 tonnes, but the static total conceals significant changes between the major producing countries. Ghana and Nigeria have declined, while Ivory Coast and Brazil have increased and now compete for the prime position. The only other country with rapidly increasing production is Malaysia.

The decline in Ghana has been due in part to a shortage of inputs for the control of certain pests. On the other hand the increase in Brazil has been due largely to increasing inputs and better control of pests and diseases.

In considering integrated plant protection methods and their application to cocoa, it is important to have a picture of the scale on which cocoa is grown and the type of farmer involved. Cocoa is largely a smallholder's crop, virtually all the cocoa in West Africa being grown on small farms of a few acres. In South America the scale of production is considerably larger but cocoa farms are largely family owned. It is only in Malaysia that cocoa is grown on a plantation basis and Malaysia is one of the few countries where pest and disease problems are tackled vigorously by the growers themselves. Elsewhere growers are dependent on Government to solve the problems arising from pests and diseases. In most cases, Government research finds the answers and Government extension services try to persuade farmers to adopt them. However this is not always so. The cutting out campaign to control swollen shoot virus will be remembered as an example when Government had to undertake the control measure.

Cocoa trees are best by pest and disease problems; an old estimate of total losses from pests and diseases amounted to 30% of the potential crop, a higher proportion than for any other crop (Padwick 1959). There are, of course, innumerable pests and diseases which attack cocoa, but some half dozen stand out as being of great importance.

Fungal Diseases

The most important fungal disease is Black Pod. The fungus causing this disease invades the pod, causing a round dark spot which enlarges rapidly to involve the whole pod surface. In immature pods the fungus spreads to the beans within, causing them to rot. This disease occurs in nearly every cocoa-growing country and it is estimated to cause the loss of at least 10% of world production i.e. 140,000 tonnes, which at current prices represents more than $500m annually.

It is probably no longer correct to consider Black Pod as one disease as it has recently been discovered that it is not caused solely by the fungus *Phytophthora palmivora,* but that two other species, *P. megakarya* and an un-named *Phytophthora,* currently known as MF4, also cause Black Pod (Brasier and Griffin, 1979). *P. megakarya* is the prime cause of Black Pod in Nigeria and Cameroon, and MF4 is the prime cause in Brazil, while *P. palmivora* occurs in most other cocoa-growing countries. With the present stage of knowledge it seems that all three species cause almost similar symptoms and lead to heavy losses of pods under conditions of high humidity. However, further research will no doubt reveal differences in epidemiology which will lead to differing methods of control.

The use of copper fungicides to control Black Pod has been advocated for many years and practiced in a few countries. It was in 1954 and 1955 that farmers in Western Nigeria started to spray their cocoa trees with carbide Bordeaux mixture following successful demonstrations, and spraying with carbide Bordeaux or alternative copper fungicides rapidly became accpted practice by Nigerian cocoa farmers.

Black Pod disease is particularly severe in Nigeria, Cameroon and Brazil and in all three countries, farmers have been advised to spray the pods with a copper fungicide at regular intervals, usually of 3 weeks, as a prophylactic measure. When properly done these measures reduce pod losses. They are, however, expensive and depend on the right equipment being used in the right way at the right time. Furthermore there are places where the climate is such that spraying becomes impossible or ineffectual and heavy losses ensue.

Alternative methods of controlling this common disease have been tried for many years. Even the first edition of van Hall's classic textbook "Cocoa" published in 1912 stressed the importance of wider spacing, the removal of dead pods, improved drainage and avoiding dense shade. These measures were designed to reduce humidity and sources of inoculum. Spraying with Bordeaux mixture was also recommended. Similar recommendations are being made today in several countries.

Black pod control is expensive and not always effective so research into improved methods of control continues. Research conducted in Nigeria during the past 6 years has not only lead to the discovery of the new species of fungus already mentioned, but has made other discoveries on the epidemiology of the diseases which will lead to better control. The control methods under trial modify the methods currently recommended by using insecticides in combination with a copper fungicide; in addition, the fungus *P. megakarya* appears to move up the trees from the soil so the removal of pods below 70 cm is being tested as an additional control.

In Jamaica, where Black Pod is the major limiting factor, recent trials have shown the value of an integrated approach to Black Pod control. These trials show that treatment of the soil with copper fungicide augmented control by spraying pods alone. The trials also showed the importance of precise timing of fungicide application for most effective control (Henry, 1977).

There are two other fungal diseases of importance: Witches' broom and *Monilia* pod rot, both of which occur only in South America.

Witches' broom is caused by *Crinipellis perniciosa*, a fungus which is specific to cocoa and other *Theobroma* species. It causes hypertrophic growth of shoots from which the disease gets its name and causes considerable pod losses.

The disease occurs in the Amazon basin and in Ecuador, Colombia and Venezuela but not in the main cocoa-growing area in Brazil.

There have been several approaches to the control of this disease. Spores are produced from mushrooms on the brooms so that removal of brooms should control the disease. However, it has never been possible to remove all the brooms and hence eradicate the disease. Prophylactic sprays with copper fungicides have similarly failed to assert an economical control. The search for resistant trees in the upper reaches of the Amazon basin which is the genetic origin of *Theobroma cacao,* gave an initial success but subsequently resistance broke down, and the possibility of there being more than one race of the fungus makes this approach more difficult.

Monilia pod rot occurs in Ecuador and neighbouring countries, that is, in a similar but smaller area than Witches' broom. It is caused by the fungus *Monilia roreri* which can infect pods at all stages of growth, producing symptoms similar to those caused by Witches' broom disease and leading to total loss of beans in young pods. As with Witches' broom, this disease cannot be controlled at present by cultural or chemical methods and in this case, no resistance has been found.

There is, however, one way by which, in Ecuador, losses from both these diseases can be reduced and that is by crop manipulation. The cocoa tree tends to flower almost continually so the crop is spread over several months, the pod maturing 5 months after pollination. In Ecuador natural pollination results in most of the crop ripening during the wet season, thereby being subject to pod diseases. By supplementary hand pollination early in the dry season it is possible to increase the proportion of the crop maturing during the dry season from 40% to 60% or more. As losses during the dry season are only 15-20% as compared with 75-80% during the wet season, the change in proportions leads to a considerable increase in actual yield, and a large return in relation to the cost of hand pollination (Edwards 1978).

Pests

The major pests of cocoa are mirid bugs. There are several species worldwide but the two most important are *Sahlbergella singularis* and *Distantiella theobroma* which attack cocoa in West Africa and are said to cause a loss of 250,000 tonnes annually. The relative importance of each species varies from country to country and to a lesser extent from season to season but the damage caused can be devastating. Feeding by a relatively small number of these bugs will kill young shoots. On young trees this causes a delay in development and repeated attacks can kill them. On mature trees the leaf canopy may be lost causing loss of crop and degeneration of the farm if attacks are repeated. These bugs were kept under control in West Africa by spraying with δ-BHC applied by mistblowers in Ghana, by knapsack sprayers in Nigeria and thermal fogging machines in Cameroon. Spraying with δ-BHC was adopted by farmers on a large scale around 1960 and was a major factor in increasing production in West Africa in the early 60s. δ-BHC was effective for many years and produced no undesirable side effects, but resistant strains of both species of mirid have appeared in Ghana and Nigeria and this necessitated a search for alternative pesticides. Aprocarb and orthobux have proved to be effective alternatives.

Much research effort has been spent on seeking alternative methods of controlling mirids. Biological control through parasites and predators has been unsuccessful, possibly because mirids in relatively small numbers (3000-5000 per hectare) can cause considerable damage. Certain species of ants predate on mirids, especially *D. theobroma*, thereby asserting a measure of control. It has been suggested that these ants (*Oecophylla longinoda*) might be conserved by avoiding spraying trees with their nests. There has been much argument about this proposal for integrated control which has not been adopted for two reasons. First there would be difficulty in training and persuading farmers to preserve a stinging ant; second the ant asserts little control over the other mirid species, *S. singularis,* the species which often predominates.

Turning to another problem in another part of the world, the first large plantings of cocoa in Sabah were heavily attacked by several species of borers and various leaf eating caterpillars. These were tackled by routine spraying with dieldrin and other persistent chlorinated hydrocarbons. The effect was to make the pest situation worse and outbreaks of other pests occurred. After 2 years it was decided to stop all spraying and as a result natural control was re-asserted over the major pests and the remaining pests were tackled by selective measures (Conway 1971). This is a prime example of the need for and success of, an integrated approach.

Integrated Control in the future

Enough has been said to prove that the principle of integrated control has been used on cocoa farms with regard to Black Pod disease for many years, but it is a principle which has to be learned and relearned time and time again.

Integrated methods will have to be regarded as the right approach, but considering the type of farmer involved in the major cocoa-growing countries, the method will need to be propagated continuously.

Looking ahead, the methods of crop manipulation already mentioned offer exciting possibilities for the control of pod diseases under certain circumstances. The method involves the removal of all developing pods at an appropriate time of year, followed by hand pollination, a task which requires a steady hand, a sharp eye and a pair of tweezers. The technique has been shown to reduce losses to Witches' broom and *Monilia* in Ecuador but it remains to be seen whether it will be adopted by farmers. This technique might be tested for control of Black Pod in places where the crop develops in the wet season. Where it can be used to avoid pod diseases it will reduce the need for spraying or eliminate it altogether. However, more research is needed before this technique can be extended to countries other than Ecuador.

Crop manipulation will only have limited application for control of Black Pod so that other techniques must be tried. The new knowledge on the species involved will no doubt be followed by a clearer understanding of the epidemiology of the different species. Much more research will be needed for this knowledge to be gained, after which the new knowledge will have to be translated into modified methods of control. New methods of control may well concentrate on the early stages of a Black Pod epidemic by trying to eliminate or contain the initial sources and the means of transmission. Another

potential development may be in the field of systemic fungicides. Ridomil, an acyl alanine, is giving encouraging results in trials in Papua New Guinea but it remains to be seen whether it will provide an economic control of Black Pod. The discovery of three *Phytophthora* species may make this more complicated.

In the control of mirids this might evolve into more selective spraying than the type of routine spraying that is currently recommended. The use of pheromones has not been investigated for these bugs and this might prove an effective approach.

In conclusion, I would emphasize that there is a great need for intensified research on the various pests and diseases mentioned besides others not mentioned in this paper. I would also emphasize the difficulty of passing on the results of research to cocoa growers. The bulk of the world's cocoa is grown by small holders whose ability to adopt the new techniques is limited. This is not to say they cannot adopt them but progress with agricultural science makes the techniques of pest control more complex and hence increasingly difficult for extension services to pass on to farmers.

It has been said—by an eminent entomologist—that "integrated control may be called common sense control. It is not a new discovery nor a panacea, but a rediscovery of the need to bring all aids to bear, including if necessary the thoughtful application (but minimal use) of toxicants, with due regard to cost" (Entwistle 1972).

The operative words are "thoughtful application" of pesticides and the value of the title "integrated control" is to remind those concerned of the need for thought.

References

Brasier, C. M. and Griffin, M. J., 1979. Taxonomy of *Phytophthora palmivora* on cocoa. *Trans. Br. mycol. Soc.* 72.1. 111-143

Conway, G. R. 1971. Pests of cocoa in Sabah and their control. Ministry of Agric. and Fisheries, Sabah.

Edwards, D. F. 1978. Studies on the manipulation of the timing of crop maturity of cocoa in Ecuador in relation to losses from pod diseases. *J. Hort. Sci.* 53. 243-254

Entwistle, P. F. 1972. Pests of Cocoa. Longman, London.

Henry, C. E. 1977. Integrated control of Black Pod disease of cocoa (*Theobroma cacao*). 6th International Cocoa Res. Conf. Venezuala, November 1977.

Padwick, G. W. 1959. Plant diseases in the colonies. *Outlook on Agriculture* 2.3. 122-6.

35.3   Integrated Sugarcane Crop Management

Dr. J. R. Orsenigo

Florida Sugar Cane League, Inc.
Clewiston, Florida 33440

Opportunities for the development of integrated crop management programs in sugarcane are abundant but poorly exploited. Plant protection methods have not been integrated fully, in general, and biological control measures have been concentrated on entomological pests.

Sugarcane is a perennial, tropical, grass plant which grows best under warm temperatures and adequate amount and distribution of rainfall or irrigation and isolation. The commercial crop is represented by interspecific hybrids of the genus *Saccharun.* which store sucrose in the stalk during maturation. The active growth cycle may range from less than 10 months (Louisiana) to 24 months (Hawaii). Most producing areas harvest after 12 to 20 months growth. Sugarcane is grown commercially in subtropical and frost-hazardous areas at about 25°S and 30°N latitude (Argentina-Brazil, and Louisiana, respectively). Commercial sugarcane is produced on media ranging from fine sands through loams, clay loams, and heavy soils to organic or muck soils.

Planting systems range from nearly flat culture in Florida to "seed-in-furrow, ridged interrow" in Taiwan to seed planted on high beds in Louisiana. Ridges are levelled mechanically before layby in Taiwan but soil compaction or settling and rainfall level slight ridges on Florida's organic soil. Planting dates vary throughout the world. Sugarcane varieties differ widely in growth habit, especially in the number, amount and position of green leaves and retention or shedding of dead, lower leaves. Many varieties grow erectly in well-defined rows during the plant crop, but form diffuse, wide rows in subsequent ratoons. Only one harvest may be taken per planting before fields are plowed and replanted. Two to six ratoon crops may be grown at other locations. The economic feasibility and success of ratooning is related to stubble damage at harvest, moisture and temperature regimes, fertilization, soil and foliar pests, weeds and herbicide usage.

A sugarcane management program should include practices for adequate plant stand and optimum growth during favorable environmental periods to enhance crop competition and to maximize recoverable sugar yields. Individual practices applicable to integrated crop management programs for sugarcane have developed naturally or have been established deliberately. Successful integration of these practices may be elusive.

The sugarcane pest spectrum is broad and economic impact may be associated with mammalian and avian pests such as rodents, feral hogs, and wallabys, and the sulfur-crested cockatoo as well as the more conventional pests such as diseases, insects, nematodes, viruses, and weeds.

Practices available to control the individual pests noted above may be: biological, chemical, cultural, and mechanical, as well as combinations of these. The text which follows will provide examples of sugarcane pests controlled by the various practices.

Biological pest controls are employed in production of many commercial crops and these practices have been developed to a greater extent in control of insects by parasitic and predacious insects. More than 1500 insect species representing both native and exotic groups reportedly feed on the sugarcane plant [Box, H. E. 1953. List of sugar cane insects. London Commonwealth Institute]. [See also, Long, W. H. and S. D. Hensley, 1972. Ann. Rev. Entomology. 17: 149-176.] A recent world review of introduced parasites and predators of arthropod pests and weeds includes more than 12 major sugarcane insect pest target species of commercial importance [Clausen, Curtis P. 1978. Agric. Handbook No. 480, U.S. Dept. Agric.]. The following parasites and predators apparently have been effective to the extent noted at one or more locations: outstanding or excellent control—sugarcane leafhopper and pink sugarcane mealybug; good control—gray sugarcane mealybug, armyworm, nutgrass armyworm, New Guinea sugarcane weevil, and sugarcane oryctes; and, questionable control—sugarcane aphid, yellow sugarcane aphid, sugarcane borer, and sugarcane grub. Many of the above pests are controlled partially by native parasites and predators, but these and other uncited insects are controlled largely through complexes of native and exotic insects.

An historical account listed 56 biological agents introduced into the United States in the years 1915-1969 as potential biological controls for insects injurious to sugarcane [Charpentier, L. J., *et al.* 1972. Proc. ISSCT 14: 466-476]. Thirty-six of the agents were introduced with the sugarcane borer as the target insect. Five species have become established, three attacking the sugarcane borer and two attacking the West Indian sugarcane fulgorid.

The so-called stalk borers or moth borers of sugarcane have been studied intensively for many years and have been the object of numerous biological and chemical control strategies. The sugarcane borer, in particular, has survived these efforts in many areas.

Before the mid-1960's the sugarcane borer (*Diatraea saccharalis* (F.), SCB) was held in check in Florida by a complex of native parasites and predators, especially *Trichogramma* sp. and ants, as well as several minor native and exotic larval parasites which had been imported by governmental agencies and private growers. Chemical control treatments were rare exceptions. The SCB has increased in spatial distribution and in intensity. The increased impact of the SCB coincides with but has not been directly attributable to increased flooding of fallow sugarcane fields. (Inundation provides wireworm and grub control, especially, but also decimates ant populations.) Species of the genera *Agathis, Apantales,* and *Lixophaga* have provided effective, albeit sometimes only partial, control of the SCB. A recent program to evaluate the potential of strategic, massive releases of the larval parasite *Lixophaga diatraeae* (Townsend), (Cuban Fly) in SCB control was based on simulated population models [Knipling, B. F. 1972. Environ. Ent. 1: 1-6]. The parasite was produced successfully on a large scale using an unnatural host, the greater wax moth [King, E. G., *et al.* 1979. USDA-SEA AAT-S-3]. Flies hatched from puparia developed in rearing facilities on both natural and the above unnatural host were released in Florida sugarcane fields in 1973 and 1974. The Tachinid parasitized 20 to 78% of SCB larvae within three weeks after release. The released flies did not appear to migrate from the release area, but their progeny eventually dispersed up to 3 km [Summers, T. E. *et al.* 1976. Entomophaga 21: 359-366]. Despite an apparent technical potential for parasite supplementation, this program has lain dormant since 1976. The economic potential of Cuban Fly supplementation has not been evaluated.

Rational control of the SCB must be based on an understanding of an interrelated complex of factors which probably are not equal wherever cane is grown and the SCB exists. An effective, and economical, control program will be based on periodic monitoring to determine the extent and intensity of economically damaging populations of the SCB during the crop cycle thereby utilizing biological control mechanisms until the economic threshold is reached.

Insecticides used post-threshold should be those which minimize damage to parasites and predators.

The toad, *Bufo marinus* L., apparently has been of some value in controlling above ground and adult stages of some insects affecting sugarcane in Puerto Rico. Adults of the genera *Phyllophaga* and *Diaprepes* appeared to be the major insects ingested by the amphibian [Dexter, Raquel R. 1932. Int'l Soc. Sugar Cane Tech. Bull. No. 74, 6 pp.].

Subterranean larvae of sugarcane pests are undoubtedly attacked by parasitic and predatory insects but the literature fails to detail this aspect of biological control in general. Mortality factors operating against *Melolonthids* attacking sugarcane in Tanganyika have been enumerated and the potential of these factors at various stages of the life cycle have been estimated [Jepson, W. F. 1956. Bull. Ent. Res. 47: 377-397]. The estimates suggest that desiccation causes greater population losses than parasites or predators. Jepson [*op. cit.*] estimated that diseases were the major biological controls of third-instar larvae. Bacterial diseases of subterranean grubs, especially the milky spore bacterium, are known biological control agents. *Bacillus popilliae* Dutky has been an effective, if occasional, natural control of *Cyclocephala* larvae in pastures in Florida. Spore powder of the milky spore bacterium can be purchased commercially in the United States [Mellinger's Inc., North Lima, OH. U.S.A.] but there is no evidence that this organism has been assayed or used commercially in sugarcane.

Biological control mechanisms for animals, birds, diseases, and viruses have not been noted in literature. Biological suppression of the root-knot nematode appears possible with pure stands of pangolagrass [Winchester, J. A., and N. C. Hayslip. 1960. Proc. Florida State Hort. Soc. 73: 100-104]. This practice has been used commercially in tomato culture on sandy soils, but it probably would be inapplicable to sugarcane since pangolagrass harbors other nematodes while simultaneously suppressing the root-knot nematode.

Numerous weeds have been controlled biologically, primarily through natural and exotic insects but these insect-weed combinations are not common to sugarcane. The ubiquitous weed pest, nutsedge, *Cyperus rotundus* L., is attacked by *Bactra truculenta* but there is little evidence of successful field control, [Clausen, Curtis P. 1978. *op. cit.* and Frick, K. E., *et al.* 1979. Weed Sci. 27: 178-183].

Plant pathogens have been evaluated and some organisms have been used commercially for weed control but few of the situations reported in the literature are applicable to sugarcane-weed complexes. An overview by Wilson outlines some of the major terrestrial weed-pathogen combinations [Wilson, Charles L. 1969. Ann. Rev. Plant Path. 7: 441-434]. The expense of a biological control program for weeds was estimated by tabulating Canadian efforts in terms of scientist-man-years by weed target, control agent and study type [Harris, P. 1979. Weed Sci. 27: 242-250]. The analysis suggested some 18.8 to 23.7 scientist-years and expenditures of $1.2 to 1.5 million current dollars for a complete biological control program. It is obvious that only major economic weeds can be targeted for such efforts.

Chemical pest control practices are in general commercial use in sugarcane and deserve little specific mention here. Rodents represent the only, or the major, animal pest in many sugarcane production areas. Chemicals and application methods available at present offer only temporary population suppression rather than sustained control. Numerous insects causing economic damage in sugarcane may be controlled chemically. The most efficient practices are those in which insect populations are visible and can be monitored readily; in which life histories have been defined well; in which the spectrum of biological controls has been determined; and, in which application can be targeted well. Control opportunities are diminished markedly for subterranean pests which cannot be scouted readily and which are not treated easily. Insect species controlled by in-furrow chemical application at planting may infest sugarcane ratoon crops. Nematodes may be associated with sugarcane wherever the crop is grown but nematodes do not cause economic damage in all cropping areas. Economic loss has occurred in sugarcane grown on sandy soils in Florida but loss has not been demonstrated on organic soils. The value of chemical controls varies with the producing area. Qualitative and quantitative changes in nematode populations follow chemical treatment but prolonged control is unlikely. Little opportunity exists for chemical control of the virus pests of sugarcane but non-chemical methods have been developed for planting stock infected with some viruses. Chemical control of weeds is a wide spread and economical practice in sugarcane. Both preemergence and postemergence application techniques offer opportunity to establish and maintain control of most problem species until the crop itself controls weeds through shading.

Cultural methods of pest control are discussed thoroughly in the several volumes of the United States National Academy of Sciences series entitled, "Principles of Plant and Animal Pest Control."

Certain techniques are used in sugarcane production and have nearly universal applicability. Destruction and incorporation of crop residues usually aid in reducing soil pest and some foliar pest pressure but incomplete comminution of sugarcane stalk pieces can harbor injurious insects. Sanitation measures used in the field and in equipment operations further impede pest inroads on the crop. Crop rotation may have limited value in many sugarcane situations since the rotational crop is usually short-lived when compared to the perennial sugarcane. Crops such as corn may maintain soil pests, wireworms and grubs especially, which are injurious to sugarcane.

Land preparation practices offer real opportunity for pest control, especially of diseases, insects, and nematodes. Where feasible, and not soil-destructive, drying of the surface soil serves to diminish populations of the foregoing pests and reduces the seed source for some weeds. The opposite practice, flooding, can provide major benefits when water supplies are ample. Flooding controls numerous soil-borne pathogens, insects, and nematodes, and eliminates rodent habitat. Flooding controls some weed species but flood waters may distribute weeds more widely within and among fields. The following steps are essential to maximize the potential of field flooding in sugarcane pest control: destroy all vegetation within the field and on field borders; inundate the field completely and maintain a flooded condition for a minimum of four weeks; remove flood waters and allow the upper soil horizons to dry out for about two weeks; reflood the field and maintain inundated for an additional four weeks. [Genung, William G. 1976. Tall Timbers Conf. Ecol. An. Contr. Habitat Mgt. 6: 165-172.] An additional cycle of drying and flooding are desirable if time and water supplies permit. Wet-and-dry alternation avoids survival problems associated with encysting and diapause and appears more effective than continuous inundation. A non-beneficial result of flooding is the temporary loss of ant populations which are predacious on SCB eggs and larvae.

The crop itself offers effective cultural control of some pests. Pest resistant varieties and healthy planting stock must be used. Time of planting practices may provide some pest control in local situations but generalizations cannot be made. Use of proper seedcane planting rates helps establish crop stands which can compete effectively by shading-out weeds.

Control of weed infestants within the crop along with non-crop weed sanitation will influence host reservoirs and pest populations. Nutsedge (*Cyperus sp.*) is a preferred host for the lesser corn stalk borer

(*Elasmopalpus lignosellus* Ziller), and attacks by this insect on young cane shoots are minimized when the sedge host is controlled.

Preharvest burning of standing sugarcane fields yielding 100 or more tonnes cane per hectare can develop internal stalk temperatures sufficient to kill sugarcane borer larvae [Questel, D. D. and T. Bregger. 1970. Proc. ISSCT XV: 871-883]. This harvest practice has achieved SCB reduction in Australia as well as in Florida.

Mechanical, or physical, methods of pest control appear to have limited utility within the sugarcane crop. Trapping may be useful in monitoring pests with motility such as disease spores, insect adults, and rodents but physical collection of insects and rodents is impractical. Roguing of sugarcane with demonstrable symptoms of virus disease may be a practical means of reducing inoculum. Roguing may be suitable also for limited infections of foliar diseases but this practice would be an uncommon means of insect control. Control of plant diseases, including the viruses, is best achieved through development of genetically resistant or tolerant sugarcane varieties, but the ratoon stunt disease virus (RSD) may be controlled by hot water or hot air treatment of sugarcane planting material. Internal stalk temperatures of 50° to 54°C for about two hours are sufficient for commercial RSD control in seed stock. Cultivation, the most commonly practiced form of mechanical pest control, is effective against many annual and perennial weed species growing in sugarcane. Timeliness of cultivation with the proper equipment has served well and continues a major means of eliminating weed plant competition.

An example of the facets of integrated crop management related to biological chemical, and cultural controls is offered for the SCB in Florida sugarcane. Major SCB host plant species common to the Florida Sugarcane area are: sugarcane and corn while minor hosts may be rice, wild and cultivated sorghum species, and wild grasses. The life cycle of egg, larval, pupal and adult stages requires about 30 days and some four generations are reared during the major period of economic susceptibility in the crop cycle, July through September. Land flooded during the fallow period to control white grubs and wireworms usually is used for the plant cane crop. Ants, efficient SCB egg and first instar larval predators, are decimated by flooding and offer little protection in the first and subsequent ratoon crops.

The grower's decisions are complex. If monitoring demonstrates economic threshold populations of the SCB in June, the grower must evaluate the consequences of an insecticide application which is likely to depress beneficial parasites and predators and thereby ensure a continued periodic need for chemical control measures. Chemical control treatments at low infestation levels may be justified for high tonnage fields early in the season under high sugar prices. But the initial application may commit the grower to four or five successive applications to maintain control. Economic justification for treatment requires higher infestation levels for low tonnage fields late in the season under low sugar prices. [Alvarez, Jose and Gerald Kidder. 1979. Proc. 9th Ann. Mtg. Amer. Soc. Sugar Cane Tech. In Press.] If commercial sources for parasite supplementation do not exist, the grower must consider how an adequate supply of SCB host larvae can be fostered to maintain a natural reservoir of *Agathis, Apantales,* and *Lixophaga* parasites sufficient to control both early and continuing borer infestations. If *Lixophaga* puparia are available commercially, the grower must know the threshold levels of the SCB which require and which will sustain supplementation. Finally, the grower needs information based on actual chemical control measures for all pests.

Commercial production of sugarcane, and of any other crop, depends on reliable cost-effective inputs. Most producers of sugarcane employ some elements of integrated crop management. But, combining the methods reviewed for the entire sugarcane pest spectrum requires greater and more detailed knowledge, in principle and in practice, than is now available. Lack of specific biological information on specific pests and their control mechanisms, and lack of information on combining control measures in production programs are major impediments to greater utilization of integrated crop management in sugarcane.

35.4 Integrated Plant Protection in Citrus

R. F. Brooks and J. L. Knapp

University of Florida, Institute of Food and Agricultural Sciences
Agricultural Research and Education Center, Lake Alfred, Florida 33850

Citrus is a perennial tree crop grown commercially in various parts of the world in a belt betwen 40° north and 40° south latitude where temperatures rarely fall below the freezing point. These latitudes represent a wide range of climatic conditions. These conditions range from the equatorial tropics with constant high temperatures inducing several crops per year with continual pest pressure to the subtropical zones with cool winters, one annual crop and a more seasonal type of pest development. There is also a wide range in annual rainfall within these temperature zones.

In order for this report to more accurately reflect differences in pest species complexes that occur within this belt, a questionnaire was prepared and sent to researchers representing all of the major production areas. The questionnaire asked each respondent to list (1) the 10 most economically important pests (disease, insect, mite, nematode, and weed) in their zone; (2) amount of land devoted to citrus with principal cultivars and production; (3) indicate important components of their pest control programs (chemical, biological, and cultural); and (4) to what extent growers are adapting integrated pest management programs.

Of the 27 questionnaires sent out, 17 or approximately two-thirds of the questionnaires were returned which represents approximately 72% of the world's known production of citrus. Some respondents included more than one citrus-growing area in their response. Each of the citrus-growing areas represented in the questionnaire are listed below. Because the USA is such a large producer of citrus (39% of the world's supply) the states of California, Texas, and Florida are listed separately. Common names of pests used in this report, where possible, were either those found in the Common Names of Insects and Related Organisms (1978) or Subtropical Fruit Pests (1959). However, if not listed in either of these texts, those attached by the individual respondents were used.

Australia—(Queensland Area) (2700 hectares)

Respondent: Daniel Smith, Senior Entomologist, Nambour, Queensland.

Production is 63.8 million kilos (1% of world's total)

Principal Cultivars: 33% in Ellendale and Imperial Mandarins, 25% in Washington Navel Orange, 25% in Valencia Orange; remainder in lemons and grapefruit.

Pests*: Insects:* California Red Scale; Citrus Snow Scale (white louse); Pink Wax Scale, *Ceroplastes rubens* Maskell; Queensland Fruit Fly, *Dacus tryoni* (Frogatt); Larger Horned Citrus bug, *Biprorulis bibax* Breddin.

Acarina: Brown Citrus Rust Mite, *Tegolphus australis;* Citrus Rust Mite; Broad Mite.

Diseases: Black Spot, *Guignardia citricarpa;* Melanose. Control: Chemical control remains most important with biological control gaining in importance. Excellent control of *Chrysomphalus ficus* with *Aphytis holoxanthus* DeBach and good parasitism of Calif. Red Scale with *Aphytis lingnanensis* Compere and *Comperiella bifasciata* Howard. Parasitic introductions have been recently made for Pink Wax Scale and the Citrus Snow Scale. In the Queensland area, integrated pest management has been introduced in the groves of their largest growers with other remaining growers awaiting results.

Brazil—(400,000 hectares in Sao Paulo state only) (80 million trees)

Respondents: Victoria Rossetti, Director, Instituto Biologica, Sao Paulo.

Production is 6,528 million kilos (Sao Paulo state only) (15.2% of world's total)

Principal Cultivars: Pera, Natal and Valencia orange account for 60% of total crop. Other cultivars including Washington Navel, tangerines, Murcott and mandarins make up the remainder.

Pests: Tristeza (virus); Mediterranean Fruit Fly; Citrus Rust Mite; False Spider Mite, *Brevipalpus phoenicis* (Geijsk.) (cause Leprosis disease); Citrus Canker, *Xanthomonas citri* (Hasse) Dawson; *Orthezia* sp. (mealybug-like damage); Various viruses suspected of causing delines; Sweet Orange Scab, *Elsinoe australis* Bitancourt and Jenkins: Rubellosis (fungus); Various species of armored scale insects.

Control: Chemicals are relied upon heavily to control Brazil's citrus pests. Biological control is not considered important at the present time. Some cultural practices are directed at the control of *Orthezia* sp.

Israel—(40,000 hectares)

Respondents: David Rosen, Hebrew University of Jerusalem, Rehovot; and E. Swirski, Volcani Center—inst. Plant Protection, Bet Dagan.

Production is 1.5 million kilos (3.9% of world's total)

Principal Cultivars: Shamouti (Jaffa), Valencia, Navel oranges, Marsh grapefruit, Eureka lemon and various mandarins.

Pests: Mediterranean Fruit Fly, *Ceratitis capitata* (Wied.); Calif. Red Scale; Black Scale, *Saissetia oleae* (Bern.); Wax Scales, *Ceroplastes floridanus* Comst. & *C. rusci* (L.); Citrus Flower Moth; Citrus Rust Mite; Chaff Scale; *P. cinerea* Hadden; Citrus aphids, *Aphis spiraecola* Patch and *A. gossypii* Glover.

Control: Chemical control is still prevalent. Biological control remains important in control of armored scale insects. The citrus flower moth is being controlled by the pheromone trap method. Integraped pest management is looked on favorably by the citrus growers but overuse of organophosphorus insecticides has led to outbreaks of Florida Red Scale, Purple Scale and Citrus Rust Mite.

Jamaica—(6070 hectares)

Respondent: Jos. R. R. Suah, Crop Research Dept. Ministry of Agriculture, Kingston.

Production is 38.5 million kilos (0.1% of world's total).

Principal Cultivars: Valencia and Ortanique oranges. Marsh Seedless Grapefruit.

Pests: *Insects and Mites:* Citrus Root Weevils, *Exophthalmus* sp. and *Pachnaeus* sp.; Armored Scale Insect Complex; Black Ant, *Crematogaster brivispinosa;* Fruit Piercing Moths, *Gonodonta* sp.; Citrus Rust Mite.

Diseases: Root Rot, *Phytophthora cinnamoni* Rands; Citrus Scab, *Elsinoe fawcetti* Jenkins; Lime Knot, *Sphaeropsis tumefasciens* Hedges; Twig Dieback (Anthracnose); Psorosis (Virus).

Control: Chemical control is used widely where fruit is grown for the export market. Biological control has been carried out under government programs. Mr. Suah reports "growers are becoming interested in integrated pest management as the local market, which is not too fussy, expands and processing of citrus products increases."

Japan—(Kuchinotsu Prefecture (77,064 hectares), (Shizuoka Prefecture) (17,100 hectares), (Total Japanese acreage is 153,200 hecatares)

Respondents: Manabu Tanaka, Entomologist, Fruit Tree Research Station, Kuchinotsu-Cho, Nagasaki-Ken; and Kaichi Furahashi, Entomologist, Shizuoka Citrus Expt. Station, Shizuoka-Ken.

Production is 3,215.1 million kilos (all Japan) (10.5% of world's total).

Principal Cultivars: Satsuma mandarins (Unshu) predominent variety with some sweet oranges and Natsudaidai (summer grapefruit).

Pests: *Insects and Mites:* Arrowhead Scale, *Unaspis yanonensis* (kuwana); Citrus Red Mite; Pink Rust Mite, *Aculus pelekassi* Keifer; Citrus Leaf Miner, *Phyllocnistis citrella* Stainton; Spirea Aphid and Brown Citrus Aphid; Citrus Whitefly.

Diseases: Lemon Scab; Melanose; Satsuma Dwarf (Virus).

Control: Chemical control remains the main method of pest control. Some success has been achieved with the introduction of *Aphytis lingnanensis* Compere for control of the arrowhead scale. Tanaka reports about 10% of the Japanese citrus growers are utilizing the integrated pest management program.

Rhodesia—(1000 hectares)

Respondent: Keith R. Pyle, Estate Entomologist, Mazoe Citrus Estates, Rhodesia, Zimbabwe.

Production is 25.4 million kilos (0.5% of world's total).

Principal Cultivars: Valencia, Premier and Navel orange, lemons, limes.

Pests (order of importance): High altitude (3900 ft). Citrus psylla, *Trioza erythreae* (Del Guercio); Grey Mite, *Cacacarus citrifoli* Keifer; Citrus Bud Mite; Southern African Citrus Thrips, *Scirtothrips aurantii* Faure; False Codling Moth, *Cryptophlebia leucotreta* Meyr.; *American bollworm, *Heliothis armigera* (Hubn.); *Tortrix rindworm, *Tortrix capeusana* Wlk.; Citrus Flat Mite, *Brevipalpus lewsi* McGregor; Calif. Red Scale; Long Tailed Mealybug, *Pseudococcus adonidum* (L.).

Control: Chemicals are being used in a selective program. Oil sprays directed against the key pest Calif. Red Scale have minimum impact on beneficial insects. Biological control agents used to suppress populations of Calif. Red Scale, Brown Soft Scale, wax scale, mealybugs, cottony-cushion scale, citrus rust mite and citrus red mite. Black spot, a fungus disease, has not yet developed into a serious threat. *Cynodon dactylon* is the most important weed among young trees. The key pest on low altitude (1500 ft) Rhodesian citrus is California red scale with oil sprays replacing organophosphorus insecticides because of insecticide resistance.

*Sporadic pests

Union of South Africa—(43,000 hectares)

Respondents: E. C. G. Bedord, Citrus and Subtropical Fruit Research Institute, Nelspruit; and

M. B. Georgala, Outspan Citrus Centre, Nelspruit.

Production is 675 million kilos (1.9% of world's total).

Principal Cultivars: Valencia, Navel and mid-season oranges, lemons, naartjes (mandarin), tangerines and various other mandarins.

Pests: *Calif. Red Scale; *South African Citrus Thrips; *3 Specias of Ants; Black spot; Citrus psylla; False Codling Moth; Citrus Rust Mite; Citrus Bud Mite; American Boll Worm; Mediterranean Fruit Fly and The Natal Fruit Fly, *C. rosa* Karsh.

Control: Chemicals still the main part of the program. Search continues for more selective material. Biological control of Calif. Red Scale essential to an integrated program. Resistant scale has forced many growers to the use of oil spray and parasites. Skirting trees and weed control are important cultural practices in suppressing obnoxious ant species.

Union of Soviet Socialist Republics (Russia)— (20,000 hectares)

Respondents: A. I. Smetnik, Central Research Plant—Quarantine, Moscow; and I. A. Shestopalov, Trade Representation, Washington, D.C.

Production is 344.7 million kilos (1.0% of world's total).

Cultivars: Mostly mandarin, some lemons.

Pests: Citrus Whitefly; Japanese Long Scale, *Lopholeucaspis japonica* CK11; Wax Scales, *Ceroplastes japonicus* Green and *C. sinensis* Guer.; Yellow Scale; Orange Scale, *Chloropulvinaria aurantii* CK11; Brown Soft Scale; Citricola Scale; Cottony Cushion Scale; Citrus Red Mite; Citrus Rust Mite.

Control: Chemical control utilizing petroleum oils with insecticides and miticides generally accepted practice. Whitewashing tree trunks aids in control of *L. japonica* CK11. Soviet citrus plantations have reduced reliance on chemicals with introduction of entophages, acariphages and entophathogene fungi.

USA—California—(113, 312 hectares)

Respondents: G. E. Carman and L. A. Riehl, University of California, Riverside; and H. S. Elmer, Lindcove Field Station, Exeter.

Production is 2,404.9 million kilos (5.8% of world's oranges, 6.1% of world's grapefruit, 11.2% of world's lemons).

Principal Cultivars: Washington Navel, Valencia oranges, lemons, grapefuit, various tangerines and tangelos.

Pests: California Red Scale (potentially most serious); Citrus Red Mite; Citrus Thrips; European Brown Snail, *Helix aspersa* Muller; Citrus Nematode, *Tylenchulus semipenetrans* Cobb; Argentine Ant, *Iridomyrmex humilis* (Mayr); Brown Rot, *Phytophthora citrophthora* (Smith & Smith) Leonian; Citrus Rust Mite; Citricola Scale; Citrus Bud Mite.

Control: Chemicals remain extremely important but trend is to play more of a supportive role to biological control. In the San Joaquin Valley, growers are timing their sprays to avoid upsetting the few beneficial organisms available. However, the citrus growers of the traditional heartland of southern California are becoming interested in IPM and are modifying their approach by lessening their dependence on chemicals.

USA—Florida—(336,390 hectares)

Respondents: J. L. Knapp and R. F. Brooks, Agricultural Research and Education Center, Lake Alfred, FL.

Production is 9,388.5 million kilos (this represents 24.6% of world's orange production, 54.7% of world's grapefruit production, 27.3% of world's total citrus production).

Principal Cultivars: Valencia (32.6% of acreage), Pineapple and Hamlin oranges. Grapefruit, tangerines, various tangelos, limes and lemons.

Pests: *Key Pests:* Greasy Spot Disease, *Mycosphaerella citria* Whiteside; Citrus Rust Mite.

Other Pests: Citrus snow Scale; Texas Citrus Mite, *Eutetranychus banksi* (McGregor); Citrus Root Weevil species complex; Fuller Rose Beetle; Citrus Root Weevil; Blue Green Orange Weevil, *Pachnaeus opalus* (Olivier); Little Leaf Notcher, *Artipus floridanus; Diaprepes abbreviatus* (L.); Caribbean Fruit Fly.

Control: Marketing of Florida citrus fruit plays an important part in formulating integrated pest management strategies. Less emphasis is placed on cosmetic appearance because 96% of the crop is processed. Chemicals remain important for the control of key pests such as greasy spot disease and the citrus rust mite. There has been a strong trend towards the use of less hazardous chemicals that at the same time are less disruptive to other pest species. The use of parasitic hymenoptera, particularly from the genus *Aphytis* has markedly reduced the importance of armored scale insects. Various factors inducing tree decline are becoming increasingly more important. Young Tree Decline or Blight, the causal organism as yet undetermined, is causing loss of trees

*Key pests, Bedford. Ant species not identified

562

on rough lemon, Florida's major rootstock. Potentially damaging populations of root weevil species have been developing since the arbitrary banning of effective soil insecticides without alternate control strategies. Three species of nematodes, although not considered major problems, are causing some acreage to be treated.

USA—Texas—(30,352 hectares)

Respondents: Herbert A. Dean, Texas Agric. Expt. Station, Weslaco; and J. Victor French, Texas A & I University Citrus Center, Weslaco.

Production is 667.7 million kilos (4.8% of world's total).

Principal Cultivars: Red grapefruit (Ruby Red and Star Ruby) and Marsh Seedless grapefruit; Marrs, Hamlin, Pineapple, Parson Brown, Valencia and Navel oranges.

Pests: Citrus Rust Mite; Chaff Scale Calif. Red Scale; Melanose; Brown Soft Scale; Phytophthora Foot Rot; Citrus Mealybug; Citrus Nematode; Texas Citrus Mite; Psorosis Virus.

Control: Selective chemical control used against key pest, the citrus rust mite, which is important to the development of other pest problems. Many problems are associated with pesticide drift from adjoining cotton fields. When cotton growers, often citrus growers also, curtailed the use of parathion, many of the citrus pest problems decreased in severity. Resident Texas citrus growers show some interest in developing an IMP program. However, 50% of acreage is owned by out-of-state residents and managed by large grove care companies who feel IPM means less profit for them.

Summary

Integrated pest management programs are in varied stages of development by researchers, the Extension Service, and citrus growers in different parts of the world. Although some similarities exist between citrus-growing areas, each has its own particular set of economic pests to manage. The pest control strategy used by a particular citrus-growing area may depend on the pest complex, climate, competition with other crops, availability of effective natural enemies or the marketing objectives for the crop.

Chemicals are expected to remain the single most important pest control strategy used on most of the world's citrus crops. One reason for this is that most of the world's citrus crops are grown for the fresh market where chemicals, essentially to control fungal organisms, are relied upon to protect the fruit's cosmetic appearance. Some of the more important citrus pests, particularly plant pathogens, lack effective natural enemies. However, efforts will continue to develop chemicals that are more selective against the target species and less harmful to natural control agents, man and the environment.

Most of the scientists responding to the questionnaire indicated that the use of effective natural enemies had the potential of being the most useful of all control strategies. Several successful natural enemy introductions were reported to be effectively controlling target pests.

References

Bedford, E. C. G. 1978. Citrus Pests in the Republic of South Africa. Republic of South Africa. Science Bull. No. 391, 253 pp.

Ebeling, W. 1959. Subtropical Fruit Pests. University of California, Berkeley 43.

Hafliger, E. 1975. Citrus Tech. Monograph No. 4. Ciba-Geigy Agrochemicals, Basle, Switzerland, 88 pp.

Kock, J., H. Schmutterer and W. Koch. 1977. Diseases, Pests and Weeds in Tropical Crops. V. P. Parey, Hamburg, 66 pp.

Reuther, W. 1967. The Citrus Industry. University of California, Berkeley. Vol. I, 611 pp.

Sutherland, D. W. S. 1978. Common Names of Insects and Related Organisms. Ent. Soc. of America. Special Pub. 78-1.

Florida Crop and Livestock Reporting Service. 1978. Citrus Summary, Fla. Dept. of Agr. and Consum. Services.

35.5 Integrated Plant Protection in Bananas and Plantains

R. H. Stover and H. Eugene Ostmark

Tropical Agriculture Research Services (SIATSA)
La Lima, Honduras

Abstract

During the past decade, the use of pesticides has decreased in some areas and increased in others. This has been accompanied by more intensive cultural practices with higher yields of quality fruit. For Sigatoka disease, spraying has been reduced about one-half because of better systems of monitoring disease and improved aircraft application. For the more virulent black Sigatoka spraying frequency has had to be doubled, but level of control is now much higher with the use of chlorothalonil fungicide. For fruit spot diseases weekly spraying has been replaced by a fungicide-dusted polyethylene bag which protects the growing fruit. This bag also controls fruit insects when the resin is impregnated or coated with an insecticide. Nematicide use has increased with granular compounds replacing DBCP. Flooding controls the burrowing nematode in Surinam and the Ivory Coast. Bacterial wilt and bunchy top virus are controlled best by strict sanitation, and the enforcement of quarantine measures where the diseases are not present. The use of foliar insecticides has been reduced to less than 10% of previous spraying by a better understanding of depredators and tolerance of the plant to defoliation. Use of insecticides for *Cosmopolites* has been eliminated except in new planting following bananas. Herbicide use has increased greatly under bare ground culture, but evidence of soil erosion in high rainfall areas has made an evaluation of ground cover necessary. A long-range breeding program will eventually further reduce the use of chemicals by providing new disease-resistant, wind-tolerant varieties.

With world inflationary trends, export taxes and increasing demands from unionized plantation labor, the costs of producing a standard 18.2 kilogram box of bananas for export has more than doubled in the past 5 years. Within this period there has been an increase in production per hectare per year which now appears to have peaked at an average of between 2470 and 2964 boxes for the best plantations of Central America.

The high yields and high quality sustained on the best plantations have been maintained only by intensifying the use of improved cultural practices such as: higher and better spaced populations, good drainage, improved irrigation, more sensitive monitoring of nutritional needs by foliar and soil analyses, and better fruit protection and handling. All of these have influenced and improved integrated plant protection.

By the utilization of more sophisticated and integrated plant protection practices, costs of disease, pest, and weed control have risen less, proportionally, than other costs of production.

Disease Control

Sigatoka and black Sigatoka

All bananas for export, and plantations where black Sigatoka is present, must be sprayed by aircraft to keep leaf spot from reducing fruit quality and yield. In the case of Sigatoka, the number of sprays required annually has been reduced from 20 or more to 10-14. This has been brought about by better application techniques, better fungicide formulations and more sensitive methods of monitoring spotting levels [1, 2]. In the case of the more virulent black Sigatoka, spraying must be carried out every 9-12 days. However, with the introduction of chlorothalonil, disease control has vastly improved and petroleum oil has been eliminated from the spray formula [3]. Cultural practices such as improved drainage and weed control also have been factors in raising the standards of control over the past 5 years. Sources of resistance to both diseases have been successfully incorporated into advanced breeding lines as a long-term approach to resolving the costly spray program [4].

Fruit Spots

Weekly spraying of young fruit with dithiocarbamate fungicides during the rainy season [5] has been eliminated in some areas because of improved drainage and weed control, early bagging of fruit and prompt removal of hanging senescent leaves. In other areas, the use of fruit bags coated with a thin layer of fungicide dust has given excellent control. As a result, labor costs have been reduced drastically and much less fungicide is placed on the fruit and the ground beneath it.

Nematodes

The burrowing nematode *Radopholus similis* is the major root pathogen [5]. Lesions are invaded by fungi and bacteria, root rot follows, and the poorly anchored plant, usually with maturing fruit, topples in light winds. Nematicide use has increased in recent years with granular compounds replacing DBCP. The latter was applied through irrigation water in some areas. Granular materials are hand-applied and cases of intoxication have resulted where careless-ness occurred, although no fatalities have been reported. Granular materials have the advantage that application is localized to the base of the mat and an area extending about 40 centimeters wide around the mat. Nematode damage is aggravated by poor drainage and poor guying or propping of the mat. Improvements in drainage and the use of wooden props and overhead cables between fixed double rows to support the plant, have helped to greatly reduce losses from plants falling over as a result of nematode root rot. Protection of plants from toppling has been greatly facilitated by the use of more dwarf varieties of the Cavendish group such as Grand Nain in Central America and Umalog in the Philippines. Dwarf plants are cheaper to support and are much less prone to topple in high winds.

In Surinam and the Ivory Coast, where nemati-cides have given mediocre results on the heavy clay and muck soils, flooding for 45 days to 6 months has greatly reduced nematode injury and substantially increased yields for several years following flooding and replanting with nematode-free rhizomes [6, 7].

Bacterial Wilt

Pseudomonas solanacearum is readily transmitted by insects and tools used for pruning and harvesting [5]. It has a wide host range among the weed flora. By breaking off the male bud after all hands have emerged, insect transmission through male bud bract scars is greatly reduced. Rapid detection and eradication of diseased plants along with strict disinfection of tools keeps the disease in check. Replacing the highly susceptible Bluggoe (Chato) plantain with the resistant Pelipita in small farms adjacent to banana plantations has effectively reduced an important outside source of inoculum. Weed-free plantations have reduced another important source of inoculum. Methyl bromide fumigation is used to quickly restore an infested soil site to production but the pathogen is also eliminated naturally during 6-12 months of bare ground fallow. Bacterial wilt of bananas is confined to Central and South America and Mindanao in the Philippines. Quarantine regulations [8] should keep it out of the Caribbean Islands, Africa, and most of Asia.

Bunch Top

This serious banana-aphid transmitted virus is not present in the Western Hemisphere [8]. It is an important cause of losses in peasant farming in the Pacific Islands where strict eradication and quarantine measures cannot be enforced. Control is effective where continuous surveys leading to rapid detection and eradication of diseased plants can be enforced by statutory means. Enforcement of suggested quarantine regulations for bananas moving between continents will continue to keep this disease out of the Western Hemisphere [8].

Insect Control

Banana pests can logically be divided into three groups based on the part of the plant attacked; defoliators, corm and pseudostem borers, and fruit feeders. The following explains how biological, cultural, and chemical control is integrated in banana plantations.

Defoliators

At least 250 species of insects and mites have been recorded as leaf feeders. Most remain endemic, their numbers kept in check by a complex of biological control agents. At times, usually as a result of adverse climatic conditions or drastic changes in plantation conditions, such as new drainage or irrigation systems, biological control agents become scarce and defoliators become epidemic.

The banana plant can withstand 20% defoliation before fruit weight is reduced [9]. If the plants are defoliated after the fruit bunch is formed, defoliation up to 40% has no effect on fruit weight. However, fruit from severely defoliated plants can ripen pre-maturely.

Translated into a practical integrated control program the banana pest control manager must determine the: (1) Species of defoliators(s) present (usually in larval stage). (2) Amount of foliage each larva consumes. (3) Number of larvae present per leaf. (4) Presence or absence of biological control agents. (5) Number of defoliators required to consume 20% of a plant's leaf area. (6) The cost of control using insecticides. (7) The state of the current market for bananas.

Counts of defoliators higher than needed to consume 20% of the plant's foliage can be tolerated if there are high populations of parasites and predators. When bananas are in seasonal surplus on the market, fruit may be lost in the plantations in order to allow natural control agents to build up.

Fruit Feeders

With the world markets requiring blemish-free bananas, the most serious insect-related losses of bananas are caused by insects that feed on, scratch or oviposit into banana peel. Only two insects, the fruit fly, *Dacus musae* of Australia [10] and the lyonetiid moth, *Opogona subcervinella* of the Canary Island actually enter the fruit pulp.

The role of biological control agents is less important than the effects of cultural practices in the integrated control of fruit pests. Since each fruit bunch must be individually bagged and, in some cases, sprayed with fungicide there is ample opportunity to combine these practices with insect control.

Some peel scarring insects enter the fruit bunch just as the bunch (bud) emerges from the plant. Insects such as *Frankliniella* flower thrips [11], the corky scab thrips, *Thrips florum* [12], and the banana scab moth, *Nacoleia octasema* [13] enter the fruit in the bud stage. Most of the insects that enter buds require individual bud injections of insecticides to produce clean fruit.

Early bagging, that is, hanging an insecticide-dusted or impregnated polyethylene bag over the bud before the bracts dehisce and fall exposing the new "hands" of fruit will protect fruit from some pests. *Colaspis* beetles and *Platynota* caterpillars [14] attack exposed young fruit and usually desert the maturing bunch. Early bagging will keep fruit damage to a minimum. *Platynota* is very common following widespread insecticide applications for defoliators. Some species of Colaspis beetles and the closely related *Metachroma* beetles live on excess suckers surrounding the mother plant. An extra pruning cycle will frequently rid a plantation of these pests. The larvae of *Colaspis blakeae* feed on grass roots and herbicide programs greatly reduce this pest [15].

As the fruit grows the hands turn upward until the fingers touch. As this point the appressed fingers make ideal feeding sites for the *Chaetanaphothrips* red rust thrips. One species, *C. orchidii,* is controlled by a conventional bagging, i.e. placing the bag over 2-week old fruit just before the fingers turn upward [16]. Another species of red rust thrips, *C. signipennis,* requires insecticide-treated bags and when populations are exceptionally high, fruit, pseudostem, and suckers must be sprayed with insecticides since *signipennis* can maintain its populations without feeding on the fruit bunch.

Corm and Pseudostem Borers

The most widespread pest of bananas is the banana root borer, *Cosmopolites sordidus* [17]. This weevil crawls, rarely if ever flies, to rhizomes to lay its eggs. Although generally considered a serious pest, the root borer confines its feeding to newly planted corms and plants growing under substandard conditions. The corms of harvested mother plants are usually riddled with galleries while adjacent followers have few or no galleries. Keeping newly dug corms to be used as planting material off the ground at night when this nocturnal weevil oviposits will do much to reduce the infestation rate.

The pseudostem of bananas is the part of the plant which suffers the least from insect pests. In the Americas the stalk borer, *Castniomera humboldti* [18] can become epidemic following destruction of the predaceous ants which keep it in check. In the Far East the pseudostem weevil, *Odoiporus longicollis,* invades pseudostems mainly by laying eggs in cut and injured tissues. Care in pruning suckers and harvesting as well as chopping up harvested pseudostem aid in keeping populations of this weevil in check.

Weed Control

Herbicides have replaced the cutlass in weed control with large savings in labor. Weed control within plantations has been facilitated by more shade resulting from an increase in plant populations from around 1235 to 1730 production units per hectare. In most areas a "bare ground" approach is followed. However, where rainfall is high and intense, erosion has become evident within the plantation and a low non-competing ground cover was sought that would reduce movement of soil away from plants that has spread slowly naturally in some herbicide-treated areas has been *Geophila repens.* Unfortunately, this plant does not reproduce rapidly from seed and vegetative propagation is slow.

As a result of the bare ground practice, interest has been revived in mulching around the base of the plant using some of the large amount of decaying banana vegetation that is left after the fruit is removed. This has been reported to increase yield as well as reduce erosion around the plant.

Discussion

Many important diseases and pests of bananas are confined to one hemisphere or continental land mass. Quarantine measures have been described that will continue to restrict movement to other areas if adhered to [8].

Breeding bananas [4] and plantains [19] is slow and difficult. Much progress has been made in the last 10 years, however. Dwarfism and resistance to leaf spots and the burrowing nematode have been introduced into advanced male diploid lines. This approach offers the only long range change of reducing the large quantities of chemicals required at present to control these diseases.

With the introduction of systemic fungicides and nematicides, disease control has been improved. Unfortunately, the problem of tolerance has arisen in the case of benzimidazole fungicides necessitating their replacement in some but not all areas. The use of systemics has resulted in less chemical being applied along with improved control.

With insects, a better understanding of depredators and tolerance of the plant to defoliation has greatly reduced and in some areas eliminated the application of insecticides to foliage and soil. Biological control measures can usually maintain populations at economically tolerable levels except in isolated instances and in localized areas where an insecticide must be introduced briefly for one or two applications.

Higher populations of dwarf varieties has increased shade preventing growth of all except shade tolerant weeds. These necessitate spot applications of herbicide, but on a much reduced scale compared with past cultural practices.

References

1. Stover, R. H. A proposed international scale for estimating intensity of banana leaf spot (*Mycosphaerella musicola* Leach). Tropical Agric., (Trinidad), 1971, 48 185-196.
2. Stover, R. H. Fungicidal control of plant diseases in the tropics. In Siegel, M. R. and H. D. Sisler, Eds. "Antifungal Compounds Vol. I", Marcel Dekker, Inc. New York and Basel, 1977; pp. 353-369.
3. Stover, R. H.; W. R. Slabaugh; M. D. Grove. Effect of chlorothalonil on a severe outbreak of banana leafspot caused by benomyl tolerant *Mycosphaerella fijiensis* var. *difformis*. Abstract. *Phytopathology News*, 1978, 12, #12.
4. Rowe, P. R.; Richardson, D. L. Breeding bananas for disease resistance, fruit quality and yield. Tropical Agriculture Research Services (SIATSA), 1975, Bulletin #2, La Lima, Honduras.
5. Stover, R. H. "Banana, Plantain and Abaca Diseases"; Commonwealth Mycological Institute, Kew, England, 1972; pp. 316.
6. Lassoudiere, A. Croissance et developpment du bananier 'Poyo' en Cote D'Ivoire. Thése. L'Université Nationale de Cote D'Ivoire. 1977.
7. Stover, R. H. Flooding of soil for disease control. In Mulder, D., Ed. *Soil Disinfection;* Elsevier Publishing Co. Amsterdarm, 1979; pp. 19-28.
8. Stover, R. H. Bananas (*Musa* spp.). In Hewitt, W. R. and L. Chiarappa, Eds. *Plant Health and Quarantine in International Transfer of Genetic Resources;* CRC Press Inc. Cleveland, 1977; pp. 71-79.
9. Ostmark, H. E. Economic insect pests of bananas. *Ann. Rev. Ent.*, 1974, 19, 161-176.
10. Saunders, G. W.; R. J. Elder. Sterilization of banana fruit infested with banana fruit fly. *Queensland J. Agr. Anim. Sci.*, 1966, 23 81-85.
11. Harrison, J. O. Notes on the biology of the banana flower thrips, *Frankliniella parvula* in the Dominican Republic. *Ann. Ent. Soc. Am.*, 1963, 56, 664-666.
12. Swaine, G.; R. J. Corcoran. Corky scab of bananas. *Queensland Agr. J.*, 1970, 96 474-475.
13. Paine, R. W. The banana scab moth *Nacoleia octasema* (Meyrick): its distribution, ecology, and control. *Tech. Pap. S. Pac. Comm.*, 1964, 145, 70 pp.
14. Bullock, R. C.; F. S. Roberts. *Platynota rostrana* (Walker), a peel-feeding pest of bananas. Tropical Agric. (Trinidad), 1961, 38, 337-341.
15. Salt, G. A study of *Colaspis hypochlora* Lefevre. Bull. Ent. Res., 1928, 19, 295-308.
16. Lachenaud, J. L. Protection contre le thrips de la roville par gainage de régime de bananes. *Fruits*, 1972, 27, 17-19.
17. Vilardebo, A. Les insectes et nématodes des bananeraies d'Equateur. Inst. Franco-Ecuator. Invest. Agrn. (IFEIA) A.N.B.E.-I.F.A.C., 1960, 75 pp.
18. Lara, E. F. The banana stalk borer, *Castniomera humboldti* (Boisduval) in La Estrella Valley, Costa Rica. V. Cultural control. *Turrialba*, 1966, 16, 136-138.
19. Rowe, P. R. Possibilités d'amélioration génétique des rendements de plantain. *Fruits*, 1976, 31, 531-586.

36.1   Implementing Pest Management on Deciduous Tree Fruits

James P. Tette and Turner B. Sutton

*Department of Entomology, N.Y.S. Agricultural Experiment Station
Geneva, New York and Department of Plant Pathology
N.C. State University, Raleigh, North Carolina*

Deciduous tree fruit pest management programs have been under development in the United States for the past seven years. Programs were first begun in California, Michigan, New York, Pennsylvania, and Washington [1,2] and more recently in North Carolina, Vermont, Massachusetts, and New Hampshire. Management strategies have been developed for mites and many insects and diseases, and are being developed for weeds, voles, and cultural practices, such as pruning and thinning. Implementation of these strategies has been brought about on both an area-wide basis and at the orchard level. The tools and techniques used at either level are similar, but the strategies become more refined and specific at the orchard level.

In an area-wide program, generalized information on pest incidence, population levels, disease infection periods, action thresholds and control information are provided to aid growers in making management decisions. For example, the first emergence of codling moth can be noted in pheromone traps in a given area and serve to alert other areas within a state [3]. In addition, the information can be used to determine the timing of insecticide applications to control this pest.

Another example of IPM information useful on an area-wide basis is provided by monitoring apple scab ascospore maturity. Perithecia are picked from leaves on a weekly basis, examined microscopically, and ranked for the presence of the five developmental stages [4]. By using information and strategies provided through an area-wide program, growers can determine when to initiate a scab spray program, have some idea of the intensity of spore discharge that might be expected, and be told when the primary scab season is over.

In an area-wide program, IPM strategies can be further refined and implemented by monitoring weather and routinely inspecting commercial and abandoned orchards. For example, one of the key components of many apple scab management programs is the duration of leaf wetness. It, along with temperature, is used in the pre-bloom period to determine the occurrence of infection periods [5].

Ideally, temperature and leaf wetness should be measured in each orchard. This is not practical because of instrument costs and maintenance. However, because wetting periods in the spring are generally widespread within certain geographical limits, several strategically located monitoring stations can effectively cover most orchards within a county.

Through frequent, routine inspections of commercial and abandoned orchards, a comprehensive picture of the current status of pests can be assembled. This information is essential in developing timely alerts and strategies for an area-wide program.

Although area-wide programs provide valuable information to assist in pest management decision-making, IPM programs at the orchard level provide the ultimate refinement in IPM decision-making. At this level, techniques and tools can be fully utilized in developing specific management strategies for individual blocks. IPM practices are usually carried out by either growers or crop protection consultants. Most states have attempted to develop programs in which growers utilize the various tools and techniques in carrying out IPM themselves. In many instances, however, the grower willingly purchases this type of expertise through a crop protection consultant. There are several different types of consultants. They may be affiliated with the agricultural chemical industry or with the processing or marketing industry. These consultants are often unable to utilize every technique and tool, but they do manage to assist growers in making sound crop protection decisions. Two other types of consultants, the private consultant and co-op consultant, usually offer a more comprehensive approach to IPM. The livelihood of private consultants is based on a fee-per-acre charge [2]. In making IPM decisions, they attempt to utilize every practical IPM tool and technique available. The same is true of co-op consultants. They are employed by grower co-operatives to gather information on crop protection, including the current status of pests in the cooperative orchards. Co-op consultants frequently are appearing as an outgrowth of state supported programs as these programs are gradually turned over to the private sector.

In an attempt to illustrate how IPM is being implemented at the farm level, we have chosen to examine some of the tools and techniques associated with many of the programs.

Tools and Techniques

A. Orchard Mapping

Orchard maps along with information on tree age, size, spacing, and variety play an important role in IPM programs [6]. Control strategies can be planned for problem areas, blocks, or varieties rather than the entire farm. In addition, accurate measurements of farm acreage and block sizes lead to better crop estimates, serve as a check on efficiency of spray rigs, and aid in communication between the decision-maker (IPM consultant) and the spray applicator (grower).

B. Strategy Sessions

A useful practice in an IPM program is to have pre-season planning sessions with each grower [6]. These sessions are used to alert growers to potential pest problems based on observations made during the previous year. They also serve to determine the intended market for each block of fruit and the grower's preference for spray materials. In some programs, a crop protection plan is worked out with the grower at this time along with preliminary estimates of pesticide requirements.

C. Sprayer Calibration

Proper sprayer calibration is an important component of fruit IPM programs. Uniform coverage is extremely essential as rates are reduced and spray intervals extended. In the New York program, every sprayer is calibrated before the season begins [6]. In North Carolina, willingness to calibrate and renozzle sprayers is a prerequisite to participation in the program.

D. Disease Monitoring

Monitoring for most tree fruit diseases is difficult because once symptoms are present, infection has already occurred. Techniques have been developed for assessing the potential threat of only a few diseases. Apple scab ascospore maturity can be followed by looking at crushed perithecia on a weekly basis and rating asci development [4]. Potential discharges can be evaluated by a 'spore tower' [7] and actual discharges by spore traps [8]. In New York, fungicide residues are monitored on newly developed or exposed leaf and fruit parts by a bioassay technique utilizing innoculated media plates. Residue levels can be determined within 24 hours, usually in time to make spray decisions. In California, the need to apply bactericides for fire blight control on pears is determined by monitoring bacterial populations on flower and foliage parts and by the use of threshold temperature relationships [9]. Monitoring techniques for other diseases, such as powdery mildew, cedar apple rust, and the summer rot diseases, are being developed and evaluated [10,11].

E. Pheromone and Bait Traps

Sex pheromone traps are used in all IPM programs to monitor flight periods of many lepidoptera that threaten tree fruit. They are used to follow pest populations, time spray applications, determine the effectiveness of control measures, and predict population development. Some programs directly interpret trap catches in making spray decisions (2). Others use the relationship between catches in commercial orchards to catches in abandoned orchards as a guide for making spray decisions.

In some IPM programs, sticky traps and boards are used to monitor adult fruit fly populations (12). States which contend with fruit fly problems, such as apple maggot, cherry fruit fly, and black cherry fruit fly, find these boards indispensible. In most instances, these pests are not found in commercial orchards, but annually migrate into commerical blocks from neighboring wild and abandoned trees. Migration can be detected by strategically placing bait boards around the perimeters of orchards. Key components in the use of these traps are frequent checking (every 3 days) and knowledge of the types of trees surrounding the orchards monitored.

F. Orchard Inspections

One of the essential practices of all IPM programs is a periodic routine for orchard inspection [2]. In most programs, this is carried out on a weekly basis. Many programs have ratings for determining the seriousness of insects and diseases. Some have developed criteria for decision-making based on sampling procedures [2,13]. Other components of individual programs include advice on growth regulator applications, tree pruning, tree planting, site selection, and bee hive placement. In the New York program, growers are able to request nematode sampling, and leaf and soil analysis.

G. Beneficial Insects and Predatory Mites

All of the IPM programs attempt to encourage and use beneficial insects and predatory mites [14]. Each apple growing area has identified and researched the biology of one or two predominent mite predators. In addition, the toxic effects of many of the commonly used pesticides on these predators have been determined [15,16,17] and are provided to growers through state pesticide recommendations. Integrated mite control programs have become a major component of all apple IPM programs. Some programs use mite brushing and counting techniques in conjunction with computer programs to aid growers in decision-making [14]. Other programs use visual inspections or field counts to make decisions [6]. Wherever possible, most IPM programs make every effort to conserve beneficial arthropods by careful selection and timing of pesticide applications.

H. Data Management

Computer programs have been extensively used in most IPM programs for organizing and analyzing data gathered in the field. Computer programs have been designed to organize insect trap information, grower spray records, fruit harvest records, orchard inspection information, and weather data. This has allowed extension and research specialists to peruse the numerous factors that influence pest development and relate these factors to proper control practices [18].

I. Weather Monitoring

Environmental monitoring is an essential component of all fruit IPM programs. These data are necessary for utilizing pest development models and following key events, such as leaf wetness as it relates to apple scab infection periods. Most projects maintain their own weather networks, but some also make use of existing NOAA weather stations [2,19]. Environmental variables most commonly monitored are rainfall, temperature, relative humidity, leaf wetness, and wind speed. During the growing season, these stations are monitored daily, including weekends.

J. Vegetation Monitoring

Vegetation monitoring and management is a component of several of the fruit IPM projects. In the North Carolina project, recommendations on herbicide use are based on spring and fall monitoring of species types and dominance within each orchard. In several states, the interaction of weeds with other pests (or predators) is an important consideration. For instance, in North Carolina, weed control is an important component of the vole management program, and in Michigan, the maintenance of ground cover around the base of the tree is important in conserving predator mite populations [20].

K. Rodent Monitoring

The pine vole and meadow vole are important pests in many orchards in the eastern United States. In North Carolina, vole populations are being followed by visually monitoring rodent activity in the fall (shortly after harvest) and spring and through live trapping in the winter. Advice on rodenticide application is based on fall population levels, species present, weather, availability of materials, and grower preference. Winter trapping is used to monitor rodenticide effectiveness.

L. Pesticides and Application Methods

Pesticides are the most important tool in orchard IPM. The selection of the proper material, rate, and timing is critical to the success of any program. Continuous evaluation of pesticide performance at the orchard level is also an important component of all programs [6,21]. The development of resistance to some of the more commonly used pesticides emphasises the need for careful pesticide selection.

Application methods vary within the IPM programs, depending on their appropriateness to a particular pest or situation. The SAT treatment [22] and Mill's after-infection program [5] are examples of application methods used for scab control instead of the standard calendar protective program. Alternate row middle spraying, perimeter spraying, and individual block application are also utilized in several programs to combat various insect problems [6].

M. Fruit Harvest Records

In most IPM projects, random samples of fruit are used to help evaluate the programs. Fruit sampled at harvest are rated for insect and disease incidence and severity as well as physiological problems, such as russet, bitter pit, etc. In some IPM programs, fruit size, color, sugar content, seed number, etc. are also evaluated. In the North Carolina project, yield records are being used to help evaluate various grower management practices. The fruit harvest record data are also useful in pointing out program weaknesses and are an aid in planning next year's programs.

Although the concepts just discussed are currently used in IPM programs, new tools and practices are incorporated each year. As they become available, IPM demonstration programs will be critical to their evaluation. In fact, some states include a continuous demonstration program as part of their overall implementation efforts in IPM.

N. Communication

As IPM programs develop, effective communication systems are essential for updating existing strategies and for testing and validating new ones. Although traditional methods of communication such as information notes, spray guides, newsletters, and Code-A-Phones, provide support to IPM systems, real-time information is needed to effectively utilize IPM strategies [2]. Michigan and New York are developing computerized communication systems for rapid assembly and delivery of IPM information. Michigan pioneered this area through the development of PMEX [18]. In New York, the SCAMP system is providing IPM information to county extension specialists, consultants, and growers [19]. Essential elements of SCAMP are 1) current weather forecasts and degree-day summaries, 2) field reports of events as they occur in the various regions of the state, 3) a summary or interpretation of various events supplied by research and extension specialists at the University, 4) strategy programs containing the latest crop protection suggestions based on field activity, 5) pest forecast programs, and 6) data assembly programs for storage and retrieval of field data.

Benefits

Because apple orchards often remain in production for 30 to 50 years, the impact of IPM programs on such factors as yield, tree growth, tree survival, variety selection, and pruning has yet to be fully realized. The most obvious and immediate benefit to growers has been through a reduction in overall pesticide use. Figure 1 shows a 4-year comparison of pesticide used between growers participating in the New York IPM demonstration program and non-IPM growers [6]. Pesticide use is expressed in dosage equivalents which are calculated from the actual rate of pesticide applied, divided by the rate recommended in the current *New York State Pesticide Registrations and Recommendations*. Dosage equivalents were calculated for each spray and totaled. The data used in Figure 1 represents the average dosage equivalent from 12 IPM farms and from 23 non-IPM farms. Fruit harvest records taken on both IPM and non-IPM farms each year show no differences in the fruit quality and quantity of either grower type. These data together with other data taken in the New York program show that growers participating in the IPM program use less pesticide

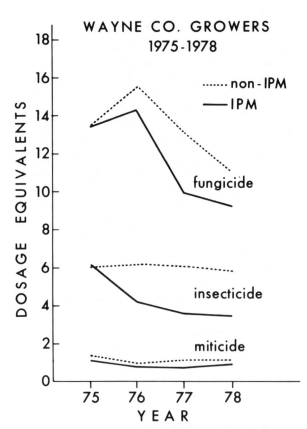

Figure 1. Average dosage equivalents for 12 IPM farms and 23 non-IPM farms in Wayne County, New York. The 12 IPM farms joined the program in 1976 [6].

and spend fewer crop protection dollars than non-IPM growers.

In summary, IPM programs are beginning to have an impact on deciduous tree fruit production. Continued development and expansion of these programs will depend upon the availability of knowledgeable individuals operating at the farm level, utilizing all of the tools and techniques available.

References

1. GOOD, J. M. 1977. Progress report pest management pilot projects. ANR-5-21, Extension Service, USDA, Washington, D.C., 20250.
2. LEEPER, J. R. and J. P. TETTE. Pest management systems for apple insects. Handbook Series in Agriculture, Section D: Pest Management, Pimentel, D., ed., CRC Press; in press.
3. RIEDL, H., B. A. CROFT, and A. J. HOWETT. 1976. Forecasting codling moth phenology based on pheromone trap catches and physiological time models. Can. Entomol. *108*:449.
4. SZKOLNIK, M. 1969. Maturation and discharge of ascospores of *Venturia inaequalis*. Plant Dis. Reptr. *53*:534.
5. MILLS, W. D. 1944. Efficient use of sulfur dusts and sprays during rain to control apple scab. Cornell Univ. Ext. Bull. *630*:4p.

6. TETTE, J. P., E. H. GLASS, D. BRUNO, and D. WAY. New York State tree fruit pest management project 1973-78. (In preparation).

7. GILPATRICK, J. D., C. A. SMITH, and R. D. BLOWERS. 1972. A method of collecting ascospores of *Venturia inaequalis* for spore germination studies. Plant Dis. Reptr. 56:39.

8. SUTTON, T. B. and A. L. JONES. 1976. Evaluation of four spore traps for monitoring discharge of ascospores of *Venturia inaequalis*. Phytopathology. 66:453.

9. THOMSON, S. R., M. N. SCHROTH, W. J. MOLLER, W. O. REIL, J. A. BEUTEL, and C. S. DAVIS. 1977. Pesticide applications can be reduced by forecasting the occurrence of fire blight bacteria. Calif. Agri. 31:12.

10. BUTT, D. J. 1977. The role of disease surveys in the management of apple mildew. Rep. E. Malling Res. Sta. for 1976. (1977):191.

11. SUTTON, T. (Unpublished)

12. LEEPER, J. R. 1978. Using sticky traps to monitor fruit flies in apple and cherry orchards. New York's Food and Life Sciences Bulletin No. 70, 1978; Plant Sciences, Entomology (Geneva).

13. BETHELL, R. S. Pear pest management. Agricultural Sciences Publications, University of California, 1422 Harbour Way South, Richmond, CA., 94804.

14. CROFT, B. A. 1975. Integrated control of orchard pests in the U.S.A. *In* Proceedings 5th Symposium on Integrated Control in Orchards. 109.

15. CROFT, B. A. and E. E. NELSON. 1972. Toxicity of apple orchard pesticides to Michigan populations of *Amblyseius fallacis*. Environ. Entomol. 1:576.

16. LIENK, S. E., C. WATVE, and J. MINNS. 1976. Susceptibility of the European red mite and its predators to chemical treatments. New York's Food and Life Sciences Bulletin No. 62, June 1976; Plant Sciences, Entomology (Geneva).

17. CROFT, B. A. and L. R. JEPPSON. 1970. Comparative studies on four strains of *Typhlodromus occidentalis*, II. Laboratory toxicity of ten compounds common to apple pest control. J. Econ. Entomol. 63:1528.

18. CROFT, B. A., J. L. HOWES, and J. M. WELCH. 1976. A computer-based, extension pest management delivery system. Environ. Ent. 5:20.

19. SEEM, R. 1979. Computers in IPM. New York's Food and Life Sciences Quarterly. 12:9.

20. CROFT, B. A. and D. L. MCGROARTY. 1977. The role of *Amblyseius fallacis* (Acarina: Phytoseiidae) in Michigan Apple Orchards. Research Report No. 333, Farm Science. Michigan State University Agricultural Experiment Station, East Lansing. 24 p.

21. SUTTON, T. B. 1978. Failure of combinations of benomyl with reduced rates of non-benzimidazole fungicides to control *Venturia inaequalis* resistant to benomyl and the spread of resistant strains in North Carolina. Plant Dis. Reptr. 62:830.

22. GILPATRICK, J. D. 1978. Single application treatments for apple scab control. p. 30-32. *In* Proceedings apple and pear scab workshop. Special Report No. ´28. N.Y. State Agric. Expt. Sta., Geneva.

36.2 Implementation of Integrated Pest Management (IPM) on Deciduous Fruits in Europe

G. Mathys

EPPO, 1 rue Le Notre, 75016 Paris (France)

Introduction

In Europe, research on integrated control in fruit growing started in the early fifties when it became evident that red spider mites had developed resistance to most of the available organophosphorus compounds and rapidly acquired the status of key pests creating difficult problems.

Another pest, the San José scale, *Quadraspidiotus perniciosus* Comst., which was of considerable concern in international trade and had resisted extensive and costly eradication campaigns also called for alternative means of containment.

In depth studies revealed the potentials of natural enemies for maintaining both pest populations below control other harmful organisms had no or little economic thresholds providing the chemicals used to disruptive effects on the bio-control agents involved. Any practical solution had thus logically to be considered in the framework of a general integrated protection scheme, a situation with which any integrated control at the crop level is normally confronted. Rosen [1974] reported on such situations in citrus groves in Israel and Bravenboer [1975] in glasshouses.

For evident practical reasons the approach towards integrated control in orchards had to be a stepwise procedure whereby the steadily increased knowledge of the ecosystems and complicated interrelationships allowed to constantly improve the system. The following 3 major phases leading to total system's management were observed:

Supervised control: taking into consideration threshold values of pests, involving the use of selective pesticides as far as applicable and safeguarding beneficial organisms.

Integrated control: in addition to the practice observed in supervised control: deliberate use of biological and biotechnical means.

Integrated production: in addition to integrated control practices: taking advantage of all positive production elements concurring to pest containment and securing optimal quality of the crop.

In several European countries, supervised and integrated control have been implemented and attempts towards introducing integrated production are currently made in France [Thiault, 1977] and Switzerland [Baggiolini, 1978]. The proportion of operational integrated control systems remains low as compared to the traditional calendar spraying which itself is however steadily improved and moving towards supervised control.

The research on past management has been largely encouraged by FAO and the International Organization for Biological Control (IOBC) to the principles of which most advanced research institutes have adhered. In fruit growing various working groups belonging to the West Palaearctic Regional Section (WPRS) of IOBC have greatly contributed to these developments.

The integration of research results obtained by specific IOBC-WPRS bodies such as the San José scale, the *Carpocapsa* and the Pheromon Groups have been secured by the Working Party on Integrated Control in Orchards. In addition, an IOBC Commission established in 1977 is in charge of promoting integrated production and quality optimization.

Advances in Integrated Control of Deciduous Fruits as Compared to Other Crops; Economic Advantages and Bottle-Necks

A survey based on an inquiry made in 9 selected European countries for the FAO/UNEP[1] Panel on Integrated Control (EPPO, 1978) has revealed the situation summarized in table 1. The data showing advances made in 11 major crops refer to the Federal Republic of Germany, France, Great Britain, Hungary, Italy, Netherlands, Norway, Spain and Switzerland.

1. UNEP: United Nations Environment Programme.

Table 1. Status of pest management in Europe. (+) means : successful for single species only.

	Research	Field experiments	Partly applied	Supervised systems	Economic interest	Integrated systems
Deciduous fruit trees	+	+	+	+	+	20,000 ha (FR, DE, CH)
Citrus	+	+	+	+	+	--
Grapevine	+	+	+	+	+	--
Olives	+	+	+	-		--
Wheat	+	+	+	[+]	[+]	--
Maize	+	+	+	+	+	10,000 ha (CH)
Potatoes	+	+	-	-	-	--
Rape	+	+	-	-	-	--
Legumes	+	+	+	-	-	--
Forests (Mediterranean)	+	+	+	+	+	+
Glasshouse crops	+	+	+	+	+	30-40% (NL) 45-60% (GB)

After some 30 years of studies, integrated pest management in fruit growing appears to offer the most solid technological background. Yet the acreage of orchards where true integrated control is practised does not exceed 20,000 ha and remains mainly confined to France, the Federal Republic of Germany and Switzerland. However, for orchard protection, the 8 mentioned countries have extensive research, field experiments and practical experimentation which led to supervised systems on extended areas where clear economic advantages are registered.

Except for corn and glasshouse crops no integrated systems are in operation in Europe, but research is actively pursued with the objective of introducing true management in other crops. In certain Italian pine forests biological control performed by the deliberate dissemination of nests of representatives of the *Formica rufa* group has become a successful practice [Pavan, 1978].

The basic principal to which the West Palearctic Regional Section of IOBC has adhered consists in strictly observing the general market requirements for quality and grading for fruit produced according to integrated systems. There are thus no extra thresholds or tolerance for infestation degrees and economic investigations reflect accordingly figures which are comparable to those obtained by traditional means.

A recent comparison of this kind has been made in the Federal Republic of Germany [Steiner, 1978; table 2], showing that the general reduction of the

Table 2. Average number of treatments in apple orchards of the Stuttgart area (1977) and reduction in costs of treatment in integrated systems (Steiner, 1978).

	Conventional	Integrated	Reduction in costs of treatment
Insecticides & acaricides	4.2	1.9	55%
Scab	11.0	8.7	26%
Mildew	1.8	1.2	64%
Total treatments	12.1	9.1	40%

costs of treatment in integrated systems reached 55% for insecticides and acaricides, 26% for scab control and 64% for mildew control. The 2 last figures indicate that thanks to regular surveillance of the orchards the frequence of fungicide treatments could also be reduced.

Another calculation performed in 1977 by Steiner [1978] in the Federal Republic of Germany revealed that the comparison of respectively 12 farms practising integrated control and 10 farms following the traditional control pattern turned in favour of the integrated system bringing about an extra income of $2330/ha (Golden Delicious, average yield of 40 t/ha; 30 cts/kg).

Yet even in the fact of such figures which might not be every year and everywhere so striking, the expansion of the integrated technology is slow. The technology transfer remains a major obstacle which can only be removed by a considerable official teaching effort and general support.

Developments Towards Integrated Fruit Production

Encouraged by the practicability of pest management systems and its various ecological and economical advantages, the West Palaearctic Regional Section of IOBC has studied the means of establishing production systems based on a whole management basis. Investigations towards this end started in 1974 jointly with representatives from OECD[2] and EEC[3].

After several meetings the WPRS decided in 1976 (IOBC, 1977) to establish a commission to study the means of developing integrated production and evaluating crops produced under optimal conditions. The commission was entrusted in particular with the establishment of 3 types of guidelines:

1. General regional guidelines on integrated pest control (IPC) in agriculture;
2. General regional guidelines on IPC for specific crops;
3. Detailed guidelines on IPC for a specific crop in a restricted area including production practices leading to optimal quality.

The groups of farmers or farmers' associations interested in producing such recognized crops of intrinsic quality had to incorporate these guidelines into a detailed code of practice which had to be submitted for approval to a national committee linked with IOBC. Such a code had to include rules on integrated protection, fertilization, pruning, weed control, conservation techniques of fruits, etc. Providing that adequate surveillance through official experts could be secured the IOBC commission could grant a special WPRS label certifying top quality.

These develoments are currently taking place in France and Switzerland. An IOBC label has been tentatively introduced in 1978 by a farmers' association in an area bordering the Swiss side of the Lake of Geneva. For the moment being, this label remains at the gross market level and experience has already shown that there is a greater demand for such fruits; it is expected that in future higher prices will be offered for them.

Conclusions

After 30 years of research and experimentation mainly in apple and peach growing, the IPM technology is sufficiently advanced to be generally introduced in most European countries. These efforts have brought about considerable improvements in traditionally controlled orchards where techniques of supervised control are increasingly applied.

The various advantages inherent in integrated control systems have been largely recognized but there remains the bottle-neck of the technology transfer which can only be removed by a considerable official encouragement and effort in the teaching component. It is expected that the West Palaearctic Regional Section of IOBC which is now backed also by the Common Market[4] will succeed in securing the breakthrough as advocated by FAO, UNEP and the Codex Alimentarius Committee on Pesticides Residues. Producing an optimal fruit quality under best protection conditions is a challenging IOBC objective.

2. OECE: Organization for Economic Cooperation and Development.
3. EEC: European Economic Community.

4. European Economic Community.

Bibliography

Baggiolini, M. (1978). La valorisation qualitative de la production agricole. *Revue Suisse Vitic. Arboric. Hortic. 10* (2):51-57.

Bravenboer, L. (1975). Biological control in protected cultivation. Semaine d'étude agriculture et hygiène des plantes, Gembloux (BE):283-296.

Eppo (1978). Biological agents used in integrated pest control. Work. Doc. FAO/UNEP Panel on Int. Contr.

IOBC (1977). An approach towards integrated agricultural production through integrated plant protection. *WPRS Bull.* 4, 163 pp.

Paven, M. (1976). Utilisation des fourmis du Groupe *Formica rufa* pour la défense biologique des forets. Min. Agr. For. Roma; *Collona Verde 39*:417-442.

Rosen, D. (1974). Current status of integrated control of Citrus pests in Israel. *EPPO Bull. 4* (3):363-368.

Steiner, H. (1978). Pers. comm.

Thiault, J. (1977). Tendances actuelles sur la vente des fruits et légumes de qualité définie. C. R. réunion d'Avignon: "Méthode de certification et de valorisation de la qualité" (OILB/SROP; multicopie).

36.3 Implementation of Integrated Pest Management on Deciduous Fruits in South America

Fausto H. Cisneros

Universidad Nacional Agraria
Lima, Peru

Temperate climatic conditions, suitable for deciduous fruit production occur only in a minor part of South America. Most of the sub-continent is tropical and sub-tropical. South to parallel 30° S (roughly equivalent to latitudes of the United States and part of Canada, in the northern hemisphere) lies only one fifth of the area. This "temperate" region has milder temperatures than their counterparts in the northern hemisphere; summer is not so warm and winter is not so cold. Rainfall distribution restricts further the relative abundance of deciduous tree fruits.

Argentina is the main producer of apples, pears and peaches, followed by Chile, Peru and Uruguay. Brazil is second in peach production. Apples are cultivated as close to the Equator as 8° S in the coastal strip west to the Andes and in the interandean valleys. The total amount of apple, pear and peach production is about 7 per cent of citrus plus banana production.

1. General Characteristics of Deciduous Fruit Ecosystems in S.A.

Climate. Apple producing areas in Argentina, Chile and Uruguay have the characteristic seasonal variations of temperate regions which influence plant phenology and occurrence of pest and diseases. In the Chilean producing areas rainfall occurs in winter rather than in summer. Peru lies within tropical latitudes but "temperate" conditions occur in the interandean valleys due to the high altitudes. The coastal strip where apple trees are also grown has very mild climate. Lowest temperature in winter is around 12°C and highest temperature in summer is around 28°C; so that the "resting stage" of the plant is not induced by winter temperature but by restraining irrigation at any time of the year. It is common to obtain 3 productive seasons each two years.

Plant Species. Apple, pear and peach trees are exotic plant species to South America, not related to the native flora. First introductions took place shortly after the Spanish conquest in the XVI century. New selected cultivars have been introduced to Chile and Argentina and to a lesser extent, or more recently, to other countries.

Orchard Management. Management systems in apple orchards are fair to good. Holdings are predominantly small to medium in size. Orchards in Argentina and Uruguay are kept under rainfall conditions while in Chile and Peru they are under irrigation (in Chile rainfall occurs mainly in winter and in Peru rainfall in the coastal area is very scarce).

Pest and Diseases. Most key pests and disease producing organisms are exotic species: codling moth, oriental fruit moth, Mediterranean fruit-fly, European redmite, the powdery mildew and the apple scab. Most secondary pests are also introduced species: apple woolly aphid, San Jose scale, oystershell scale and the apple leaf curling midge. Most of the native species are phytophagous insects with no-economic importance except for the South-american fruit-fly and the polyphagous mites, the carmin mite and the two-spotted spider mite. Chile is the only southamerican country free of both the South-american and Mediterranean Fruit flies.

Natural Enemies. Being most of the deciduous fruit pests of exotic origen important natural enemies were also imported. Introduction of natural enemies started during the first decades of the century. The most notorious examples are *Aphelinus mali* Hald, parasite of the apple woolly aphid; several species of *Aphytis* and *Prospaltella* parasites of diaspidid scales; species of *Metaphycus* and *Coccophagus* and *Scutellista cyanea* parasites of coccid scales; several general predators of aphids and related insects (Chrysopidae, Coccinellidae and Sirphidae) are native. There is an important native complex of predatory mites, studied particularly in Chile, which under favorable conditions is able to keep spider mite populations under control. *Prospaltella berlesei* How gives only partial control of the white peach scale in Peru. The codling moth, the oriental fruit moth, and the fruit flies are free of efficient natural enemies.

2. Old and Current Status of Deciduous Fruit Pest Control

The older methods for pest and disease control in deciduous fruits included: lead arsenate against the codling moth, other lepidoterous larvae and chewing insects; emulsifiable oils against scale insects and eggs of several species of insects and mites; lime-sulfur and DNOC against scale insects, eggs of various insects and mites as well as diseases; and Bourdeaux mixture for the control of diseases. Winter oil and Bourdeaux mixture is still in use against apple scab in Uruguay.

The discovery of DDT and other modern insecticides and fungicides improved the control of the codling moth and other pests as well as diseases. Unfortunately, as it occurred in other parts of the world and with other crops too, new problems arose as a consequence of the new methods. The most general and common phenomenon was the increase of mite populations; as a result of the destruction of predatory species; and the development of resistance of the mite populations to many products. The decimation of other beneficial insects produces commonly increases of several other pests as the apple woolly aphid, scale insects, the apple leaf curling midge, and even native species as the leaf roller, Eulia sp. in Argentina.

The current system of Chemical control in Argentina and Chile involves 8 to 16 treatments per year, including 6 applications of insecticides against the codling moth and other insects, 5-7 applications of acaricides and 6-8 applications of fungicides.

3. Status of the Implementation of IPM in deciduous Fruits in S.A.

The implementation of IPM in apple orchards in S. A. is in a phase of development in limited areas of Chile and Uruguay. It is not of general qpplication as is the IPM in cotton fields and to a lesser extent in citrus and olive orchards, in Peru. Very little has been done in other countries or in pear and peach orchards.

In the experimental areas of Chile, it has been possible to reduce the normal 6 treatments of insecticides to 3-4 applications; and from 5-7 treatments of acaricides to 3 applications (González, 1976). This was possible through a better timing of the applications based on field assessment of pest densities, instead of the fixed schedule of sprays, and the utilization of selective products or selective dosages.

Seasonal Monitoring and Population Assessment of Key Pests.

Seasonal appearance of codling moth and oriental fruit moth is detected by using bait, black light, or pheromone traps; the last being the most efficient. The Mediterranean and the South-american fruit flies populations are assessed by using bait traps for both species, or specifically pheromone traps for the medfly. *Panonychus ulmi* is sampled by taking leaves and counting the number of mites.

Selective use of Pesticides. Much attention is given to reduce the effects of pesticides against beneficial insects and mites. It is considered the degree of plant coverage, the timing of the applications, and the use of selective products or selective dosages.

The recommended selective pesticides are: acaricide-fungicide products: triamiphos, Benomyl and Dinocap; acaricide products: tetradifon, chloropropilate and Plictram; insecticide products: mineral oils; low dosages of dimethoate, azinphos-methyl and chlordimeform; not more than twice applications a year of naled, trichlorphon, tetra-chlorvinfos and malathion. Excluded are the broad spectrum products parathion, Binapacryl, dicofol, Metidation and others.

Partial tree coverage (upper part of the tree) is

recommended for the control of the oriental fruit moth; and only one fifth of the foliage for the baits sprayed against fruit flies.

A winter treatment of an acaricide-ovicide (Lambrol M-2060) is recommended in Chile against mites and aphids. A winter oil and Bourdeaux mixtures is used in Uruguay against the apple scab. Pre-bloom treatments are recommended for apple orchards in Chile (González, 1976).

DNOC is used in Peru as ovicide and apple defoliant to complement the effect of water restriction to produce the resting period of the plant and stimulate the subsequent blooming.

In the demonstrative orchards in Uruguay, carbaryl and azinphosmethyl have been discarded for the control of the codling moth; and TEPP and

parathion, for the control of the brown mite. Imidan, which does not destroy predatory mites, is recommended against the codling moth. A winter oil spray, with an optional partial application of a summer oil is used against the San Jose scale and other scales, with little effect on beneficial insects. Apple scab is controlled by the winter oil spray plus Bourdeaux mixture. Herbicides, like paraquat, which decimate the predatory mite *Cidnodromus* are discarded. The management of the pest populations are complemented with cultural method such as keeping the canopy of trees open by adequate pruning and limiting the Nitrogen fertilization not to exceed 6 kg. N/ha/MT of fruit. Both measures restricts the favourability of the plants to insect and mite increase.

36.4   Current Status of Biological Control in Deciduous Fruit Orchards in Australia

J. L. Readshaw

The Co-operative Research Programme on Orchard Pest Management (CRP) by Lloyd *et al.* [1970] showed that in apple orchards at least theoretically there was scope for implementing biologically centered control methods for some of the pests involved. In order of probable responsiveness to this approach were tetranychid mites using coccinellid predators (*Stethorus* spp.); woolly apple aphid using parasites (*Aphelinus mali*); and San Jose scale using coccinellid predators (*Rhizobius lindii*) and parasites (*Prospaltella* spp). Less certainly perhaps light brown apple moth (LBAM) was considered responsive to the appraoch but not so plague thrips, dimple bug or codling moth. Codling moth was clearly the major fly in the ointment since its control under Australian conditions was shown to require repeated applications of broad-spectrum pesticides.

Further work by the CRP found that both codling moth and LBAM could be attacked with specific non-inclusion granulosis viruses but because of problems associated with production, formulation, application and residues neither virus has been developed to the stage of being a commercial proposition [Morris, 1972; Geier & Briese, 1979].

We are left with the fact that commercial orchards still have to be sprayed almost routinely with various broad-spectrum insecticides in order to produce a

saleable crop. In this circumstance it is virtually impossible to gain any benefits from the normal complex of beneficial species. From November to March the commercial orchard is almost a biological desert, the only resident animals being those few individuals that survive the barrage of sprays. These tend to be either sheltered from the the sprays at some time during their life-cycles (e.g. codling moth and LBAM larvae and pupae) or resistant to the chemicals (e.g. tetranychid mites). For controlling the former, more frequent, better timed or more thorough spraying may suffice, but for the latter there is the constant need to develop new pesticides that are chemically unrelated to those used previously.

Thus it is clear that in practice there is little scope for conventional biological control in Australian orchards.

The major exception happens to be in my own field which concerns orchard mites. Here a few important natural enemies of the two-spotted mite (*Tetranychus urticae*) and the European red mite (*Panonychus ulmi*) have themselves developed resistance to some pesticides. They are all small predacious mites (Phytoseiidae) and the main species are *Typhlo dromus occidentalis, T. pyri* and *Amblyseius fallacis*.

T. occidentalis was introduced from North America in 1972; *A. fallacis* from New Zealand (*ex*

578

North America) in 1976 and *T. pyri* from New Zealand in 1976 and 1977.

Detailed work by CSIRO and State Departments has shown that *T. occidentalis* can control the two-spotted mite in apples and peaches throughout the major fruit growing areas of eastern Australia provided the background spray programme does not include chemicals that are harmful to the predator [Readshaw, 1975b; Field, 1974, 1976, 1978]. This is illustrated by data from Hauptmann's orchard in Canberra: 1972-1979.

Similar studies are in progress with *T. pyri* against the European red mite (for which *T. occidentalis* is less effective) and *A. fallacis* is also being investigated.

There is good reason to expect that the successful control of orchard mites with "resistant" predators will be commonplace throughout eastern Australia within the next few years with a consequent saving of about $4 million per annum on sprays. More importantly perhaps, is that resistance by mites to acaricides will develop less quickly than in the past, if at all.

Apart from mites the only other biological control projects in deciduous fruit that I am aware of are in peaches. Rothschild [1975] has been successful in controlling oriental fruit moth (OFM) using pheromones to confuse the male moths and thus prevent mating, and Bailey has been releasing parasites against the same species. Also, some attempt is being made to control the long-tailed mealy bug in fruit trees in S.A. using Encyrtid and Pteromalid parasites [Furness, 1977].

Future possibilities include the use of specific controls such as pheromones, viruses and insect growth regulators (IGRs) for codling moth and LBAM but there are still major problems to overcome [e.g. MacFarlane & Jameson, 1974]. Even if successful I shudder to think of what new problems lie hidden under the broad-spectrum lid of Pandora's Box.

At the other end of the spectrum we have the new pyrethroids waiting in the wings—more deadly than DDT—and probably much more disruptive to natural control. For example, like DDT, they are not very effective against tetranychid mites but their effect on the "resistant" predators is catastrophic.

The use of these compounds in orchards will certainly create a major mite problem and unless someone comes up with a pyrethroid-resistant strain of predator we will have to rely entirely on chemical control in a field where only one or two compounds are currently effective.

So I must end my song rather sadly with a warning and apologies to Jonathan Swift [1733] or August de Morgan [1806-1871] or R. R. Fielder [?], whichever version you fancy:

> Great fleas have little fleas upon their backs to bite 'em,
> And little fleas have lesser fleas,
> but not unless we fight 'em.

Useful References

Attia, F. I. 1974. Laboratory evaluations of insecticides against *Nysius vinitor* Bergroth and *Nysius clevlandensis* Evans (Hemiptera: Lygaedae). J. Aust. ent. Soc. *13*:161-4.

Edwards, B. A. B. & Hodgson, P. J. 1973. The toxicity of commonly used orchard chemicals to *Stethorus nigripes (Coleoptera: Coccinellidae). J. Aust. ent. Soc.* 12(3):222-24.

*Field, R. P. 1974. Occurrence of an Australian strain of *Typhlodromus occidentalis* (Acarina: Phytoseiidae) tolerant to parathion. J. Aust. ent. Soc. *13*(3):255.

*Field, R. P. 1976. Integrated pest control in victorian peach orchards: The role of *Typhlodromus occidentalis* Nesbitt (Acarina: Phytoseiidae). Aust. J. Zool. *24*(4)L565-72.

*Field, R. P. 1977. Reduced spray program that controls oriental fruit moth. Vic. Hort. Dig. Dept. Agriculture, Melbourne. 72:73.

*Field, R. P. 1978. Control of the two-spotted mite in a Victorian peach orchard with an introduced insecticide-resistant strain of the predatory mite *Typhlodromus occidentalis* Nesbitt (Acasrina: Phytoseiidae). Aust. J. Zool. 26:519-27.

*Furness, G. O. 1977. Apparent failure of two parasites, *Anarhopus sydneyensis* (Hymenoptera: Encyrtidae) and *Hungariella peregrina* (Hymenoptera: Pteromalidae) to establish on field populations of *Pseudococcus longispinus* (Hemiptera: Coccidae) in South Australia. J. Aust. ent. Soc. *16*(1):111-12.

Geier, P. W., Hillman, T. J. & Wilson, A. G. L. 1967. Evaluating new prospects in pest control. J. Aust. ent. Soc. 6(2):91-92.

*Geier, P. W. & Briese, D. T. 1979. The light brown apple moth, *Epiphyas postvittana* (Walker): three differences in susceptibility to a nuclear polyhedrosis virus. Aust. J. Ecol. 4:(in press).

Hogan, T. W. 1973. Integration process as it affects entomology. J. Aust. ent. Soc. *12*(4:241-7.

*Lloyd, N. C., Jones, E. L., Morris, D. S., Webster, W. J., Harris, W. B., Lower, H. F., Hudson, N. M. & Geier, P. W. 1970. Managing apple pests: A new perspective. J. Aust. Inst. agric. Sci. *36*(4):251-8.

*MacFarlane, J. R. & Jameson, G. W. 1974. Ovicidal effect of juvenile hormone analogues on *Cydia pomonella* L. and *Cydia pomonella* L. and *Cydia molesta* Busk. (Lepidoptera: Tortricidae). J. Aust. ent. Soc. *13*(1):31-6.

*Morris, D. S. 1972. Evaluation of a nuclear granulosis virus for control of codling moth. 14th International Congress of Entomology, Canberra. Abstracts: 238.

Morris, D. S. & Field, R. P. 1975. Role of predatory mites in mite control on tree fruits. Vict. Hort. Dig. *66*:10-13.

Morris, D. S. & Webster, W. J. 1971. Managing apple pests—a new perspective. Vict. Hort. Dig. *15*(3):12-16.

Readshaw, J. L. 1971. Ecological approach to the control of mites in Australian orchards. J. Aust. Inst. agric. Sci. 37(3):226-30.

Readshaw, J. L. 1972. Failure of lead arsenate in an ecological approach to the control of mites in orchards. J. Aust. Inst. agric. Sci. *38*(4):308-9.

Readshaw, J. L. 1975(a). The ecology of tetranychid mites in Australian orchards. J. Appl. Ecol. *12*:473-95.

*Readshaw, J. L. 1975(b). Biological control of orchard mites in Australia with an insecticide-resistant predator. J. Aust. Inst. agric. Sci. 41(3):213-14.

Richardson, N. L. 1971. An integrated pest management programme for oriental fruit moth (Cydia molesta) and other pests of peaches. M.Sc. Thesis, University of Adelaide.

Rose, H. A. & Hooper, G. H. S. 1969. The susceptibility to insecticides of Cydia pomonella (L.) (Lepidoptera: Tortricidae) from Queensland. J. Aust. ent. Soc. 8(1):79-86.

*Rothschild, G. H. L. 1975. Control of oriental fruit moth, Cydia molesta (Busk.) (Lepidoptera: Tortricidae) with synthetic female pheromone. Bull. ent. Res. 65:743-50.

Schica, E. 1975. Predacious mites (Acarina-Phytoseiidae) on sprayed apple trees at Bathurst (N.S.W.). J. Aust. ent. Soc. 14(3):217-19.

*Swift, J. 1733. On poetry: A Rhapsody, I. 337.

Terauds, A. 1978. Light-brown apple moth damage assessment. J. Aust. ent. Soc. 16(4):367-69.

Thwaite, W. G. 1975. Control of apple pests. Agric. Gaz. N.S.W. 86(6):15.

Unwin, B. 1971. Biology and control of the two-spotted mite Tetranychus urticae (Koch.). J. Aust. Inst. agric. Sci. 37(3):192-211.

Walters, P. J. 1973. Integrated pest control in apple orchards. Agric. Gaz. N.S.W. 84(1):16-17.

Walters, P. J. 1976. Chlordimeform: its prospects for the integrated control of apple pests at Bathurst. J. Aust. ent. Soc. 15(1):57-61.

Walters, P. J. 1976. Susceptibility of three Stethorus spp. (Coleoptera: Coccinellidae) to selected chemicals used in N.S.W. apple orchards. J. Aust. ent. Soc. 15(1):49-52.

Walters, P. J. 1976. Effect of five acaricides on Tetranychus urticae (Koch) and its predators Stethorus spp. (Coleoptera: Coccinellidae) in an apple orchard. J. Aust. ent. Soc. 15(1):53-6.

Webster, W. J. & Field, R. P. 1977. Mass release of predatory mites in Goulburn Valley orchards. Vict. Hort. Dig., Dept. Agric., Melbourne. 72:2-5.

(*Mentioned in Text)

36.5   Use of Allelopathic Cover Crops to Inhibit Weeds[1]

A. R. Putnam and Joseph DeFrank

*Pesticide Research Center, Michigan State University
East Lansing, Michigan, 48824, U.S.A.*

Introduction

The term *allelopathy* was coined by Molisch in 1937 [13]. In current use, the term refers to the detrimental effects of higher plants of one species (the donor) on the germination, growth, or development of plants of another (receptor) species. Allelopathy can be separated from other mechanisms of plant interference in that the detrimental effect is exerted through release of a chemical by the donor species. During the past two decades, numerous review articles [1, 2, 3, 5, 7, 8, 12, 15, 16, 19, 22, 24] and a book [17] have been written about allelopathy. Recent articles suggest exploitation of this phenomenon as an aid in weed control [2, 16].

Numerous phytotoxic secondary products have been isolated and identified from higher plants. Introduction of these compounds into the environment occurs by exudation of volatiles from living plants [14]. Leaching of water-soluble toxins from aerial portions [21], or subterranean tissues [6], or by release of toxins from non-living plant materials [9]. In the latter case, microbial products generated during litter decomposition may also exert action [12].

Natural products play important roles in plant resistance to insects, nematodes, and pathogens. Parallel approaches may allow the use of allelopathic chemicals to inhibit weed germination and growth, or prevent propagule production. We propose to exploit allelopathy as an additional strategy for weed suppression in several agroecosystems, including the deciduous fruit cropping system.

Experimental Methods

The residues of numerous crops were initially screened for weed suppressing activity by drilling them into rows 18 or 30 cm apart and allowing the plants to reach a height of 25 to 40 cm after which they were dessicated by the herbicides glyphosate or paraquat. In other experiments, cover crops were planted in the late summer (September 1 to 15) and were allowed to freeze. Weed density estimates were obtained at intervals of 30 to 360 days after treatment, and total weed biomass was obtained at

1. This research was supported by the Science and Education Administration of the U.S. Dept. of Agriculture under Grant No. 7800385 from the Competitive Grants Office.

the end of the growing season. Among the plants under study were sorghum *(Sorghum bicolor)*, sudangrass *(Sorghum vulgare)*, wheat *(Triticum aestivum)*, oats *(Avena sativa)*, barley *(Hordeum vulgare)*, and rye *(Secale cereale)*. Peat moss was included as an organic mulch control. Additional orchard experiments were conducted in which straw of selected species was placed under 7 year-old 'Montmorency' cherry *(Prunus cerasus)* and 'McIntosh' apple *(Malus sylvestris)* trees at rates equivalent to 4450 kg/ha to ascertain their influence on weed populations and tree growth. Terminal growth measurements were obtained from 10 lateral branches per tree. Plots contained 2 or 3 trees per replicate and treatments were replicated 4 times.

Results and Discussion

Where fall-planted cover crops were utilized, both weed populations and biomass were greatly reduced in the next growing season. "Tecumseh" wheat dessicated in spring or fall reduced weed weights 76 and 88% respectively (Table 1). Fall-killed "Balboa" rye was similar in effectiveness. In contrast, fall-killed "Garry" oats or spring-killed rye had no toxic action on weeds. In fact, in this experiment, oat residues appeared to stimulate weed germination.

The mulching experiments also indicated that certain plant residues can contribute exceptional weed control. For example, sorghum and sudangrass straw provided weed biomass reductions of approximately 90% and 85% respectively, whereas peat moss provided only a marginal, if any, reduction (Table 2). The species encountered in both series of experiments were primarily annuals and consisted of *Digitaria sanguinalis, Ambrosia artemisiifolia, Chenopodium album, Polygonum persicaria, Setaria viridis,* and *Cerastium vulgatum.* Partial control of perennial *Festuca* species was also obtained.

After one season, tree growth was not adversely affected by any of the mulches. All mulches, including peat moss, provided terminal growth equal to or better than the bare ground control (Table 3).

The promising results obtained in these experiments lead us to the conclusion that selected cover crop residues or mulches can provide excellent suppression of a number of annual weed species. To date, we have no experience on the efficacy of these residues on several problem perennial species which invade deciduous fruit ecosystems.

The use of mulches under trees has provided several additional benefits which include improved water penetration and retention [18, 20], reduced soil erosion [23], and an improved habitat for predators

Table 1. Weed suppression provided by residues of cover crops planted in September of the previous year.

Cover Crops	Time of Dessication	Weeds/M²	Kg/plot
None	–	69	18.5
"Garry" oats	Fall	102	33.0
"Balboa" rye	Spring	43	23.3
"Balboa" rye	Fall	34	2.9
"Tecumseh" wheat	Spring	7.7	4.4
"Tecumseh" wheat	Fall	14	2.2
HSD at 5% level		21	5.7

Table 2. Weed suppression by allelopathic mulches applied in early spring.

Mulch	Weed Biomass (Kg/plot)
None	21.4
"Bird-a-Boo" Sorghum	2.1
"Monarch" Sudangrass	3.3
Peat Moss	15.9
HSD at 5% level	6.4

Table 3. Growth of two species of deciduous fruit trees after early spring mulching.

Mulch	Av. Terminal Growth (cm)	
	"Montmorency" Cherry	"McIntosh" Apple
None	18.5	30.2
"Bird-a-Boo" Sorghum	25.2	44.1
"Monarch" Sudangrass	21.7	37.2
Peat Moss	24.3	31.5
HSD at 5% level	4.7	6.8

of the European red mite [4]. Cover crops planted in late summer may slow tree growth and allow adequate hardening of tissue against winter stress. A major disadvantage of mulch is that it provides favorable habitat for rodents which may cause extensive tree damage [10]. However, practical orchard experience has shown that trees must be protected from rodents with barriers or guards even in the absence of weeds or mulches.

The toxicity of water-soluble compounds leached from several of these plant species lead us to the conclusion that allelopathy is a major factor which contributes to their effectiveness. Long term studies must be conducted in orchard ecosystems to ascertain that no harmful direct or indirect effects occur on trees after repeated use of these cover crops or mulches.

Literature Cited

1. Aemisepp. A., and H. Osvald, 1962. "Influence of higher plants upon each other—allelopathy." *Nova Acta Regiae Soc. Sci. Ups.* 18:1-7.

2. Altieri, M. A. and J. D. Doll. 1978. "The potential of allelopathy as a tool for weed management in crop fields." *PANS* 24(4):495-502.

3. Bonner. J. 1950. "The role of toxic substances in the interactions of higher plants." *Bot. Rev.* 16:51-65.

4. Croft, B. A. 1975. "Integrated control of apple mites." *Mich. State Univ. Coop. Ext. Serv. Bull.* No. E-825. 12 pp.

5. Evanari, M. 1961. "Chemical influences of other plants (allelopathy)." *Handb. Pflanzenphysiol.* 16:691-736.

6. Fay, P. K. and W. B. Duke. 1977. "An assessment of allelopathic potential in *Avena* germ plasm." *Weed Sci.* 25:24-28.

7. Grummer, G. 1955. Die gegenseitige Beeinflussung hoherer Pflanzen-Allelopathie. Jena:Fischer. 162 pp.

8. Grummer, G. 1961. "The role of toxic substances in the interrelationships of higher plants." *Symp. Soc. Exp. Biol.* 15:219-28.

9. Guenzi. W. D. and T. M. McCalla. 1962. "Inhibition of germination and seedling development by crop residues." *Proc. Soil Soc. Sci. Am.* 26:456-58.

10. Lord, W. J., D. A. Marini, and E. R. Ladd. 1967. "The effectiveness of fall applications of granular simazine and dichlobenil for weed control in orchards and the influence of weed control on mouse activity." *Proc. Northeast Weed Control Conf.* 213-217.

11. Martin, P. 1957. "Die Abgabe von organischen Verbindugen insbesondere von Scopoletin, aus den Keimwurzein des Hafers." *Z. Bot.* 45:475-506.

12. McCalla, T. M. and F. A. Haskins. 1964. "Phytotoxic substances from soil microorganisms and crop residues." *Bacteriol. Rev.* 28:181-207.

13. Molsich, H. 1937. Der Einfluss einer Pflanze auf die andere—Allelopathie. Jena:Fischer.

14. Muller, C. H. 1965. Inhibitory terpenes volatilized from *Salvia* shrubs. Bull. Torrey Bot. Club 92:38-45.

15. Muller, C. H. 1966. The role of chemical inhibition (allelopathy) in vegetation composition. Bull. Torrey Bot. Club 93:32-51.

16. Putnam, A. R. and W. B. Duke. 1978. "Allelopathy in agroecosystems." *Ann. Rev. Phyto. Pathol.* 16:431-51.

17. Rice, E. L. 1974. *Allelopathy.* New York Academic. 353 pp.

18. Rom, R. C. 1972. "Herbicide effects on surface soil in orchards and rainfall infiltration." *Ark. Farm Res.* 21(4):5.

19. Swain, T. 1977. "Secondary compounds as protective agents." *Ann. Rev. Plant Physiol.* 28:479-501.

20. Toenjes, W., R. J. Higdon, and A. L. Kenworthy. 1956. "Soil moisture used by orchard sods." *Quart. Bull. Mich. Agr. Exp. Sta.* 39(2):1-20.

21. Tukey Jr., H. B. 1966. Leaching of metabolites from above ground plant parts and its implications. Bull. Torrey Bot. Club 93:385-401.

22. Tukey Jr., H. B. 1969. "Implications of allelopathy in agricultural plant science." *Bot. Rev.* 35:1-16.

23. Unger, P. W., R. R. Allen, and A. F. Wiese. 1971. "Tillage and herbicides for surface residue maintenance weed control, and water conservation." *J. of Soil and Water Conserv.* 147-150.

24. Whittaker, R. H. 1970. "The biochemical ecology of higher plants." In *Chemical Ecology.* ed. E. Sondheimer, J. B. Simeone, pp. 43-70. New York Academic. 336 pp.

36.6   Use of Pesticides in Integrated Control in Orchards

P. Gruys

Experimental Orchard De Schuilenburg TNO
4041 BK Kesteren, The Netherlands

Introduction

Orchards are subjected to many pests. This fact makes them a favorite crop for new developments in plant protection: the application of intensive chemical spray programs, soon after the advent of the modern pesticides, as well as research aimed at alternatives for intensive chemical control, at present.

Several harmful orchard arthropods are amenable to biological control by native or introduced natural enemies, including major pests such as spider mites (*Panonychus ulmi, Tetranychus urticae* and *Tetranychus mcdanieli*), coccids (*Quadraspidiotus perniciosus, Lepidosaphes ulmi*), and *Eriosoma lanigerum* (MacPhee et al. 1976; Mathys and Guignard 1967).

Furthermore, microbial control (Dickler this Symposium), mating disruption by pheromones (Arn 1979), and sterile insect releases (Proverbs et al. 1978) offer promise, particularly against codling moth and other tortricids. Breeding apple trees for resistance to diseases is beginning to yield benefits (Aldwinckle and Lamb this Symposium), microbial antagonists as possible control agents of pathogens of fruit trees are receiving increasing attention (Corke and Hunter, 1979; van den Ende 1980), and dormant season sprays may reduce the need for fungicide applications in the growing season (Hislop this Symposium).

In spite of all these developments chemical pesticides are, and will be at least for a long time, indispensible to achieve the degree of perfection in quantity and quality of harvest required at present. However, essential as pesticides are, biotic control factors are the basic element of integrated pest management (IPM) in orchards, and the chemicals applied must not violate their proper functioning.

Two main approaches to achieve this have been followed. The more recent one, prominent particularly in the USA, is based on the discovery of resistance to certain O-P and carbamate insecticides in phytoseiid mite predators of spider mites (Croft 1976). This fact allows an integration of biological spider mite control with carefully timed and applied chemical control of insects (Hoyt 1969; Croft and Brown 1975). Insecticide resistant phytoseiids also show promise in Australia and New Zealand and work with introduced O-P resistant *Typhlodromus pyri* from New Zealand has been initiated in the UK (Cranham 1979; Readshaw this Symposium). It is anticipated that certain other beneficial arthropods will become resistant with time, provided the use of pesticides is managed properly (Croft and Morse 1979).

The second approach to integrating biological and chemical control is to use selective chemicals, or apply broad-spectrum pesticides selectively. It is followed in Nova Scotia and Europe, where acquired resistance to pesticides in native beneficial arthropods has not been observed. Within this approach, there is a difference of emphasis. In Central Europe, the aim is to spare the whole complex of useful species, whereas in the Netherlands and the UK, the attention is concentrated on certain key beneficial species (Gruys 1980a).

This paper is concerned with recent progress of the latter type of integration.

Orchard Fauna under IPM

In selectively sprayed apple orchards in Holland, *Panonychus ulmi* is kept at a very stable and low level by phytoseiid mites, of which *Typhlodromus pyri* is the most important species. Characteristically, numbers fluctuate below one-tenth of the economic threshold. Furthermore, the leaf miner, *Stigmella malella,* a minor but troublesome pest where it occurs, is kept in check by its specific chalcid parasite, *Chyrsocharis prodice* (Gruys 1980b). Biotic mortality of other major pests can be considerable but it is not, or not always, sufficient to provide satifactory control. This applies to the main tortricid pest in Holland, *Adoxophyes orana,* which can be attacked heavily by the polyphagous chalcid, *Colpoclypeus florus* (Evenhuis 1974).

The other characteristic feature of selectively sprayed orchards is the more diverse complex of

Table 1. Pest Complex on Apple in Holland: Broad-Spectrum and Selective Control Compared (underlined = major, p = primary, s = secondary pest)

Order, fam.	Broad-Spectrum Control		Selective Control; Additional Pests	
LEP. Geom.	*Operophtera brumata*	p		
	Orthosia spp.	p		
Tortr.	*Laspeyresia pomonella*	p		
	Adoxophyes orana	p/s		
			Spilonota ocellana	p
			Pandemis spp.	p
			Archips podana	p
			Archips rosana	p
			Pammene rhediella	p
			(+ innocuous spp.)	
Nept.	*Stigmella malella*	s		
Hom. Aphid.	*Rhopalosiphum insertum*	p		
	Dysaphis plantaginea	p		
	Aphis pomi	p		
	Eriosoma lanigerum	p		
Cocc.			*Lepidosaphes ulmi*	
HET. Mirid.	*Lygus pabulinus*	p		
ACAR. Tetr.	*Panonychus ulmi*	s		
HYM. Tenthr.			*Hoplocampa testudinea*	p
COL. Curc.			*Anthonomus pomorum*	p
			Phyllobius oblongus	
DIPT. Cecid.			*Dasineura mali*	

Table 2. Percent Fruits Blemished by Insects under a Minimum Selective Program, 1975–1978

Caterpillars, spring*	3.0	
Codling	4.5	15.4
Lead rollers	7.5	
Minor lepidopt. pests	0.4	
Lygus	1.0	
Sawfly (*Hoplocampa*)	1.8	3.2
Aphids	0.3	
Scale (*Lepidosaphes*)	0.1	
	18.6	

*See note to Table 3

phytophagous insects, including several minor pests, as compared to orchards subjected to broad-spectrum pesticides (Table 1).

Four 0.5 hectare plots in our experimental orchard (Gruys, 1975), receiving a selective minimum program, demonstrated the effect of the minor pests. This program was limited to fungicides and occasional sprays against aphids and Lygus bugs. Crop loss due to apple sawfly, *Hoplocampa testudinea,* varied from 0 to 40%, depending on the year and cultivar. The degree of attack by *Anthonomous pomorum* and *Phyllobius oblongus* would have worried many ordinary fruit growers, and none of them would have accepted the fraction of insect-blemished fruit, in which damage by various species of caterpillar predominated (Table 2). To investigate

583

their potential for controlling this caterpillar complex, three selective insecticides were added to the minimum program, each on its own 0.5-1.0 ha plots: Dipel, Dimilin, and epofenonane.

Selective Insecticides

Dipel (*Bacillus thuringiensis*) was unsatisfactory at a reasonable dose (1.5 kg/ha), possibly because of the moderate temperatures characteristic of the average Dutch summer.

Dimilin (diflubenzuron), an ingestion poison which disturbs chitin synthesis during moults, is primarily toxic to the larvae of phytophagous insects, but it also has an ovicidal effect on some species, either directly or via the female (Grosscurt 1978). It is an effective insecticide against *Operophtera brumata* and *Orthosia* spp. Within the orchard tortricids, the susceptibility to this insecticide varies considerably. In feeding tests, the LD-50 values of the most and the least susceptible species, *Laspeyresia pomonella* and *Pandemis heparana* respectively, differed by a factor 3000 (van der Molen 1976). Besides *L. pomonella*, *Spilonota ocellana*, *Hedya nubiferana*, and *Archips podana* are susceptible and besides *P. heparana*, *Archips rosana* and *Adoxophyes orana* are tolerant to Dimilin. In the plots where Dimilin was applied, the leaf roller fauna changed accordingly: *Pandemis heparana*, *Archips rosana*, and *Adoxophyes orana* were left and the other species virtually disappeared; the damage to the fruit by *Lepidoptera* greatly diminished but was not eliminated (Table 3).

Epofenonane (Ro-10-3108/018) will probable cure some of the problems that Dimilin leaves, such as *Adoxophyes orana*. It is an insect growth regulator (IRG) with juvenile hormone activity, acting on the last instar of caterpillars and inducing morphogenetic effects resulting in death, or sterility of emerged adults. It has a long residual life on the foliage as compared to other IGRs of this type (Hangartner et al. 1976). The rationale of its

application against *Adoxophyes orana* is that the hibernating larvae of this species, in spring, do not inflict economic damage. If the population could greatly be reduced this time, and no immmigration would occur during the subsequent flights, then the numbers might remain sufficiently low during the next two injurious generations of that season. Detailed field experiments have shown that the total effect of carefully timed spring application of epofenonane can amount to 98% (Schmid et al. 1978). The 1 ha plot in our orchard to which epofenonane was applied was not very suitably situated to put this principle to a test, since *A. orana* was everywhere around it. Therefore, *Adoxophyes* control by epofenonane was included in trials on the functioning of IPM in commercial holdings, selected to provide a variation in the distance to the nearest orchard of up to 800 meters. This work was started in 1978 and the results, against *A. orana* and other leaf rollers, are promising. Regrettably, it is doubtful whether this compound will ever be marketed as an insecticide. However, a similarly acting, economically more promising compound is being developed (Anonymous 1979). We therefore feel justified to continue our large-scale trials with epofenonane as a model of the applicability of this type of insecticide.

Dimilin as well as epofenonane showed useful side-effects in suppressing some of the minor pests.

Dimilin, applied at petal fall, killed the half-grown larvae of apple sawfly. Although this does not prevent damage in the same year it may cause some 90% of the larvae to die, which keeps an already low population from rising. If needed, additional selective control of apple sawfly can be obtained before damage is inflicted by a thiophanate-methyl spray at pink bud (Predki and Profic-Alwasiak 1976). Furthermore, Dimilin very effectively controlled the weevils, *Anthonomus pomorum* and *Phyllobius oblongus*, through an ovicidal effect via the females.

Epofenonane gives good control of scale insects, including *Quadraspidiotus perniciosus* (Frischknecht and Muller 1976) and *Lepidosaphes ulmi*, and, when applied early in spring, of aphids and woolly aphid. Other satisfactory selective compounds against the latter insects are, however, already available. Furthermore, it controls pear psylla (*Psylla pyrisuga*, *P. Pyri;* Frischknecht et al. 1978) as does Dimilin, if the applications are carefully timed to the eggs and the young larvae (van Frankenhuyzen and Meinsma, 1978).

In our orchard where these insecticides are applied for the fifth and the fourth year, respectively, they have not disrupted the spider mite-phytoseiid balance, a fact confirming expectations based on small-scale tests, nor significantly reduced parasitization of

Table 3. Percent Fruit Blemished by *Lepidoptera* insects under a Schedule of Broad Spectrum Pesticides Including O-P and Carbaryl Sprays against Caterpillars and a Selective Program Including Dimilin; 1975-1978

Insects	O-P and Carbaryl 3-6 Sprays	Dimilin 3 Sprays
Caterpillars, spring*	0.2	0.6
Codling	0.1	0.1
Leaf rollers	0.5	1.7
	0.8	2.4

*Including *O. brumata*, *Orthosia* spp., *Archips rosana*, and some of the damage by *Pandemis* spp.

Adoxophyes orana. Dimilin has strongly changed the natural populations of *Stigmella malella*—against which it is very effective, without, however, affecting its specific parasite directly.

Other Pesticides

It takes 2-4 years after the initition of IPM in an orchard for the spider-mite-phytoseiid relationship to become firmly established. In this time, acaricidal treatments may be required. We found Plictran (cyhexatin) unsuitable for this purpose because it virtually exterminated the phytoseiids, due to toxicity to eggs and larvae; the harmlessness of Plictran to adult phytoseiids is deceptive (van Zon, pers. comm.). Fenbutatinoxide, benzoximate, and white oil at bud burst were not detrimental to phytoseiid populations.

The choice of fungicides is essential to the success of biological control of spider mites because of their frequent application. In large-scale applications or small field trials continued for a least one season on established phytoseiid populations consisting predominately of *Typhlodromus pyri*, we found captan, dodine, nitrothal-diisopropyl, bupirimate, and triadimefon harmless, as were one or two applications of thiophanate-methyl. On the other hand, repeated applications of mancozeb, thiram, fungicide mixtures containing 15-35% maneb, and sulphur were detrimental to phytoseiid populations. Dinocap and binapacryl, known to be toxic, were not tested.

Horticultural demands interfere with the implementation of IPM in part of the orchards in Holland. Sulfur is preferred on Golden Delicious because of its cosmetic effect on the fruit, as are Mn-containing bisdithiocarbamates which improve the quality of foliage. Moreover, chemical crop thinning is indispensable in modern apple growing, and carbaryl is the only satisfactory agent on most cultivars under our climatic conditions. Small field trials showed single applications of carbaryl to be acceptable on populations of *T. pyri*.

Final Remarks

With a limited number of new selective insecticides (diflubenzuron, epofenonane), added to some already existing ones (pirimicarb, white oil) and occasional use of special broad-spectrum materials (endosulfan, bromophos) and special fungicides (including captan, bupirimate, and thiophanate-methyl), all the main pests of orchards in Holland left over by biological control could be suppressed satisfactorily, with no quantitative loss and insect-

blemish fruits remaining under 4%. A list describing the details of this program has been published elsewhere (Gruys 1980a). Notice that the main selective insecticides in this program are far from species-specific.

Owing to their mode of action, the new selective insecticides have to be used in a more preventive way, necessitating a different approach to the management of certain pests as compared to the last-resort-type-of action that the usual contact insecticides allow. Fruit growers may find this an advantage over present supervised control procedures in Europe.

We have already better tools within reach for managing the complex of orchard pest than the old selective insecticides such as lead arsenate and Ryania. Moreover, research is building-up the apparatus for selectivity tests to which newly discovered pestidical compounds could and should be subjected. The chemical industries willing to develop these materials should be stimulated to start or continue producing them, by government research investigating, and showing the potentials of, these chemicals in large-scale pilot projects in as many countries as possible.

References

Anonymous 1979. Ro 13-5223. Insecticide. Technical Data Sheet, Dr. R. Maag Ltd. 5 pp. CH-8157 Dielsdorf Switzerland.

Arn, H. 1979. Developing insect control by disruption of sex pheronome communication: Conclusions from programs on lepidopterous pests in Switzerland. in *Chemical Ecology: odour communication in animals.* F. J. Ritter, ed. pp. 365-374. Elsevier North Holland Biomedical Press.

Corke, A. T. K. and Hunter, T. 1979. Biocontrol of *Nectria gallegenia* infection of pruning wounds on apple shoots. *J. Hort. Sci.* 54:47-55.

Cranham, J. E. 1979. Managing spider mites on fruit trees. *Span* 22:28-30

Croft, B. A. 1976. Establishing insecticide resistant phytoseiid mite predators in deciduous tree fruit orchards *Entomophagae* 21:383-399.

Croft, B. A. and Brown, A. W. A. 1975. Responses of arthropod natural enemies to insecticides. *Annu. and Rev. Entomol.* 20:285-335.

Croft, B. A. and Morse, J. G. 1979. Research advances on pesticide resistance in natural enemies. Entomophaga 24:3-11.

Ende, G. van den. 1980. Fungi from phyllosphere and carposphere in relation to integrated control in an apple orchard. In *Integrated control of pests in the Netherlands.* A. K. Minks and P. Gruys, eds. pp. 53-55. Pudoc: Wageningen.

Evenhuis, H. H. 1974. *Colpoclypeus florus* (Hymenoptera, Eulophidae), an important potential parasite of *Adoxophyes orana* (Ledioptera, Tortricidae) in apple orchards. Meded. Fak. Landbouwwet. Rijksuniv. Gent 39:769-775.

Frankenhuyzen A. v. and Meinsma, E. 1978. De werking van diflubenzuron (Dimilin) op de gewone perebladvlo (*Psylla pyri*). Gewasbescherming 9:53-59.

Frischknecht, M., Jucker, W., Baggiolini, M. and Schmid, A. 1978. Mode of action and practical possibilities of an insect growth regulator with juvenile hormone activity in pear psyllid control. A. Pflanzenkr. Pflanzenschutz 85:334-340.

Frischknecht, M. L. and Muller, P. J. 1976. The use of insect growth regulators in integrated pest control. Mitt. Schweiz. Entomol. Ges. 49:239-244.

Grosscurt, A. C. 1978. Diflubenzuron: some aspects of its ovicidal and larvicidal mode of action and an evaluation of its practical possibilities. Pestic. Sci. 9:373-386.

Gruys, P. 1975. Development and implementation of an integrated control programme for orchards in the Netherlands. Proc. 8th Brit. Insecticide and Fungicide Conf. (1975) 3:823-835.

Gruys, P. 1980a. Significance and practical application of selective pesticides. Proc. Int. Symp. Integrated Control of IOBC—WPRS, Vienna, 8-12 Oct. 1979.

Gruys, P. 1980b. Development of an integrated control program for orchards. Natural control of the leaf miner Stigmella malella in apple orchards. In: Integrated control of pests in the Netherlands. A. K. Minks and P. Gruys, eds. pp. 6-10. Pudoc: Wageningen.

Hangartner, W. W., Suchy, M., Wipf, H. K., and Zurflueh, C. 1976. Synthesis and laboratory and field evaluation of a new, highly active and stable insect growth regulator. J. Agric. Food. Chem. 24:169-175.

Hoyt, S. C. 1969. Integrated chemical control of insects and biological control of mites on apple in Washington. J. Econ. Entomol. 62:74-86.

Mac. Phee, A. W., Caltagirone, L. E., Van de Vrie, M., and Collyer, E. 1976. Biological control of pests of temperate fruits and nuts, In Theory and practice of biolgical control. C. B. Huffaker and P. S. Messenger, eds. pp. 337-358. Academic Press: New York.

Mathys, G. and Guignard, E. 1967. Quelques aspects de la lutte biologique contre le pou de San José (Quadraspidiotus perniciosus Comst.) a l'aide de l'aphélinide Prospaltella perniciosi Tow. Entomophaga 12:1967, 223-234.

Molen, J. P. van der. 1976. Inpassing van selectieve bestrijdingsmiddelen in geintegreerde betrijdingsprogramma's. Annual Report 1976 Instituut voor Plantenziektenkundig Onderzoek Wageningen: 91-93.

Predki, S. and Profic-Alwasiak, H. 1976. The effectiveness of systemic fungicides in controlling the European apple sawfly, Hoplocampa testudinea Klug. Fruit Science Reports, Skierniewice, 3:39-45.

Proverbs, M. D., Newton, J. R. and Logan, D. M. 1978. Suppression of codling moth, Laspeyresia pomonella (Olethreutidae) by release of sterile and partially sterile moths. Can. Entomol. 110:1095-1102.

Schmid, A., Molen, J. P. van der, Jucker, W., Baggiolini, M. and Antonin, Ph. 1978. The use of insect growth regulators, analogues of the juvenile hormone, against summer fruit totrix moth, Adoxophyes orana and other pests. Ent. Exp. and Appl. 24:65-82.

36.7 Use of Host Plant Resistance in Tree Fruits

Herb S. Aldwinckle and Robert C. Lamb

*Departments of Plant Pathology, and Pomology and Viticulture, Cornell University
New York State Agricultural Experiment Station, Geneva, NY 14456*

Rationale

Host plant resistance is the most desirable method of pest control on deciduous tree fruits for economic, environmental, and toxicological reasons. At present tree fruits receive the greatest amount of pesticides per hectare of any crop in the USA. Annual expenses for these chemicals and their application would be greatly reduced on pest-resistant cultivars. There would be no danger of environmental pollution, effects on non-target organisms, exposure of orchard workers to pesticides or contamination of food. Crop losses from pests that are poorly controlled at present would be eliminated on cultivars that were resistant to those pests. The undesirability of the dependence on the tree fruit industry on pesticides is magnified by the increasing prices and declining supplies of the petrochemicals from which they are manufactured, and by the development of pesticide-tolerant strains of pests.

Difficulties

Integrated pest management (IPM) on deciduous tree fruits, however, currently utilizes host plant resistance in very few instances. There are good reasons for this situation. (1) Reliable quantitative data on resistance are available for only a limited number of diseases. (2) For pollination purposes orchards frequently consist of mixed cultivars, often with different levels of resistance. (3) Although differences in resistance of present commerical cultivars may be statistically significant, they may not be sufficient to justify economizing on chemical sprays without increasing the risk of crop loss. (4) A cultivar may be resistant to one pest but not to others, for which chemical sprays must still be applied. (5) In contrast with many agronomic crops, cosmetic quality of fruit is as or more important than yield, so that blemished fruit must approach 0% in commerical production.

Thus, the prospects for using the pest resistance available in present commmercial cultivars are poor except in a few cases. There are, however, good prospects that cultivars of the major deciduous tree fruits with a high degree of pest, especially disease, resistance will be developed in the future.

Development of New Resistance Cultivars

The major problems associated with the development and effective utilization of new pest-resistant cultivars of tree fruits include: (1) The inordinate length of the breeding process which stretches into decades with these long-generation plants. (2) The necessarily slow replacement of existing orchards of suspectible cultivars in which growers have large investments. (3) The need for processors, supermarkets, and consumers to accept new cultivars. (4) The need for assurance that resistance will remain effective indefinitely in the face of pest variation. (5) Lack of information on pests that may develop in large plantings of new cultivars that are not sprayed with pesticides. (6) The need for diversity of germplasm to avoid the dangers of genetic uniformity

With sufficient long term investment in breeding and related research in plant pathology and entomology, there are no technological reasons why these problems in the development and application of pest-resistant cultivars should not be overcome.

Let us examine in detail the research and development procedures that are necessary to produce a new resistant cultivar. Apple will be used as an example because it is the most important deciduous tree fruit worldwide and because most progress has been made with it.

Factors Affecting Disease Development

It is crucial to know what factors are most favorable for development of the particular disease so that resistance can be evaluated under the most severe conditions that a cultivar will encounter and so that disease escape is minimized. (Disease escape may be a desirable attribute of a cultivar but must be specifically selected for. An example is breeding pears that do not have secondary bloom to avoid fire blight.) Environmental factors including temperature, relative humidity, leaf wetness, and soil moisture should be optimal for infection. The phenophase of the plant is also crucial. For example, apple seedlings are most susceptible to *Venturia*

inaequalis at the two leaf stage, but resistance is manifested at this stage. Resistance to *Gymnosporangium juniperi-virginianae*, however, does not develop until seedlings are several weeks old.

Screening Techniques

Using knowledge of the factors favoring infection, efficient and reproducible techniques must be developed for screening large numbers of seedlings. It is important that the resistance selected in young seedlings should also be operative when they become mature trees.

It is less important if some seedlings that are susceptible in the screening test, but which could have been resistant in the orchard are discarded, since seedlings are relatively cheap to produce compared with mature trees. Selection for resistance against pests (e.g. quince rust) that affect the fruit only must be delayed until seedlings flower. However, in the case of scab and cedar apple rust there is an excellent correlation between leaf susceptibility of young seedlings and fruit susceptibility of mature trees.

Sources of Resistance

Not surprisingly, high degrees of resistance are usually not available in commercial cultivars and must be sought in ornamental or wild relatives, in the same or different species. In apple good resistance to scab, fire blight, and mildew is being bred in from small-fruited *Malus* species, and requires several backcrosses. Resistance to cedar apple rust, on the other hand, is available in commercial cultivars such as Spartan and McIntosh. It is very desirable that diverse sources of resistance be identified and used so that genetic uniformity can eventually be avoided.

Physiologic Races

Resistant cultivars should remain resistant wherever they are grown. Therefore, all known physiologic races of the pathogen must be used in screening tests. Unfortunately, for most diseases information on physiologic races is not available and extensive research must be done to define them.

Breeding

The actual breeding consists of hybridizing parents that are selected for favorable characteristics, particularly quality and pest resistance; identifying the high quality resistant progeny; and then backcrossing for improved quality, or out-crossing for other resistance. Several such back- or out-crosses will probably be required to develop a cultivar with good quality and multiple pest resistance. Liberty, the new disease-resistant apple cultivar released by us, is the result of a cross, a sib-cross, and two modified back-crosses. Because of a discontinuity in the breeding, it took about 64 years from the initial cross to final release. However, using classical breeding and improved cultural techniques in a continuous program it would now be possible to accomplish this process in half that time. Some experimental techniques have allowed us to shorten the generation time of apple to 2 years so that five crosses could be made in 10 years. This time span would be comparable with the time required at present to develop and register a new pesticide.

Genetics of Resistance

Although it is desirable to know how the resistance is inherited, this is not necessary for conducting the breeding program. Furthermore, it is usually not feasible for long-generation crops. Nevertheless, we have learned something about the inheritance of resistance in apples from analysis of crosses in the breeding program.

It would be more important to determine the genetics of resistance if there were a constant correlation between the number of genes on which resistance depends and the durability of resistance. There is a hypothesis that monogenic resistance is race-specific and less stable than polygenic resistance, which is non-race-specific. There are, however, many exceptions to this hypothesis. Nevertheless, it is interesting that most of the resistances that are being used in apple breeding are oligogenic, or monogenic plus minor genes. Further analysis may well reveal that the resistances depend on complexes of genes.

Yield and Quality Testing

An important tenet of our philosophy is that yield and quality must not be compromised for the sake of disease resistance. Otherwise growers will prefer to grow susceptible cultivars and use pesticides for control than to grow the resistant cultivars. Furthermore, if mutant races or biotypes were to attack the resistant cultivar the result would be a poorer susceptible cultivar than those in present use. In fact, we are striving to improve the quality and yield of our resistant cultivars over those of cultivars like McIntosh, Delicious, and Golden Delicious.

Durability of Resistance

Durability is particularly important in tree fruits where cultivars take long to develop and cannot be changed frequently, and where orchards are expected to produce for decades. It is impossible to determine precisely how durable a particular resistance is unless it "breaks down". However, the longer the resistance "holds up" the more durable it is.

Nevertheless, there are certain procedures that we would recommend to increase the probability of selecting durable resistance. Seedlings should be screened against all physiologic races under optimal disease conditions. Resistance selections should then be grown in the field, preferably in widely spaced locations with different climatic conditions, and observed closely for infection. Although resistance need not be polygenic to be durable, resistance based on more than a single gene is preferred. We do not subscribe to the view that incomplete resistance that allows some infection and sporulation is more durable than complete resistance. Futhermore, the high cosmetic quality requirements for fruits would render such resistance of little value with fruit-blemishing pests.

Resistant Apple Cultivars

Nine cultivars of apple bred specifically for disease resistance have been introduced in North America and Europe. Some of the earlier cultivars have high susceptibility to certain diseases (Table 1). It is only recently that the concept of multiple disease resistance has been accepted. Some resistant cultivars are very good in quality, but whether they will be widely planted depends on their performance

Table 1. Apple Cultivars with Resistance to *Venturia inaequalis*: Their Origin and Resistance to Other Pathogens

Cultivar	Year Introduced	Source	Resistance[1] PM[2]	FB[3]	CAR[4]
Prima[5]	1970	Univ. Illinois[6]	2	2	4
Priscilla	1972	Purdue Univ.[6]	3	2	1[7]
Macfree	1974	Ottawa, Canada	–	2	3
Priam	1974	Angers, France[8]	–	–	–
Nova Easygro	1975	Nova Scotia	3	2	1
Sir Prize	1975	Purdue Univ.[6]	2	3	3[7]
Florina	1977	Angers, France	–	–	–
Liberty	1978	Geneva, NY	2	2	1
Novamac	1979	Nova Scotia	–	–	–

[1] 1 = very resistant, no control needed.
   2 = resistant, control needed only under very high disease pressure.
   3 = susceptible, controls usually needed.
   4 = very susceptible, controls always needed.
[2] *Podosphaera leucotricha* (apple powdery mildew).
[3] *Erwinia amylovora* (fire blight).
[4] *Gymnosporangium juniperi-virginianae* (cedar apple rust).
[5] Prima is susceptible to bitter pit.
[6] Introduced by the cooperative program of Purdue University, Rutgers, and the University of Illinois (PRI).
[7] Susceptible (3) to *Gymnosporangium clavipes* (quince rust).
[8] First grown at Purdue University under the PRI program, then evaluated and introduced by the INRA Station d'Arboriculture fruitière. Angers, France.

in trials by growers. It is expected that at least five active disease-resistant apple breeding programs will continue to introduce similar cultivars with improved quality and disease resistance in the next few years.

Less progress has been made in breeding apple cultivars with resistance to arthropd pests, although research is underway in several places. It is unlikely that cultivars with good resistance except to certain aphids will be developed soon.

There is a critical need for new apple rootstocks with resistance to *Erwinia amylovora* (fire blight) and *Phytophthora cactorum* (crown rot), which cannot be satisfactorily controlled by other methods. Good progress has been made by Cummins and Aldwinckle at Geneva, New York in indentifying sources of resistance to *E. amylovora, P. cactorum,* and *Eriosoma lanigerum,* the woolly apple aphid, and using them in breeding.

Table 2. Current Status of Development of Pest-Resistant Cultivars of Pear, Peach, and Cherry

Pest	Resistant Cultivars Introduced	Breeding Underway
Pear		
Erwinia amylovora	X	X
Fabraea maculata		X
Venturia pirina		X
Psylla pyricola		X
Peach		
Xanthomonas pruni	X	X
Meloidogyne spp.	X	
Cytospora spp.		X
Sanninoidea exitiosa		X
Cherry		
Pseudomonas spp.	X[1]	X
Coccomyces hiemalis		X
Phytophthora spp.	X	

[1] Cultivars will probably be released shortly.

Pears, Peaches, and Cherries

The present status of the development of pest-resistant cultivars of other important deciduous tree fruits is summarized in Table 2.

Conclusion

After decades of research and breeding, good quality apple cultivars of the other major deciduous tree fruits with resistance to specific diseases have also been introduced or are being bred. With continued investment in research and breeding programs, the prospects are good that host resistance will play a major role in the control of many of the important diseases of deciduous tree fruits by the end of this century.

Limited research on arthropod pest resistance is in progress but generally breeding is lagging behind that for disease resistance. The evidence indicates that arthropod-resistant cultivars can be bred. For progress to be made, however, greater long-term investments will be required perhaps by making shifts in our priorities for tree fruit research.

Bibliography

Day, P. R. 1974. Genetics of Host-Parasite Interaction. Freeman, San Francisco. 238 pp.

Faust, M. *et al.* 1976. (Proceedings of Fruit Breeding Symposium, July 22-24, 1975, Beltsville, MD) *Fruit Varieties J.* 30(1):1-34.

Janick, J. and J. N. Moore (eds.). 1975. *Advances in Fruit Breeding.* Purdue Univ. Press, W. Lafayette, IN. 623 pp.

Nelson, R. R. 1973. *Breeding Plants for Disease Resistance.* Pennsylvania State Univ. Press, University Park, PA. 401 pp.

Russel, G. E. 1978. *Plant Breeding for Pest and Disease Resistance.* Butterworths, London. 485 pp.

Visser, T. (ed.). 1976. Proceedings Eucarpia Meeting on Tree Fruit Breeding, Wageningen, 7-10 Sept. 1976. Institute for Horticultural Plant Breeding, Wageningen. 142 pp.

36.8   A Novel Method of Eradicating Apple Powdery Mildew and Its Relevance to Integrated Pest and Disease Control Programmes

E. C. Hislop

University of Bristol, Long Ashton Research Station
Long Ashton, Bristol, England

It is evident from the tone of this Congress that the rational use of pesticides by means of integrated or supervised schemes is an aim to which we all subscribe, and one which over the last 10 years or so has made considerable progress. Clearly we can only afford to use pesticides when we can be sure that they will have maximum effect on a target organism and minimum effect elsewhere.

Yet the majority of fruit growers throughout the world who have a problem with the control of apple mildew spray at frequent calendar intervals without any precise knowledge of the epidemiology of the pathogen; often using chemicals which adversely affect non-target species. This admission is bad enough in itself, but I suggest that for two reasons it has deeper significance. Firstly, the practice of routine spraying induces a state of mind which is not conducive to the acceptance of rational schemes. Secondly, since the cost of pesticides is only a small fraction of the total cost of spraying there is a great temptation to add other pesticides to the fungicide mix, not because they are essential but because it is apparently a cheap form of insurance.

The reasons for the present unsatisfactory situation with regard to the control of apple mildew are not difficult to understand and have been ably reviewed recently in general terms [9]. Apple mildew survives during the dormant season in apple buds so that at the beginning of the growing season infected tissues are produced near to abundant healthy and very susceptible leaves. Wind-dispersed conidia are produced in profusion, and with the exception of prolonged rainfall there are no clearly defined meteorological or biological parameters which prevent infection in a normal late spring and summer in the UK [8]. Thus the disease can develop with a rapid initial momentum and in consequence is often difficult to control. This situation contrasts dramatically with that which exists for many wetness-requiring pathogens—e.g. the apple scab fungus—where the recognition of discrete infection periods in the epidemic permits a rational approach to control by disease monitoring and forecasting procedures [22].

At this point it is important that I declare my interests. I am primarily a plant pathologist interested in fungicides *per se,* and most of the work I have done on apple mildew eradication has been done in collaboration with Derek Clifford—a chemist. Although we, and others, realized several years ago that in some of our treatments we had a tool which might permit a re-appraisal of the sledge hammer approach to mildew control we did not have the facilities to pursue this line. It was largely left to colleagues who constitute the British Working Group on Integrated Control on Fruit and Hops to examine the relevance of mildew eradication. Thus I wish to acknowledge formally the part played by the Group and in particular by R. T. Burchill, D. J. Butt, P. W. Carden, J. E. Cranham, M. E. Upstone, S. G. Evans and M. Simkin, because much of the limited information available is presently unpublished.

If you accept the explanation given above for the existing irrational approach to mildew control and all that this implies, I think you will see why attempts to eradicate the disease are of interest in the context of this Symposium. The rationale is simple. As Butt summarizes [9] "the severity of an epidemic is a function of the quantity of primary inoculum, so that on apple the amount of primary mildew in spring is correlated with the amount of secondary mildew in summer". Eradicate the primary mildew and perhaps we can reduce or avoid the use of the popular dinitrophenol-based fungicides and some benzimidazole and organophosphorus compounds which are also toxic to fruit tree red spider mites and some of their predators thus interfering with some integrated pest control schemes. In discussing sanitation I am well aware that long ago Horsfall [19] warned of the problems of using eradicant fungicides, and more recently Van Der Plank [16] has explained why this approach may be naive for the control of powdery mildew.

Since time and space for this presentation are limited I will not describe the details of mildew eradication but will give a brief summary and a comprehensive bibliography. We now know that a single spray of one of a variety of chemicals to dormant apple trees can virtually eradicate severe mildew infections [2, 5, 12, 13, 14, 16, 17]. To do this in early trials we had to use about 5% active ingredient in 2000 l of spray per hectare. This was very costly ($250/ha) and in some instances caused bud death or sometimes delayed flowering. In severely mildewed orchards this treatment was cost-effective and even some phytotoxicity was acceptable because it had to be weighed against the debilitating effects of the disease. Yields of fruit in the year of treatment were not significantly altered, but were often greatly increased in the following year, depending upon the degree of eradication achieved [17].

However, most orchards are not severely mildewed and phytotoxicity becomes a major problem. Thus a compromise was reached and a non-ionic phenol ethoxylate material coded PP.222 was test-marketed in the UK [2] for use at 3.5% a.i., and at the growers' risk. Applied with automatic spraying machines this treatment in cold spells in December to mid-February achieved 85-95% eradication of mildew without injury, although flowering was sometimes delayed by up to a week. In several instances this delay resulted in increased fruit yields due to the avoidance of frost injury.

The next significant development was the introduction of surfactant/fungicide mixtures for use in the dormant season [2, 18, 20, 21, 23, 25]. With this technique between 1 and 2% surfactant is mixed with certain mildew fungicides at concentrations similar to those used in the growing season. The mixtures are advantageous because they are cheaper and seemingly less phytotoxic than PP.222 at 3.5%, and yet of similar efficacy. Speculation that they might be usable after harvest but pre-leaf fall so that more of the spray is caught by the leaves and funnelled to buds and perhaps have an eradicant effect on scab [4, 14, 18, 23] has yet to be verified. An interesting recent observation concerns the effect of heavy rain falling soon after treatments have been applied and indicates that the mixture may penetrate buds more quickly than surfactant alone and therefore be less affected.

The vigorous vegetative shoot growth produced on formerly mildewed trees is not, as suspected at one time, resistant to secondary infection. Trials carried out at Long Ashton where there was a severe mildew infection pressure clearly showed that routine spraying could not be delayed if one was going to capitalize on the eradication achieved [16, 18]. We concluded that if there were to be economies in the fungicide programme they should be made late rather than early in the growing season [16], i.e. when the crucial periods for bud protection had passed [6, 7].

Incorporation of the mildew eradication technique into rationalized schemes for pest and disease management was subsequently taken over by Members of the Working Group on Integrated Control who had access to larger orchards where the spray programs could be tested under more realistic conditions. The pest data at one of these have been published [11], and I am indebted to P. W. Carden and M. E. Upstone for providing mildew data from an Agricultural Development and Advisory Service Development Farm Project [1]. Under this far-sighted scheme a commercial grower sprays insecticides and fungicides only with the approval of the Advisor and in return is guaranteed financial compensation if any loss of revenue results from the sale of damaged or reduced crop. In this trial with two plots, each of 2.4 ha, the grower's semisupervised program consisting of a November application of urea for scab eradication [3], four insecticide sprays, seven applications of dinocap and 8 of bupirimate were compared with an integrated program of November urea, PP. 2222 mixed with bupirimate in January, 4 insecticide sprays and only 2 of bupirimate (summer sprays). Primary mildew on blossoms was 0.6% on the grower's plot and 0.2% on the integrated plot. A feature of the trial was frequent assessment of secondary infection [10], and on the growers plot this remained between 5 and 7% throughout the year. In contrast, secondary infections increased rapidly on the integrated plot so that by mid-June they exceeded the arbitrary threshold level of 15% so that two curative sprays of bupirimate had to be applied. Levels than fell to below the threshold but rose sharply again in early August so that by 1st September over 30% leaf infection was recorded. The number of spray rounds on the grower's plot was 18 compared with 8 on the integrated plot, but because of the considerable cost of the mildew eradication treatment chemicals costs were 50% higher on the latter. This grower was, however, already practicing a degree of supervision and his costs were only 70% of a grower's program (unsupervised) on another site. Apparently the winter treatment did not affect the viability of mite or aphid eggs or the numbers of beneficial predatory mites [24].

Apple scab was virtually absent from both plots in 1978, but urea was again applied overall in November while the integrated control plot was sprayed with PP.222/bupirimate (1.75% PP. 222 + 1.4 l. bupirimate active ingredient in 2200 l per ha) in January 1979. This May there was virtually no primary mildew on the integrated plot and about 5% on the grower's plot. However, although subsequent secondary infections were few on the grower's plot a new threshold level of 10% was soon exceeded on the integrated control plot necessitating the application of four bupirimate sprays by mid July. This year May and June were wetter than normal, and despite the use of urea apple scab appeared on both plots so that four sprays of delancol have also already been applied to the integrated plot. Thus, eradication of overwintering infections of scab and mildew has not facilitated the control of these diseases with the reduced spray programs examined so far, confirming at least for mildew some previous tentative results. Data obtained this spring at Long Ashton indicate that for mildew this might be due to failure to protect new foliage in May, because when this was done the amount of secondary mildew on 18 July on a plot treated with eradicant in the winter was 12% compared with 38% on another plot not treated in the winter. Clearly an important question remaining to be answered concerns the degree of primary mildew eradication which must be achieved before epidemics can usefully be delayed. Fortunately, the Development Farm Project has several more years to run and next year the eradicative treatment will almost certainly be followed by early spring protective sprays before any attempt is made to reduce the mildew program.

In conclusion, it is obvious that there are many fundamental problems yet to be examined but from the practical point of view it is clear that we must find cheaper treatments if the technique of apple mildew eradication is to play a significant role in integrated pest and disease control schemes. In a wider context, I believe that the future of disease control by eradication of primary or overwintering infections is a challenge to everyone attending this Conference.

References

1. Anon. (1978). ADAS Ann. Rep. for 1977, 87-88.
2. Bent, K. J., Scott, P. D. and Turner, J. A. W. (1977). Proc. Br. Crop Protect. Conf.: Pests and Diseases. 331-339.
3. Burchill, R. T. and Cooke, R. T. A. (1971). In *Ecology of leaf surface micro-organisms.* T. F. Preece and C. H. Dickinson, Eds. 471-483. Academic Press.
4. Burchill, R. T., Frick, E. L., Cook, M. E. and Swait, A. A. J. (1979). *Ann. appl. Biol.* 91. 41-49.
5. Burchill, R. T. and Swait, A. A. J. (1977). *Ann. appl. Biol.* 87, 229-231.
6. Butt, D. J. (1971). *Ann. appl. Biol.* 68, 149-157.
7. Butt, D. J. (1972). *Ann. Appl. Biol.* 72, 239-248.
8. Butt, D. J. (1975). In "Climate and the Orchard". (H. C. Pereira, Ed.), 125-126. Commonwealth Agricultural Bureaux, England.
9. Butt, D. J. (1978). In *The powdery mildews.* (D. M. Spencer, Ed.), 51-81. Academic Press.
10. Butt, D. J. (1979). Rep. E. Malling Res. Stn for 1978, 211-214.
11. Carden, P. W. (1977). Proc. Br. Crop Prot. Conf.: Pests and Diseases. 259-367.
12. Clifford, D. R. and Hislop, E. C. (1975). *Pestic. Sci.* 6, 409-418.
13. Frick, E. C. and Burchill, R. T. (1972). Pl. Dis. Reptr. 56, 770-772.
14. Hislop, E. C., Barnaby, V. M., and Burchill, R. T. (1977). *Ann. appl. Biol.* 87, 29-39.
15. Hislop, E. C. and Clifford, D. R. (1974). Plant Dis. Reptr. 58, 949-951.
16. Hislop, E. C. and Clifford, D. R. (1977). Rep. Long Ashton Res. Stn for 1976, 177-182.
17. Hislop, E. C., Clifford, D. R., Hosgate, M. E., and Gendle, P. (1978). Pestic. Sci. 9, 12-21.
18. Hislop, E. C. and Harper, C. W. (1978). Rep Long Ashton Res. Stn for 1977, 104-105.
19. Horsfall, J. G. (1956). "Principles of fungicidal action". Chronica Botanica Co.
20. Hunter, L. D., Blake, P. S. and Swait, A. A. J. (1977). Rep. E. Malling Res. Stn for 1976, 165.
21. Hunter, L. D., Blake, P. S. and Swait, A. A. J. (1978). Rep. E. Malling Res. Stn for 1977, 103-104.
22. Jones, A. L. (1978). Proc. apple and pear scab Workshop, 1976. Rep. No. 28. 19-21.
23. Mercer, R. T., Stevens, C. C., Beech, B. G. W. and Paul, G. C. Diseases. 369-374.
24. Solomon, M. G. (1979). Rep. E. Malling Res. Stn for 1978, 123.
25. Swait, A. A. J., Souter, R. D. and Butt, D. J. (1979). Rep. E. Malling Res. Stn for 1978, 87.
26. Van Der Plank, J. E. (1963). *Plant Diseases: Epidemics and Control.* Academic Press.

36.9   Simulation Models of Apple and Associated Arthropods[1]

B. A. Croft

Department of Entomology and Pesticide Research Center
Michigan State University, East Lansing, MI 48824

Simulation models describing the biology of several species associated with apple integrated pest management (IPM) have been developed as research tools (R) and in some cases are used for implementing (I) plant protection practices (Table 1). They include two apple tree models (R), phenology or timing models (R, I) for the codling moth, oriental fruit moth, red-banded leafroller, tufted apple bud moth, apple maggot, tentiform leafminer, white apple leafhopper and San Jose scale and a population dynamics model (R) of the codling moth. Four predator-prey models (R, I) of plant-feeding mites and their natural enemies also have been developed.

Several of these models are used for decision making in the field and are implemented via charts, nomograms and in some cases by computer-based processor units which operate as stand-alone devices. These models in some cases may be accessed by computer-based, extension plant-protection delivery systems. In this paper, several of these species or

multispecies simulation models are discussed. Emphasis is on verbal rather than detailed mathematical descriptions of each model.

Models of perennial crops or plants are rare in existing horticultural or agronomic literature. For apple, several components of tree physiology and growth have been modeled including canopy photosynthesis, canopy structure, allocation of photosynthate, shoot growth, tree water balance and flowering phenology; reviewed in Elfving et al. [1979]. Only 2 attempts have been made at modeling total tree growth [Lansberg et al. 1975, Elfving et al. 1978]. The Lansberg model was designed to describe apple production in England relative to weather and tree growth; research on this program has been discontinued recently. The model of Elfving and co-workers was specifically designed to integrate species models of pests and their feeding effects on fruit tree physiology and growth. This model describes the whole tree processes of assimilate production, utilization and storage, as well as the growth of the various organ systems of the tree (i.e., all leaves, extension shoot biomass, stem biomass, root

[1]Published as Journal Article *9050* of the Michigan Agricultural Experiment Station, East Lansing, MI 48824

Table 1.   Plant and arthropod simulation models developed for apple plant protection.

Species		Model Classification	Deterministic/ Stochastic	Discrete/ Continuous	Research/ Implementation	Reference
Malus	(1)	Whole Tree	D		R	Landsberg et al. 1977
	(2)	Whole Tree	D	D	R	Elfving et al. 1979
Laspeyresia pomonella	(1)	population-dynamics	D	C	R	Berryman et al. 1973
						Brown et al. 1978
	(2)	developmental	D	C	R,I	Welch et al. 1978
	(3)	developmental	D	D	R,I	Falcon et al. 1976
Grapholitha molesta		developmental	D	C	R,I	Welch et al. 1978
						Croft et al. 1980
Argyrotaenia velutinana		developmental	D	C	R,I	Welch et al. 1978
Playtnota idaeusalis		developmental	D	C	R,I	Welch et al. 1978
Rhagoletis pomonella		developmental	D	C	R,I	Welch et al. 1978
Typhlocyba pomaria		developmental	D	C	R,I	Michels, unpublished
Quadraspidiotus perniciosus		developmental	D	C	R,I	Michels, unpublished
Lithocoletus blancardella		developmental	D	C	R,I	Michels, unpublished
Panonychus ulmi - Amblyseius fallacis		predator-prey	D,S	D	R,I	Dover et al. 1979
Panonychus ulmi - Stethorus punctum		predator-prey	D	D	R,I	Mowry et al. 1975
Panonychus ulmi - Amblyseius potentillae		predator-prey	D	D	R	Rabbinge 1976
Tetranychus mcdanieli - Typhlodromus occidentalis		predator-prey	D	D	R	Logan 1977
						Wollkind and Logan 1978

biomass, fruit biomass, stored reserve carbohydrate). Growth is expressed in terms of dry weight changes. Linkage of the tree to the environment via weather data provides the external means of driving the tree model. The model simulates growth for trees having a variety of tree sizes and canopy shapes. As yet, an internal hormonal control component for regulating tree growth is not included in the model. At present, the greatest use of the model relative to plant protection has been in more precise definition of economic damage thresholds for foliage feeding pests of apple (e.g., mites, aphids, leafhoppers, leaf rollers). As this model is further developed and refined, it will greatly assist researchers in defining causal interactions between pests and fruit tree physiology and growth more precisely [see identification of more specific areas in Croft et al. 1979b].

Phenology or developmental models for some 8 pests species associated with apple in the midwestern and northeastern regions of North America (Table 1) have been developed in connection with a computer based, phenology or timing modeling system under the acronym of PETE [*Predictive Extension Timing Estimator* system; Welch at al. 1978]. PETE employs a generalized, continuous time model based on the - *K*-th order distributed delay process reported by Manetsch and Park [1974] for each species. Species models are constructed with a minimum set of biological parameters including developmental rates and temperature thresholds, initial maturity distributions, oviposition functions and population variance components for generation to generation development. Models using this approach are easily

developed by research personnel without extensive computer programming background. These timing models are then used by extension personnel to schedule plant protection activities such as sampling, spraying, parasite release etc. The PETE system has recently been proposed as a model for a national apple pest forecasting system in the United States [Gilpatrick and Croft 1978]. PETE models for the major species of the pest complex attacking this crop are being developed at various institutions and validation on a national scale is underway for several of these models developed to date (e.g., codling moth, Oriental fruit moth).

In Figure 1, validation comparisons between the simulated development of the Oriental fruit moth using the PETE system vrs trap catch data taken from pheromone and bait traps during the period 1972-78 in California peach orchards are given [Croft et al. 1980]. As can be seen, model outputs and actual developmental data are in reasonable agreement. When appropriate biofix (biological fixpoint indicator) monitoring assessments are taken from the actual population (e.g., 1st pheromone trap catch of males) and are then used to synchronize these models, rather accurate predictions of pest development are possible. As independent phenology model for predicting codling moth development under California conditions has been developed and preliminarily validated by Falcon et al. [1976].

Whereas the codling moth models discussed previously are mostly designed to provide short-term, within season estimates of the temporal biology of this species, researchers in the western

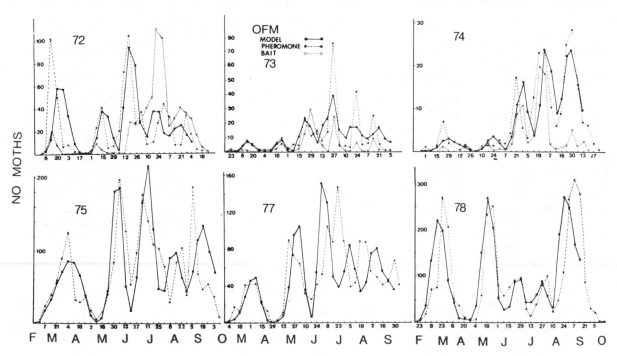

United States have developed a series of population dynamics models designed to evaluate control strategies and tactics for this pest (i.e., sterile male application, insecticidal control, pheromone control) when applied over a period of 10 yrs or more [Berryman 1973, Brown et al. 1978]. The most recent version is a state dynamics model using transition equations arranged in a modified Leslie matrix. Outputs from this model (CODMOTH) have been validated against field data and show acceptable agreement with population behavior over at least a 5 year period. Model robustness or its capacity to compensate for errors in parameter estimates is indicated by the fact that sensitivity analysis studies have shown only a 10% error in model output when varying parameter values as much as 45% from the true value. The significance of various density-dependent components associated with each life stage of the codling moth have been examined. Feedback processes were found to be most important relative to fecundity of adults and 1st instar larval survival. Larval survival within apples (i.e., beyond 1st instar) and survival to cacooning sites were intermediately affected by density dependent feedback processes while sex ratio and voltinism effects were least affected by these type of factors.

Four prey-predator models for tetranychid mites and their natural enemies have been developed and preliminarily validated [reviewed in Welch 1979]. They include the species *Panonychus ulmi: Amblyseius fallacis* in Michigan, [Dover et al. 1979], *P. ulmi: Stethorus punctum* in Pennsylvania [Mowry et al. 1975], *Tetranychus mcdanieli: Typhlodromus occidentalis* in Washington state [Logan 1978, Wollkind and Logan 1978] and *P. ulmi: Amblyseius potentillae* in the Netherlands [Rabbinge 1976]. Each of these models has a similar structure including features of development, consumption, oviposition and mortality, but varies considerably in relation to the level of modeling detail and the emphasis given to validation and use. The Washington and Dutch models are primarily research tools, while the Pennsylvania and Michigan models are used to assess the potential for biological control in grower orchards and the need to readjust predator:prey ratios in favor of the predator so that effective natural control is achieved thereafter.

Each of these predator-prey mite models have brought increased understanding of how to develop and use these type of models for IPM. The model of Dover et al. [1979] is unique in showing the need to include spatial variations between leaf and tree components in the simulation before one can reasonably forecast the temporal and numerical dynamics of these mites in the orchard. The work of Mowry et al. [1975] demonstrates that excessively detail biological data are not always necessary to establish models which can be functionally useful under field conditions. The model of Logan [1978], Wollkind and Logan [1978] emphasizes the influence of temperature dependent rate phenomenon in explaining such basic processes as reproduction, survival, predation etc. Lastly, the work of Rabbinge [1976] by use of a micro-weather submodel showed that for these species at least, microhabitat weather differences are less important than small hypothetical changes in biological parameters (e.g., oviposition rates, survival) in describing biological control results. He also clearly points out the need to carry out model validation at each level of spatial complexity in the crop ecosystem (e.g., leaf branch, tree, orchard) in a systematic manner.

Implementation of computer models in crop protection systems including apple is being investigated in several of the major agricultural states in the USA [reviewed in Croft et al. 1979a]. For apple systems, scientists have developed a pest management executive system (PMEX) housed in a large main-frame computer for accomplishing several IPM implementation delivery tasks including communications, information processing, environmental and biological monitoring and model applications [Croft et al. 1976]. Similar approaches using a smaller processor are also under evaluation [Lellevik et al. 1977, Jones et al. 1980] for use in regional centers or for in-field management. These processors can function as stand-alone units or may be linked to other computers. In this regard, most recently an attempt to implement timing aspects of apple plant protection via a hierarchial-distributed processor based system has been initiated [Croft et al. 1979a]. Prototype feasibility and economic evaluations of such a system are currently being conducted with 4 linked processors located at statewide, regional, area and on-farm sites and the PETE extension timing system model described previously to better time IPM practices at each of these levels of plant protection operation.

References Cited

Berryman, A. A., T. P. Bogyo and L. C. Dickman. 1973. Computer models and the application of the sterile male technique. IAEA Rept. Vienna Aust. (1973):31-43.

Brown, C. M., A. A. Berryman and T. P. Bogyo. 1978. Simulating codling moth population dynamics: model development, validation and sensitivity. Environ. Entomol. 7:219-227.

Croft, B. A., J. L. Howes and S. M. Welch. 1976. A computer-based, extension pest management delivery system. Environ. Entomol. 2:20-34.

Croft, B. A., S. M. Welch, D. J. Miller and M. L. Marino. 1979a.

Developments in computer-based, IPM extension delivery and biological monitoring systems design. p. 223-50 Cpt. 12 in Pest Management of Tree Fruits and Nut Crops. Plenum Press. 256 pp.

Croft, B. A., D. C. Elfving and J. A. Flore. 1979b. Pest management and fruit tree physiology and growth. Cpt. 3 In Integrated Pest Management of Pests of Pome and Stone Fruits Eds. B. A. Croft and S. C. Hoyt. Wiley Intersci. (in press).

Croft, B. A., M. F. Michels and R. E. Rice. 1980. Validation of a PETE timing model for the Oriental fruit moth in Michigan and California. Environ. Entomol. (in press).

Dover, M. J., B. A. Croft, S. M. Welch and R. L. Tummala. 1979. Biological control of Panonychus ulmi (Acarina:Tetronychidae) by Amblyseius fallacis (Acarion:Phytoseiidae) on apple: A prey:predator model. Environ. Entomol. 8:

Elfving, D. C., R. Bagley, P. Elliott, S. Friday, T. Oren and J. Smith. 1978. A computer simulation of apple tree growth. Mich. St. Univ. Syst. Sci. Rept. 62 pp.

Falcon, L. A., C. Pickel and J. White. 1976. Computerizing the codling moth. Furit Grower. 96:8-14.

Gilpatrick, J. D. and B. A. Croft. 1978. A national program for the development of a comprehensive, unified, economical and environmentally sound system of a integrated pest management: Apple subproject. EPA/USDA Grant Proposal RF-78-222. 48 pp.

Jones, A. L., S. L. Lillevick, P. D. Fisher and T. C. Stebbins. 1980. Development and validation of a microcomputer-based instrument to predict primary apple scab infection periods. Plant Disease. (in press).

Landsberg, J. J., M. R. Thorpe, R. L. Watson, W. M. Huxley and G. Heion. 1977. Model of apple tree growth. Long Ashton Res. Sta. Ann. Rept. 1976:117-119.

Lellevik, S. L., A. L. Jones, and P. D. Fisher. 1977. A predictive CMOS-based instrument for agriculture. p. 275-86 In Micro-Computer Design and Application ed. S. L. Lee. Academic Press, NY.

Logan, J. A. 1977. Population model of the association of Tetranychus mcdanieli (Acarina:Tetranychidae) with Metaseiulus occidentalis (Acarina:Phytoseiidae) in the apple ecosystem. Ph.D. Diss. Wash. St. Univ. 107 pp.

Manetsch, T. J. and G. L. Park. 1974. Systems analysis and simulation with applications to economic and social systems. Part II Publ. Dept. Elect. Eng. and Syst. Sci. Mich. St. Univ. 214 pp.

Mowry, Paul D., Dean Asquith and William M. Bode. 1975. Computer simulation for predicting the number of stethorus punctum needed to control the European red mite in Pennsylvania apple trees. J. Econ. Ent. 68:250-254.

Rabbinge, R. 1976. Biological control of the fruit tree red spider mite. Centre Ag. Pub. Doc., Wageningen, The Netherlands. 288 pp.

Welch, S. M., B. A. Croft, J. F. Brunner and M. F. Michels. 1978. PETE: an extension phenology modeling system for management of a multispecies pest complex. Environ. Entomol. 7:482-94.

Welch, S. M. 1979. Application of simulation models to mite pest management. Proc. V. Intern. Cong. Acarology. 5:(in press).

Wollkind, D. J. and J. A. Logan. 1978. Temperature-dependent predator-prey mite ecosystem on apple tree foliage. J. Math. Biol. 6:265-83.

36.10 Forecasting and Modeling Pests in Apple Orchard Ecosystems: Diseases

R.C. Seem and A.L. Jones

Department of Plant Pathology, New York State Agricultural Experiment Station
Cornell University, Geneva 14456, and
Department of Botany and Plant Pathology Michigan State Unversity, East Lansing, 48823, respectively

Our entomologist colleagues have aptly demonstrated the uses and benefits of forecast and model systems in the management of insect and mite pests. Obviously, the general model systems of the plant pathogens are similar to those for insects and mites but because plant pathogens are made of simpler life forms than insects and mites, their behavior is more closely associated with their immediate environment. As a result, pathogen forecasts and models must utilize more environmental parameters, and this added complexity has forced most models to describe only small portions of the disease system. A simple "infection period" model may utilize temperature and hours of wetting to determine if infection has occurred, yet the model will not address such important components as amount of available inoculum, or the amount of susceptible host tissue. Nevertheless, plant disease models play a significant role in the management of the orchard agro-ecosystem.

It is our intent to describe the types of forecast and model systems currently being used to manage orchard diseases. In doing so, we will also describe new modeling endeavors and explain how they contribute to a total management system.

The complex nature of most plant diseases dictates a well structured management system. The basic system is often illustrated as having a monitoring component, a decision making component and an action component all linked together with a delivery system. General disease models work well in this system by aiding in the decision process. Models designed to determine when conditions are favorable for infection are the most common. However, diseases create two serious problems in the monitoring and delivery components of the system. First, pathogens are very difficult to monitor. Macroscopic observations of the disease only tells us of pathogen activity initiated days or weeks earlier. Other techniques, such as direct spore counting have been used but require both skill and time. Time is the second problem. As we said, most pathogens respond directly to their immediate environment. A rapid change in environmental conditions usually causes an equally rapid response by the pathogen and thus reduces the effective delivery time of the management system. To lessen these problems, models now play an increasingly important role in all components of the management system.

Infection Models

W.D. Mills [6] must be considered one of the first orchard ecosystem disease modelers. His intuitive model of temperature and moisture conditions that affect the occurrence and intensity of infection by the apple scab fungus (*Venturia inaequalis*) has served as a world-wide infection model for more than 35 years.

Recent adaptations of this sort of infection model by Pearson and coworkers [1, 7] have been developed for basidiospore infection of the cedar-apple rust fungus (*Gymnosporangium juniperi-virginianae*). Several refinements have been made to the general infection model. Prior to infection of apple, the fungus must germinate from its resting spore (teliospore) to produce the basidiospore which in turn will germinate to infect the apple. Pearson et al. [7] have added a teliospore germination model and basidiospore formation model to the infection model (which includes basidiospore germination) and thus have developed a forecast model which can account for conditions unexplained by a simple infection model. For example, the simple infection model might forecast infection at 4 °C, but the addition of the teliospore germination model would nullify the infection prediction because the teliospore will not germinate at this temperature.

Infection models have great potential in pest management systems as shown by the longevity and widespread application of the Mills model. We must, however, point out several limitations of these models: (1) They are effective only if after-infection control strategies are available (usually eradicant fungicides). (2) They generally are built for constant temperature and continuous wetting conditions and therefore, do not account for wide temperature fluctuations or so called "split wetting periods". (3) They do not allow much time for an action program to be implemented (often a matter of hours).

Inoculum Models

As mentioned earlier, monitoring the microscopic pathogen is a substantial task, particularly in the time required by most action programs. Models can play a significant role in disease management systems by determining either when full scale monitoring should be initiated and terminated, or how frequently monitor samples should be taken. Massie and Szkolnik [5] used a regression model based on accumulated heat degree units and precipitation to forecast the level of mature ascospores (*V. inaequalis*) available for release and subsequent infection of apple. This system is ideally suited as a trigger of actual ascospore monitoring. Fungi with similar modes of overwintering such as the causal agents of black knot of plum (*Dibotryon morbosum*) or cherry leaf spot (*Coccomyces hiemalis*), are likewise amenable to ascospore maturity models.

Eventually models will not only aid inoculum monitoring, but will help us forecast seasonal intensities of inoculum by predicting the spore load based on level of intensity, numbers available and environmental conditions associated with each release period. Pearson and coworkers have prepared such a model for cedar apple rust using hours RH 85%, and initial temperature, relative humidity and rainfall ($R^2 = 69\%$). Similar models can be developed for apple scab ascospores and conidia as well as powdery mildew (*Podosphaera leucotricha*) conidia since much of the necessary data has already been collected [9, 10]. Models can also be developed to equate inoculum level to actual disease intensity.

Inoculum models are important because with them, decisions can be made prior to a disease event, thus allowing true disease forecasts and full utilization of management strategies.

Nonpest Models

This is an exclusive category of ecosystem models directly associated with the pest but not categorized as such. They often have broader discipline interest and thus, may be touched upon in previous or subsequent papers. There are two non-pest models of direct interest to us.

One is the apple phenology model under continuing development by Szkolnik and Seem [8]. It is a growth model designed to simulate the first 90 days of seasonal growth from the break of dormancy. The model passes through a detailed set of phenophases for both floral and vegatative development. It was designed to provide a distribution of phenophases at any given time based on environmental parameters accumulated from an initial trigger (usually break of dormancy). The model is used to determine the types and age distribution of apple tissues susceptible to various diseases. It also has the capability of incorporating cultivar differences in phenological development and tissue susceptibility. A model for predicting leaf emergence of sour cherry from degree day accumulation has been developed in Michigan [2]. It predicts the number of leaves on fruiting spurs and terminal shoots, and has been coupled with a leaf expansion model to predict leaf area.

The second nonpest model is a pesticide degradation and redistribution model. This model is under early stages of development by workers in New York. Pest control strategies utilizing repeated applications of pesticide may cause excessive and wasteful residue build up which could be avoided by the use of a pesticide residue model. Using weather conditions, leaf or fruit surface characteristics and pesticide attributes as parameters, the model will determine effective surface concentration of the chemical by calculating degradative decline due to washoff, abrassive removal, volatilization, photolysis, hygrolysis, biolysis (including pest uptake), and surface binding as well as redistributive decline due to host growth. The pesticide residue model comes closer to a true simulation model than any previously described model because we are attempting to account for all known factors that affect residue. However, it remains to be seen if the utility of a residue prediction can outweigh the difficulty in implementing such a complex model in the field.

Model Implementation

We emphasized earlier that effective use of models required a rapid delivery system. The most efficient system is one in which the grower does his own monitoring, decision making and control action (as he has usually done in the past). The development and implementation of models, usually meant the concurrent development of a large-scale delivery system. However, a microprocessor-based delivery system has been developed by Fisher and coworkers [3,4] which is ideally suited for disease models that require localized weather data. The programmable device frequently monitors several environmental parameters (temperature, humidity, leaf wetness, and rainfall) and uses these data to determine if an apple scab infection period has started. If a period has started, the device begins to store pertinent information and forecasts the hours remaining until a low level of infection will be reached. After a low, moderate, or severe level of infection has occurred, the instrument forecasts the number of hours remaining to achieve eradicant control by each of several different fungicides.

The infection period forecasts are based on a modified Mills model. Periods favorable for infection to occur but not lasting long enough for a low level of infection, can be extended when leaves are dry if the humidity remains above 90% or if the period when humidity is $< 90\%$ does not exceed 8 hours. Jones et al [3] have reported on the ability of this device to reduce the number of sprays normally applied to control scab. We are continuing field studies to "fine tune" both the device and the model. It is expected that other models can be added to the instrument as it will be able to process several models and develop forecast simultaneously.

Summary

Forecast and model systems used in the management of orchard diseases have usually been designed to address one particular aspect of the disease (inoculum or infection). This approach seems to work well given our current level of understanding of the disease systems, in spite of the fact there are still areas where we lack information necessary to make present models more meaningful. Significant advances can also be made in the areas of inoculum models to aid the monitoring component of our management systems as well as non-disease models to aid in a better understanding and therefore, management of the total orchard ecosystem

Literature Cited

1. Aldwinckle, H. S., R. C. Pearson, and R. C. Seem. 1980. Infection periods of *Gymnosporangium juniperi-virginianae* on apple. *Phytopathology* 70:1070-1073.
2. Eisensmith, S. P., and J. A. Flore. 1980. Predicting leaf emergence of sour cherry (*Prunus cerasus* L. 'Montmorency') from degree day accumulation. *J. Am. Soc. Hort. Sci.* 105: (in press).
3. Jones, A. L., S. L. Lillevik, P. D. Fisher, and T. C. Stebbins. 1980. A microcomputer-based instrument to predict primary apple scab infection periods. *Plant Dis.* 64:69-72.
4. Lillevik, S. L., P. D. Fisher, and A. L. Jones. 1977. A predictive CMOS-based instrument for agriculture. Pages 275-286 in S. C. Lee, (ed). *Microcomputer Design and Applications.* Academic Press, New York.
5. Massie, L. B. and M. Szkolnik. 1974. Prediction of ascospore maturity of *Venturia inaequalis* utilizing cummulative degree-days. *Proc. Amer. Phytopathol. Soc.* 1:140 (Abstr.).

6. Mills, W. D. 1944. Efficient use of sulfur dusts and sprays during rain to control apple scab. N.Y. Agric. Exp. Stn. (Ithaca) Ext. Bull. 630 4 pp.

7. Pearson, R. C., H. S. Aldwinckle, and R. C. Seem. 1977. Teliospore germination and basidiospore formation in *Gymnosporangium juniperi-virginiae*: a regression model of temperature and time effects. *Can. J. Bot.* 55:2832-2837.

8. Seem, R. C., and M. Szkolnik. 1978. Phenological develop-

ment of apple treeş. Vermont Agric. Exp. Stn. Bull. 684:16-20.

9. Sutton, T. B., and A. L. Jones. 1979. Analysis of factors affecting dispersal of *Podosphaera leucotricha* conidia. *Phytopathology* 69:380-383.

10. Sutton, T. B., A. L. Jones, and L. A. Nelson. 1976. Factors affecting dispersal of conidia of the apple scab fungus. *Phytopathology* 66:1313-1317.

36.11 The Granulosis Virus of Codling Moth: Its Use as a Biological Agent in an Integrated Control Program

Dr. Erich Dickler

Biologische Bundesanstalt fuer Land- und Forstwirtschaft,
Institut fuer Pflanzenschutz im Obstbau,
Postfach 73, 6901 Dossenheim ueber Heidelberg, Federal Republic of Germany.

Abstract

In commercial apple orchards in Southern Germany codling moth is the most serious fruit pest and the limiting factor in integrated control programs. The efficacy of a granulosis virus against codling moth was tested in the field over five growing seasons at Dossenheim, Germany. Granulosis virus sprayed at a concentration of ca. 10^{11} capsules/liter was as effective or even more effective than organophosphorus insecticides. The virus used is highly virulent and in an apple biocoenosis species specific to the codling moth. A report will be given about a long term project started at Dossenheim in order to study the effect of codling moth granulosis virus on the apple fauna, on beneficial arthropods as well as on other pest species. Omitting organophosphorus insecticides and using granulosis virus against codling moth led to an increase of fruit injuries by leafrollers, whereas the density of winter eggs of the European red mite was decreased. Field trials were started to develop an integrated control program combining the control of tortricids by using different species specific viruses.

The codling moth (CM), *Laspeyresia pomonella*, is a serious apple pest in Southern Germany. The larvae cause fruit loss of 10% or more each year where no chemical insecticides are sprayed. In the sunny and hot summer of 1976, in unsprayed orchards, damage reached 80%. In the upper Rhine valley, where CM is bivoltine, up to 4- applications each season are necessary to protect the apples against *L. pomonella*. Most farmers use broad-spectrum organophosphorus insecticides which have undesirable effects on beneficial and other innocuous arthropods. Therefore CM is the limiting factor in integrated control programs, and for this reason the development of a specific CM insecticide which is nontoxic to beneficial and our nontarget arthropods is desirable and has much potential.

Since the discovery of a codling moth granulosis virus (CMGV) by Tanada [1964] orchard tests elsewhere [Falcon et al., 1968; Morris, 1972; and Keller, 1973] have indicated that this virus may be a useful microbiological control agent against L. pomonella. In these first field trials GV was found to be highly virulent and very specific to the CM. Until now apart from L. Pomonella, only 2 other species, the Oriental fruit moth, Grapholitha molesta, and the European pine shoot moth, Rhyacionia buoliana, are known to be susceptible to CMGV.

In 1974 field experiments were conducted at Dossenheim Germany to study the efficiency of the GV against CM in comparison with organophosphorus insecticides, and to test the potential of GV as an applied agent for control of L. pomonella.

Materials and methods

The virus was propagated in vivo on CM larvae by Dr. Huber at the Institut fuer biologische Schaedlingsbekaempfung of the Biologische Bundesanstalt, Darmstadt.

Most of the field experiments were carried out in apple orchards of the experimental field of the Institut fuer Pflanzenschutz in Obstbau of the Biologische Bundesanstalt, Dossenheim and a few field trials were accomplished in commercial apple orchards in the Rhine Valley. Materials and methods, the study orchards, application technique, and virus propagation, are described in previous papers [Dickler, 1977; Huber and Dickler, 1975, 1977], and only a brief description will be given here. The main study orchard on the experimental field is 11 years old and consists of 25 rows with 25 small trees on rootstock M 9. Each row contains 4 times 5 cultivars. The results given in this paper are limited to the cultivar 'Golden Delicious' on plots with green covered soil.

Results and Discussion

The first year of experiments was extremely cool and wet and 4 GV applications were as effective against CM damage as the same number of treatments with chemical insecticides (Table 1). 8 GV sprays resulted only in a slightly higher damage reduction which was nonsignificant. In the following 2 years, when summers were sunny and hot the 1974 results could be proved and a similar good protection of the fruit against CM damage was found in all the years of experiments with different weather conditions. Throughout the experiment, corrugated cardboard bands were placed around the trunks of the trees to collect fifth- instar larvae.

As shown by the results of the trap bands (table 2) the total reduction of CM population by GV was much higher than the reduction of fruit damage, reaching 100% in 1975 [Huber and Dickler, 1977].

In other experiments carried out at Dossenheim a better efficiency of the GV could be observed over green covered soil compared to orchards with clean cultivated soil [Dickler, 1977; Huber and Dickler, 1977]. This effect might be partly due to beneficial arthropods which are not killed by GV and which probably find more suitable conditions in the green cover.

In 1976 three different virus concentrations of 10^9, 10^{10}, and 10^{11} capsules/liters were sprayed in order to test the possibility of reducing the concentration of the virus suspension in the field [Huber and Dickler, 1978]. A GV concentration of 10^{11} capsules/liter was most effective and in some trials better than chemical insecticides.

Even in orchards with high fruit damage above 70%, GV treatment led to a damage reduction of more than 80% (Huber and Wundermann 1978). In the same heavily infested orchard, the CM damage was differentiated in "deep entries" and "stings" [s. Jaques et al., 1977]. No significant difference could be found in apples with "stings only" between GV, chemical, and untreated trees. Thus, under the conditions of the Upper Rhine Valley, virus treatment didn't lead to an increase of stings done by larvae which died after having already injured the fruit.

In all field trials the activity of the GV deposit was bioasseyed in the laboratory [Huber and Dickler, 1977].

In several other european countries where GV has given very good selective control of the CM, similar results were obtained [Charmillot, 1978; Fischer-Colbrie, 1977].

In 1975 a long-term project* was started in the orchard of the experiment field in Dossenheim, as described before, to study the effect of CMGV on the

Table 1. Effect of granulosis virus and chemical insecticides on codling moth infestation.

Treatment	Total of apples	Apples infested by codling moth (%)	Reduction by treatment (%)
Results 1974			
virus 8x	1601	0,62a	85,9
virus 4x	3466	0,75a	82,9
insecticides 4x	3121	0,74a	83,2
untreated	684	4,39b	–
Results 1975			
virus 4x	2218	0,72a	89,0
insecticides 4x	2962	1,79b	72,8
untreated	3403	6,58c	–

a Figures from the same year followed by the same letter do not differ significantly at the 5% level of probability (contingency table χ^2 test).

Table 2. Number of diapausing codling moth larvae collected in trap bands on virus-, insecticide-, and untreated trees.

Treatment	Diapausing larvae in trap bands No. collected	per 1000 apples	Reduction by treatment (%)
Results 1974			
virus	4	0,35	97,7
untreated	45	15,27	–
Restults 1975			
virus	0	0	100
insecticides	13	2,31	85,1
untreated	95	15,54	–

Table 3. Leafroller infestation on apples from an orchard treated with granulosis virus and chemical insecticides against codling moth,

Treatment	Results 1977 Total of apples	Apples infested by leafrollers (%)	Results 1978 Total of apples	Apples infested by leafrollers (%)
virus	2912	23,6a	435	17,2a
insecticides	3025	2,8b	593	1,2b
untreated	2173	22,7a	146	13,7a

a see footnote table 1

arthropod fauna, especially on beneficial arthropods and other major apple pests.

As indicated in table 3, omitting organophosphorus insecticides and using the host specific granulosis virus led to an increase of fruit injuries by leafrollers, whereas the density of winter eggs of the European red mite was decreased. In the 4 years of experiments significantly lower leafroller damage was found in the trees sprayed with chemical insecticides as against those treated with GV and the untreated check.

Several leafroller species, *Adoxophyes reticulana, Archips podana, Pandemis heparana* and others, cause the fruit damage, but the summer fruit tortrix seems to be the species responsible for the main damage done.

Baculoviruses for some of these tortricids are available and it will be the objective of our further research work to develop an integrated program which combines the control of tortricids by using different microbiological methods especially species specific viruses.

Similar pilot trials under different ecological conditions were started in England, France, and the Netherlands. These research activities are coordinated in an "EC-Workshop on integrated and biological control in apple orchards" and are partly supported by EC-funds.

*This research is supported in part by the German Federal Ministry of Research and Technology (BMFT).

36.12   The Current Status of Codling Moth Granulosis Virus in the USA

Louis A. Falcon

Department of Entomology-Parasitology, 333 Hilgard Hall
University of California, Berkeley, California 94720

The granulosis virus of the codling moth was first isolated by Gorden Marsh in 1963 in the then Division of Invertebrate Pathology, University of California, Berkeley. The virus was obtained from diapaused codling moth larvae collected by L. E. Caltagirone on apple or pear trees near Valle de Allende, Chihuahua, Mexico, September 12, 1963 [Tanada, 1964].

Subsequent to the Berkeley isolation, granulosis virus was isolated from codling moth cultures maintained at the USDA laboratory, Yakima, Washington [S. R. Dutky correspondance to B. A. Butt dated March 16, 1964]. Later isolations included: Entomology Laboratory, Summerland, British Columbia, 1966; Colorado State University, Orchard Mesa Research Center, Grand Junction, 1968; USSR 1972, and Hungary 1975.

Electron microscope studies of the virus showed that the average size of the outer capsule (=granule) is 394 × 208 millimicrons and for the inner single rod 314 × 51 mu [Tanada 1964]. The virus is very pathogenic for codling moth larvae. Only 4-5 capsules *per os* larva gives an LD_{50} for first instar [Falcon, 1971; Sheppard and Stairs, 1977], and about 49 per larva produces the same results in 5th instar [Sheppard and Stairs, 1977]. It is a polyorganotropic disease and produces pathologies in the fat body, hypodermis, tracheal matrix and malpighian tubules [Tanada and Leutenegger 1968].

The first field studies with the granulosis virus were conducted in California beginning in 1966 [Falcon, Kane and Bethell, 1968]. The granulosis virus was applied in water to apple using a sprayer equipped with two handguns. The field dosages tested were: 3.4×10^{11} granulosis virus capsules per gallon of finished spray in 1966; 1.2×10^{11} in 1967; and 9.5×10^9 for 1968. The dosages were similarly effective 0 days after application with the higher dosages providing longer residual effectiveness. Repeated spray applications at 7-day intervals in 1968 provided the greatest degree of control for the lowest cost. Residual effectiveness was extended markedly with the addition of Bio-film®, crude molasses and skim milk to the virus-water mixture in the spray tank.

The granulosis virus caused heavy larval mortality in the field population of codling moth. The majority of larvae perished soon after feeding on the epidermis of granulosis virus treated fruits (80%). Of the 20% or so that entered the fruit about 3/4ths died within.

601

The remainder succumbed during the prepupal or pupal stages in cocooning sites. Overall population reductions averaged 98%.

Similar results were obtained by other workers including D. S. Morris in Australia [1969, 1972]; S. Keller, Switzerland [1973], R. E. Sheppard and G. R. Stairs, USA [1976]; J. Huber and E. Dickler, Federal Republic of Germany [1975, 76, 77, 78] and R. P. Jaques et al., Canada [1977].

In 1976 an Experimental Use Permit for granulosis virus against codling moth was issued by the Federal Health Office of the Federal Republic of Germany to the Federal Biological Facility for Agriculture and Forestry.

In the USA a few research workers and one industrial firm are interested and working to develop a registration for the granulosis virus of codling moth. The major objectives are: (1) establish a central production facility to provide experimental quantities of a standardized granulosis virus preparation; (2) distribute the experimental preparation to interested workers for field testing; (3) develop and refine the virus to provide consistent, effective and economical control of codling moth; (4) compare results and establish use patterns; (5) register the granulosis virus with EPA and develop commercial products; and (6) establish the use of the granulosis virus for control of codling moth in apple, pear and walnut production.

The potential for the granulosis virus of codling moth is great. It has been shown to be highly infective for codling moth and can reduce population levels as effectively as chemical insecticides used for control of this pest [Dickler and Huber]. The virus appears to be specific for codling moth and other closely related genera (e.g. it infects oriental fruit moth and pine shoot moth larvae). It can be produced relatively easily in codling moth larvae. The virus can be formulated and stored until needed and be applied with conventional spray equipment.

The granulosis virus of codling moth is the first biological control agent that totally, on its own, can effectively reduce codling moth populations below economic threshold levels. The industrialization and commercialization of this virus could revolutionize orchard pest management for the future in the same way DDT did in the middle of this century.

Literature Cited

Falcon, L. A. 1971. Microbial control as a tool in integrated control programs. P. 346-64. *In* C. B. Huffaker (ed.), Biological Control. Plenum Press, N.Y./London. 511 pp.

Falcon, L. A., W. R. Kane and R. S. Bethell. 1968. Preliminary evaluation of a granulosis virus for control of codling moth. J. J. Econ. Entomol. *61*, 1208-13.

Huber, J. E. and E. Dickler. 1975. Freilandversuche zur Bekämpfung des Apfelwicklers, *Laspeyresia pomonella* (L.), mit granuloseviren. Z. Planzen-Krankh. Pflanzensch. *82*, 540-546.

Huber, J. and E. Dickler. 1976. Das granulosevirus des Apfelwicklers: seine Erpobung als biologisches schädlingsbekämpfungsmitel. Z. Angew. Entomol. *82*, 143-147.

Keller, S. 1973. Mikrobiologische Bekämpfung des Apfelwicklers (*Laspeyresia pomonella* (L.)) (= *Carpocapsa pomonella*) mit spezifischem Granulosisvirus. Zeitschrift fur Angewandte Entomologie. *73*, 137-181.

Morris, D. S. 1972. A cooperative programme of research into the management of pomefruit pests in southeastern Australia. III. Evaluation of a nuclear granulosis virus for control of codling moth. Abstr. 145h Int. Congr. Entomol. Aug. 22-30, Canberra, 238 pp.

Morris, D. S. and R. van Baer. 1969. Codling moth granulosis virus. Victorial Plant Research Institute Report No. 5, 27.

Sheppard, R. F. and G. R. Stairs. 1976. Effects of dissemination of low dosage levels of a granulosis virus in populations of the codling moth. J. Econ. Entomol. *69*, 583-86.

Sheppard, R. F. and G. R. Stairs. 1977. Dosage-mortality and time-mortality studies of a granulosis virus in a laboratory strain of the codling moth, *Laspeyresia pomonella*. J. Invertebr. Pathol. *29*, 216-221.

Tanada, Y. 1964. A granulosis virus of codling moth, *Carpocapsa pomonella* (Linnaeus) (Olethreutidae, Lepidoptera). J. Insect Pathol. *6*, 378-380.

Tanada, Y. and R. Leutenegger. 1968. Histopathology of a granulosis-virus disease of the codling moth, *Carpocapsa pomonella*. J. Invertebr. Pathol. *10*, 39-47.

36.13   Nematodes and the Replant Problem in Fruits

Eldon I. Zehr

Department of Plant Pathology and Physiology
Clemson University, Clemson, SC 29631

"The replant problem" for fruits implies any difficulty that is encountered when orchard or vineyard sites are replanted with the same or another fruit crop. Causes of such problems are numerous—such as root rots caused by various fungi, crown gall caused by *Agrobacterium tumefaciens,* or stem-pitting of peach caused by tomato ring spot virus. However, discussions of "the replant problem" of fruits in the United States usually refer to a disease complex that is characterized by poor growth of trees, reduced yields, and sometimes premature death resulting from cold injury of other factors. This particular replant problem is the subject of this discussion. At least two excellent reviews of this subject have been published [4, 14]. Factors in addition to nematodes usually are involved in this problem, but in this limited time frame I shall confine my remarks to the role of nematodes in the replant problem.

Generally, two types of replant problems are recognized. Specific replant problems are those in which difficulties in replanting orchard sites appear only when the crop being planted is of the same or closely related species as that which preceded it. Best known of specific replant problems is the "specific apple replant disease" (Sard), which is common in Great Britain and other European countries [4, 14]. Nonspecific replant problems are those which are not confined to the plant species which preceded in the replant situation. Cherries may suffer from the disease syndrome when following apples, or pears may be affected when following peaches in replanted orchards.

Some research has implicated nematodes in both specific and nonspecific replant problems [4, 12, 14]. However, the most recent research information indicates that nematodes probably are not the primary cause of specific replant disorders. Fungi, such as *Thielaviopsis basicola* in cherry [16] and *Pythium sylvaticum* in apple [15] have been implicated as primary pathogens in specific replant diseases. Plant-parasitic nematodes also are not always associated with specific replant problems [4]. Therefore, we probably are talking about the role of nematodes in *nonspecific* replant diseases.

As you probably have notices, I have been talking about replant *diseases*—not replant disease. This terminology stems from my belief that the nature and causes of replant diseases are diverse on the same fruit crop, and perhaps even in the same general geographic location. This is an important point if for no other reason than identification of the steps needed to control the problems. As examples, I would like to discuss two nonspecific replant diseases—the fruit replant problem in New York [7, 8, 9] and the peach tree short life syndrome in South Carolina and elsewhere in the southeastern United States [17].

Peach tree short life is a disease of complex origin characterized by sudden collapse of trees, or portions thereof, before or just after bloom [17, 18]. The replant problem in New York is a disease of fruit trees characterized by poor growth, necrosis or absence of tertiary roots, and premature death [7]. Symptoms of the two disorders are similar in many respects. However, the spectacular tree losses associated with short life usually are not seen in the Northeast, while the notably poor tree growth observed in replanted New York orchards is less evident in peach orchards in South Carolina.

Is peach tree short life a replant disease? Characterisitcs of PTSL in common with replant diseases are, first, the site-related nature of the problem. Peach tree short life occurs earlier and is more severe on replanted sites than on those where peaches or other stone fruits have not grown previously. Second, the general symptom pattern is similar to that of replant diseases—reduced growth, yield, and vigor; poor root development; and predisposition to cold injury and other factors. Third, the severity of PTSL can be corrected in part by soil fumigation; while yields of trees in fumigated soil also increase.

However, certain characteristics of peach tree short life are not shared by replant diseases. One is that PTSL is not confined to replanted orchard sites. Peach orchards in sites where peaches and other stone fruits have not grown before sometimes are severely affected by PTSL. Usually, when this occurs, losses occur later in the life of the orchard than in replanted orchards. Second, the response of trees to soil fumigation is different than for most replant problems. Trees in replanted orchard sites where PTSL might be expected to occur do not respond to soil fumigation if plant-parasitic nematodes are absent. No changes in growth are evident, and no tree loss occurs. Third, post-plant nematicide applications, usually at 2-year intervals, are required for satisfactory extension of tree life.

Speaking to the point, what is the role of nematodes in these two disorders? The lesion nematode *Pratylenchus penetrans* is very important for initiating the root injury that characterizes the replant disease in New York [7]. However, the involvement of other biological agents is implicated by the differential response of tree growth to different nematicides. Parasitism by *Pratylenchus* also appears to predispose cherry trees to cold injury in New York [3, 7].

In the southeastern United States, the direct involvement of nematodes in peach tree short life also is evident. Severe necrosis of feeder roots precedes or accompanies tree losses due to short life. Feeder root necrosis can be corrected by soil fumigation. The nematodes in this case are ring nematodes *(Macroposthonia xenoplax)* and root-knot *(Meloidogyne* spp.). Lownsbery and coworkers [5, 6, 10, 11] have shown the potential for ring nematodes to predispose peach and plum trees to *Pseudomonas syringae,* which together with cold injury is directly responsible for the early death of peach trees in the PTSL syndrome. Although Nesmith and Dowler [13] have found that peach trees in ring nematode-infested soil were less cold hardy than those in fumigated soil, ring nematodes *per se* have not been shown to predispose peach trees to cold injury. However, they do interfere with growth factors which regulate dormancy in peach trees, notably indole-3-acetic acid and abscisic acid [2, Nyczepir and Lewis, *personal communication*]. Peach trees parasitized by ring nematodes have elevated IAA levels in the fall and winter months [2], which suggests that they are not fully dormant and therefore might be subject to cold injury. Evidence, largely circumstantial at this point, indicates that nematodes, by interference with growth regulators, predispose peach trees to cold injury, and thus to peach tree short life. Other factors, notable fall pruning of trees [13], may have similar effects.

Replant problems, whether specific or nonspecific, seem to hold several factors in common. First, replant problems basically are root disease problems. Whether they are specific or nonspecific may depend largely on the host susceptibility to micro-organisms or adverse physical factors present in the soil environment. Nematodes in some instances may be the dominant factor in root disease problems, while in other situations they have a more passive role. Second, soil fumigation often is of significant value in correcting replant problems. The nature and spectrum of the fumigant selected may be very important, especially when biologic agents other than or in addition to nematodes cause the problem. Therefore, it is important to identify all of the contributing factors involved in each replant situation. Third, physical factors in the root environment may be very important. Any event which acts to injure roots or interfere with their development may contribute to replant problems. A total management program must therefore be considered when attempting to combat replant problems. The management program for peach tree short life consists of 10 steps which growers are urged to use for effective control [1]. Most of these are designed to develop and maintain healthy root systems for peach trees. By so doing, productivity increases even when short life does not become a problem.

Literature Cited

1. Brittain, J. A. and R. W. Miller, eds. 1978. Managing peach tree short life in the Southeast. Clemson University Circ. 585. 19 pp.

2. Carter, George E., Jr. 1976. Effect of soil fumigation and pruning date on the indoleacetic acid content of peach trees in a short life site. *HortScience* 11:594-595.

3. Edgerton, L. J. and D. G. Parker. 1958. Cold hardiness of Montmorency cherry affected by nematode damage. *Farm Research 24(6):12.*

4. Hoestra, H. 1968. Replant diseases of apple in the Netherlands. Meded. Landb. Wageningen 68-13. 105 pp.

5. Lownsbery, B. F., H. English, E. H. Moody, and F. Schick. 1973. *Criconemoides xenoplax* experimentally associated with a disease of peach. *Phytopathology* 63:994-997.

6. Lownsbery, B. F., H. English, G. R. Noel, and F. J. Schick. 1977. Influence of Nemaguard and Lovell rootstocks and *Macroposthonia xenoplax* on bacterial canker of peach. *J. Nematol.* 9:221-224.

7. Mai, W. F. and K. G. Parker. 1967. Root diseases of fruit trees in New York State. I. Populations of *Pratylenchus penetrans* and growth of cherry in response to soil treatment with nematicides. Plant Dis. Rept. 51:398-401.

8. Mai, W. F. and K. G. Parker. 1972. Root diseases of fruit trees in New York State. IV. Influence of preplant treatment with a nematicide, charcoal from burned brush, and complete mineral nutrition on growth and yields of sour cherry trees and numbers of *Pratylenchus penetrans.* Plant Dis. Reptr. 56:141-145.

9. Mai, W. F., K. G. Parker, and K. D. Hickey. 1970. Root diseases of fruit trees in New York State. II. Populations of *Pratylenchus penetrans* and growth of apple in response to soil treatment with nematicides. Plant Dis. Reptr. 54:792-795.

10. Mojtahedi, H. and B. F. Lownsbery. 1975. Pathogenicity of *Criconemoides xenoplax* to prune and plum rootstocks. *J. Nematol.* 7:114-119.

11. Mojtahedi, H., B. F. Lownsbery, and E. H. Moody. 1975. Ring nematodes increase development of bacterial cankers in plums. *Phytopathology 65:556-559.*

12. Mountain, W. B. and Z. Patrick. 1959. The peach replant problem in Ontario. VII. The pathogenicity of *Pratylenchus penetrans* (Cobb, 1917) Filip. and Stek. 1941. *Canad. J. Bot.* 37:459-470.

13. Nesmith, W. C. and W. M. Dowler. 1975. Soil fumigation and fall pruning related to peach tree short life. *Phytopathology* 65:277-280.

14. Savory, B. M. 1966. Specific replant diseases. Research Review No. 1, Commonwealth Bur. Hort. and Plant. Crops, E. Malling. 64 pp.

15. Sewell, G. W. F. and Stephanie C. Fleck. 1979. Effect of *Pythium* ssp. on growth of apple seedlings. Rept. East Malling Res. Stn., 1978. p. 92.

16. Sewell, G. W. F., J. F. Wilson, and Christine M. Blake. 1976. Specific replant disease of cherry. Rept. East Malling Res. Stn. 1975. p. 117.

17. Taylor, Jack, J. A. Biesbrock, F. F. Hendrix, Jr., W. M. Powell, J. W. Daniell, and F. L. Crosby. 1972. Peach tree decline in Georgia. Univ. Ga. Agric. Exp. Sta. Res. Bull. 77. 42 pp.

18. Zehr, E. I., R. W. Miller, and F. H. Smith. 1976. Soil fumigation and peach rootstocks for protection against peach tree short life. *Phytopathology* 66:689-694.

37.1 Integrated Plant Protection in *Pinus radiata* in Australia and New Zealand

J. W. Ray, M. J. Nuttall and P. D. Gadgil

Forest Research Inst., Private Bag, Rotorua, New Zealand

D. W. Edwards

Forestry Commission of New South Wales
P.O. Box 100, Beecroft., N.S.W., Australia

F. G. Neumann

Forests Commission
Victoria, Box 4018 GPO, Melbourne, Australia

Introduction

Both Australia and New Zealand rely heavily on exotic conifers (mainly *Pinus radiata* D. Don) to meet their softwood needs. Plantations of *P. radiata* occupy ca. about 432,000 ha in Australia [1], mostly on low-productivity eucalypt forest sites in areas receiving around 800 mm of annual rainfall but occasionally affected by prolonged severe droughts. New Zealand has ca. 525,000 ha under *P. radiata* [2] and most of the plantations are on the central North Island volcanic plateau with an evenly distributed annual rainfall of about 1500 mm. Current annual roundwood production is ca. 3 million m^3 in Australia [1] and ca. 8 million m^3 in New Zealand [3].

Although established plantations of *P. radiata* in both countries are simple monocultures with mostly even-aged stands, they have generally experienced little damage from insects (and diseases until recently) despite the adaptation to pines of indigenous defoliators.

Harmful Insects

In Australia, economically damaging outbreaks have occurred of the North American bark beetle *Ips grandicollis* Eichhoff (Scolytidae); the wood wasp *Sirex noctilio* Fabricius (Siricidae) which has caused up to 40% mortality after several years in a plantation in Tasmania [4] and up to 80% in some stands in Victoria; the indigenous undescribed Lepidopterous defoliators *Lichenaula* sp. and *Procometis* sp. (Xyloryctidae); and some native acridids in young plantations [5].

Sirex noctilio is the only potentially serious pest of *P. radiata* in New Zealand. It killed about one-third of the trees in the heavily overstocked plantations in the central North Island between 1946 and 1950 after severe summer droughts [6]. The bark beetle *Hylastes ater* (Paykull) (Scolytidae), the ambrosia beetle *Xleborus saxeseni* (Ratzeburg) (Scolytidae), the pit weevils *Psepholax* spp. (Curculionidae), and the longhorn *Arhopalus ferus* (Mulsant) (Cerambycidae) all attack logs soon after felling, making such logs unacceptable for export.

Diseases

The most important foliar pathogen in both countries is *Dothistroma pini* Hulbary (=*D. septospora* (Doroquin) Morelet), which can cause severe growth loss in young plantations; *Pinus radiata* becomes resistant to attack by *D. pini* at about 15 years of age. The fungus was first identified in New Zealand in 1964 [7] and in Australia in 1975 [8]. It is now present in most of the *P. radiata* plantations in New Zealand except for eastern South Island. In Australia it is found in New South Wales, southern Queensland, the Australian Capital Territory, and north-eastern Victoria. In New Zealand it has been found that losses in volume become significant when more than 25% of the green crown is infected [9]. Two other needle-cast fungi, *Lophodermium pinastri* (Schrader ex Fries) Chevalier and *Naemacyclus niveus* (Persoon ex Fries) Saccardo, are widespread in Australia and cause occasional severe epidemics of needle cast in Victoria and New South Wales although the subsequent increment loss appears to be minor. In New Zealand *Naemacyclus minor* Butin causes chronic shedding of older (greater than 1-year-old) needles and preliminary work indicates that growth losses in affected trees could be significant. Dieback of terminals caused by *Diplodia pinea* (Desmazieres) Kickx is of some importance in Australia [1] but it is only locally serious in New Zealand. *Armillariella limonea* Stevenson and *Armillariella novae-zelandiae* Stevenson have caused up to 27% mortality in the first 2 years after planting in New Zealand on sites which formerly carried native forests [11]. In Australia, damage from *Armillaria luteobubalina* Watling & Kile is minor; *Phytophthora cinnamoni* Rands and *Phytophthora cryptogea* Pethybridge & Lafferty cause some damage on very wet sites.

The Pest and Disease Management System

Quarantine

The general absence of serious harmful insects and diseases in the Australian and New Zealand 'high risk' *P. radiata* plantations reflects the geographic location of the two countries, away from other pine-growing areas, and also the effectiveness of quarantine—an essential element of a successful pest and disease management system. In both countries, quarantine officers inspect all imported timber and timber products, with special emphasis on detection of the prohibited bark-bearing material as bark beetles are regarded as the greatest threat. Any products showing signs of infestation or fungal infection are either burnt or fumigated. Imports of plants of most conifer species are prohibited and, if permitted, entry is allowed only after a period in quarantine. Internal quarantine is practiced with respect to *S. noctilio* in Australia; transport is prohibited of sawn pine timber (unless milled to below 2.5 cm thickness or kiln dried) and of untreated and treated roundwood derived from unhealthy trees, from infested areas to *Sirex*-free areas during the insect's pupal or flight period from late-spring to autumn. In New Zealand, transport of nursery stock of the *D. pini*-susceptible genera—*Pinus, Picea, Pseudotsuga,* and *Larix*—from areas infected by *D. pini* to uninfected areas is totally prohibited. These internal quarantine measures are aimed at containing the insect or the fungus within the existing range.

Surveillance

Another major component of pest and disease management is routine surveillance. In Australia, district forest staff carry out this function backed, when necessary, by specialist ground and aerial photographic surveys with normal color or infrared film. In New Zealand, officers of the Forest Biology Survey are responsible for inspecting forests and forest nurseries and they pay particular attention to areas of forest near ports. The use of district forest staff and local Forest Biology Survey officers instead of occasional visits by itinerant specialists for early detection of pests and diseases has been successful. In both countries, the detection services are backed by research institutes. Surveys are also carried out for monitoring current outbreaks such as those of *S. noctilio* in south-eastern Victoria and of *D. pini* in Australia and New Zealand. For *D. pini*, disease assessment is done from the ground in Australia; in New Zealand, where the areas of infected plantations are much bigger, aerial assessment followed by ground checks is relied upon. Neither aerial photography nor earth satellite imagery has proved successful in assessment of *D. pini* infection.

Preventative Control

In preventative control, the aim has been to create conditions which are unfavorable for the development of harmful insects and fungi. Use of high quality stock for planting, careful establishment, effective fire prevention, and attention to forest hygiene should lead to stand improvement and lessen damage. Selective thinning for the early removal of suppressed trees and for minimizing stress by reducing between-tree competition is of major importance in preventing attack by *S. noctilio* [12] and may reduce dieback of terminals, caused by *Diplodia pinea* in New Zealand [13]. The rejection of export logs through infestation by insects can be reduced by rapid removal of logs from the forest during the flight season, by keeping the skid sites free from logging debris, and by stockpiling logs on skids rather than on the ground. Fumigation of logs is a last resort. Damage from *Armillaria* root rot can be avoided by not planting sites which have recently carried indigenous forest [11]. There appears to be strongly heritable variation in the susceptibility of individual trees to attack by *Naemacyclus minor* [14] and in New Zealand highly susceptible trees are not used in the breeding program. A clone bank of trees selected for *D. pini*-resistance is maintained in New Zealand although it has not yet been used in the breeding program.

Curative Control

Curative control involves the application of silvicultural, biological, or chemical control measures. Silvicultural control consisting of salvage felling and burning of infected bark and slash has been successful for *Ips grandicollis* in South

Australia [15]. *Sirex noctilio* populations in Victoria have been kept low by the felling and burning of dying trees during winter. As physiologically stressed trees are most susceptible to attack by *S. noctilio* [16] control can be improved by timely thinning, by avoiding damage to standing trees during thinning, and by thinning only during winter so that the slash dries out and becomes unsuitable for oviposition before the insect flight season. Pruning of trees infected by *D. pini* reduces the amount of infected foliage and checks the build-up of infection.

Biological control of *S. noctilio* has been attempted in Tasmania, Victoria, and New Zealand. The egg and early instar larval parasitoids *Ibalia leucospoides leucospoides* (Hochenwarth) and *Iblia leucospoides ensiger* Norton (Ibaliidae) and the late larval instar parasitoids *Rhyssa persuasoria* (Linnaeus) and *Megarhyssa nortoni nortoni* (Cresson) (Ichneumonidae) have been introduced into both Australia [4] and New Zealand. In addition, Australia has introduced *(Ibalia rufipes drewseni* Borries and *Schletterius cinctipes* (Cresson) (Stephenidae). The major success in the field of biological control has been brought about by the nematode *Deladenus siricidicola* Bedding (Neotylenchidae) [17], introduced accidentally into Zealand and deliberately into Australia. This nematode sterilizes females without impairing general vigor and is spread each season by infected females during oviposition [18]. In the moist Tasmanian and New Zealand plantations, *D. siricidicola* and the parasitoids have been associated with the reduction of *S. noctilio* to very low populations. In Victoria, however, these control agents alone cannot check *S. noctilio* in drought-stressed plantations of intermediate age that have remained unthinned.

Apart from the occasional use of insecticides as an emergency measure against locust swarms in young plantations in Australia, insecticides are generally not used in both countries, except in nurseries. *Dothistroma* needle-blight is the only disease in both Australia and New Zealand against which chemicals are routinely and successfully used. Copper oxychloride (2.2 kg in 50 litres of water/ha) is the usual fungicide [7]. Many plantations may not require fungicidal control because disease development is not favored by either the climate or the nutritional status or both but in New Zealand an average of 26,000 ha have been treated annually since 1967. Current costs are ca. $A16/ha in Australia and $NZ12/ha in New Zealand. Fixed-wing aircraft are usually used for the spraying work but helicopters are employed when many small plantations in one locality have to be treated. Timing is crucial to the success of the treatment [19] and, in New Zealand, a single spray applied in November/December (late spring/summer summer) gives very good control. In New Zeland, fungicide is applied when the average level of infection of the unsuppressed crowns in a stand has reached 25%. The spray program is coordinated by a committee of state and private forest owners. This committee buys the fungicide in bulk and organizes flying contracts for the whole country, thus reducing costs. In New Zealand, three to four applications during a rotation are usually enough to keep infection at an acceptably low level, but in Australia it is expected that more frequent spraying may be needed for satisfactory control in plantations on sulfur-deficient soils in areas with an evenly distributed annual rainfall in excess of 1200 mm.

Intregration of Nutritional Problems and Disease Control

One of the most fruitful areas for integrated plant protection is the linking of information on the nutritional status of future pine plantations with the likely response to fungal pathogens of tree planted on such sites. In New South Wales many of the soils derived from igneous rocks have low sulfur levels [20] and all the plantations infected by *D. pini* in Australia are on such sites. It is suggested that if the climate favors the development of *Dothistroma* needle-blight, far more severe epidemics can be expected on S-deficient soils than on soils with an adequate S-supply. A very similar situation occurs with certain types of dieback caused by *Diplodia pinea*. One type affects large numbers of trees on S-deficient soils within some New South Wales stands. This link between S-deficiency and *Diplodia* infection is often sharply delineated by soil type boundaries. In Victoria, S-defiency has not necessarily been linked with *Diplodia* outbreaks. In New South Wales it is proposed to use a combination of land classification system based on soil and foliar nutrient analysis together with climatic information and disease surveys.

Weed Control

Application of herbicides to control weeds such as bracken, gorse, blackberry, and wattle is often needed for the successful establishment of plantations in both Australia and New Zealand. It is theoretically possible to combine herbicide applications with either liquid fertilizer to correct nutrient deficiencies or with compatible insecticides or fungicides for the control of insects and diseases. No systematic work on this aspect of integrated control has been done in either country.

Future Outlook

Current forest policy in both Australia and New Zealand is to expand the total area planted under *P. radiata*. This expansion will place increasingly heavy demands on the resources available for routine surveillance, monitoring, and control of harmful insects and diseases as well as research into the ecology, population dynamics, and management of exotic and indigenous pests and diseases. This challenge might best be met if planned plantations are established with genetically improved stock on suitable sites and are properly managed for sustained health and vigor. Furthermore, the current Australian practice of retaining substantial areas of native vegetation near or within plantations is expected to result in greater diversity of fauna and flora in pine forests, a trend which could contribute to ecological balance among destructive and other organisms within plantations.

References

1. Cameron, R. J. 1978: Year Book, Australia No. 62, *Aust. Bur. Stat. 1977-1978*. Canberra.
2. New Zealand Forest Service 1977: The 1977 National Forestry Planning Model—a report for the Forestry Council. *NZFS, For. Econ. Div.*, Wellington.
3. New Zealand Forest Service 1978: Statistics of the forests and forest industries of New Zealand to 1977. *NZFS Inf. Series No. 33*.
4. Taylor, K. L. 1976: The introduction and establishment of insect parasitoids to control *Sirex noctilio* in Australia. *Entomophaga* 21:429-440.
5. Neumann, F. G. and Marks, G. C. 1976. A synopsis of important pests and diseases in Australian forests and forest nurseries. *Aust. For.* 39:83—102.
6. Rawlings, G. B. 1955: Epidemics in *Pinus radiata* Forests in New Zealand. *N.Z. J. For.* 7:53-55.
7. Gilmour, J. W. 1967: Distribution and significance of the needle blight of pines caused by *Dothistroma pini* in New Zealand. *Pl. Dis. Rep.* 51:727-730.
8. Edwards, D. W. and Walker, J. 1978: *Dothistroma* needle blight in Australia. *Aust. For. Res. 8:125-137.*
9. Whyte, A. G. D. 1978: Spraying pine plantations with fungicides: the manager's dilemma. *For. Ecol. Manag.* 1:7-19.
10. Wright, J. P. and Marks, G. C. 1970: Loss of merchantable wood in radiata pine associated with infection by *Dipolida pinea*. *Aust. For* 34:107-119.
11. Shaw, C. G. and Calderon, S. 1977: Impact of *Armallaria* root rot in plantations of *Pinus radiata* established on sites converted from indigenous forest. *N.Z. J. For Sci.* 7:359-373.
12. Morgan, F. D. and Stewart, N. C. 1966: The biology and behaviour of the wood wasp *Sirex noctilio* in New Zealand. *Trans. Roy. Soc. N.Z. Zool.* 7:195-204.
13. Birch, T. T. C. 1936: *Diplodia pinea* in New Zealand. *N.Z. State For. Serv. Bull. No. 8.*
14. Wilcox, M. D., Shelbourne, C. J. A. and Firth, A. 1975: General and specific combining ability in eight selected clones of radiata pine. *N.Z. J. For. Sci.* 5:219-225.
15. Morgan, F. D. 1967: *Ips grandicollis* is South Australia. *Aust. For.* 31:137-155.
16. Madden, J. L. 1968: Physiological aspects of host tree favourability for the wood wasp, *Sirex noctilio* F. *Proc. ecol. Soc. Aust.* 3:147-149.
17. Zondag, R. 1969: A nematode infection of *Sirex noctilio* (F) in New Zealand. *N.Z. J. Sci.* 12:732-747.
18. Bedding, R. A. and Akhurst, R. J. 1974: Use of the nematode *Deladenus siricidicola* in biological control of *Sirex noctilio* in Australia. *J. Aust. Ent. Soc.* 13:129-135.
19. Gilmour, J. W. and Noorderhaven, A. 1971: Influence of time of application of cuprous oxide on control of *Dothistroma* needle blight. *N.Z. J. For Sci.* 1:160-166.
20. Lambert, M. J. and Turner, J. 1977: Dieback in high site quality *Pinus radiata* strands—the role of boron and sulphur deficiencies. *N.Z. J. For Sci.* 7:333-348.

37.2 Prospects for Integrated Pest Management in Slash Pine Ecosystems in Florida

R. A. Schmidt and R. C. Wilkinson

Professor of Forest Pathology, School of Forest and Conservation and Professor of Forest Entomology, Department of Entomology and Nematology, respectively, University of Florida, Gainesville, 32611

Introduction

Slash pine is one of the four major pine species in the southeastern USA. This pine occurs abundantly throughout much of the Lower Coastal Plain from South Carolina south to Florida and west to Louisiana (15). The slash-longleaf pine ecosystem is bounded on the north first by the loblolly pine (*P. taeda L.*) then by the shortleaf pine (*P. echinata* Mill.) ecosystems.

Slash pine is a subclimax species with moderate fire-resistance and intermediate shade-tolerance. In the absence of wildfire or site treatments, which control competition, this pine is succeeded naturally by hardwoods, especially oak on the drier sites. In Florida's natural forests the fire- and rust-resistant longleaf pine (*P. palustris* Mill) was abundant on these drier sites and slash pine occurred naturally in pure, even-aged stands on flatwood sites typified by poorly drained soils.

Because of North Florida's favorable climate for pine growth—[e.g., in Alachua County, Florida [7] average January and July temperatures are 13°C and 26°C, respectively; average frost-free days number 295 and average rainfall is 137 cm]—and an abundance of soils often too infertile or wet for agricultural crops, forestry is of prime economic importance. Of the state's 14.3 million ha, at least 6.2 million ha are in commercial forest land; approximately 2.2 million ha of the latter are owned by forest industries [16]. Much of this industry-owned land is planted with slash pine and intensively managed for pulpwood production. Each year in Florida approximately 50,000 ha are planted to pine.

Because slash pine is indigenous and easily regenerated artificially and grows rapidly on appropriate sites, it has been favored and planted extensively in North Florida, even beyond the flatwoods sites where slash pine occurs naturally. Intensive management includes: producing improved seed in open-pollinated orchards of selected, progeny-tested parents; planting vigorous seedlings from nurseries onto accessible sites intensively prepared by chopping, burning and bedding; fertilizing and harvesting by clearcutting at 20-30 years of age. Normally, intermediate thinnings have been avoided in these short pulpwood rotations.

Florida's warm, wet climate is also favorable for forest pests and these cause significant losses to tree crops. Insect and disease problems are of special concern in the slash pine monoculture which possesses minimal functional diversity [12] to limit pest outbreaks in time and space. Such outbreaks and their associated losses must be avoided if increased demands upon the southern pine resource are to be satisfied. Hopefully this potential vulnerability to pests can be countered with the increased opportunities for pest management which can accompany intensive forest management. This paper identifies, via chronological management units, some important pests of slash pine in Florida and considers some opportunities for integrated pest management.

Some Important Pests and Their Current Management in Slash Pine

Pests of slash pine are numerous, but easily grouped since most are associated with specific age-classes of the host. Accordingly, Table 1 lists for operational management units some important insect and disease pests on slash pine together with the damage and current pest management practice.

In addition to these major pests of slash pine, there are many minor insect and disease problems which are not the object of control programs, but which could become major problems in the future. Also, birds and rodents may consume pine seeds during direct seeding operations and, on occasion, chemical repellants are utilized for their control. Control of plants that compete with pine for light, soil moisture and nutrients is primarily by mechanical means during site preparation or by prescribed fires prior to or after plantation establishment. Chemical control of competitive vegetation is used sparingly in pine plantation management.

Despite these pests and their host being endemic species, sporadic and chronic infestations of insect pests and steadily worsening epidemics of several diseases are occurring in the intensively managed slash pine forest. Increased pest incidence and damage occurs as susceptible host tissue is increased and concentrated, carrying capacities are exceeded and functional diversity is decreased.

For example, fusiform rust, a most damaging pest of slash pine, was rare in the natural forests at the turn of this century, but now occurs in epidemic proportions in Florida and over large areas of the South [4, 8]. Likewise pitch canker, initially of sporadic occurrence, has increased in frequency and severity within the last 10 years and is widely distributed in the slash pine ecosystem [10]. Chronic infestations of *Ips* and black turpentine beetles and sporadic outbreaks of sawflies and weevils [13] also occur. Then too, pests of little importance in the natural ecosystem may present unique problems in orchards and nurseries necessary for intensive forest management.

Some Additional Prospects for Integrated Pest Management

Pest management within perennial forest crops is complex. Forest managers must attempt to minimize losses due to pests while optimizing diverse, multiple-use resources within relatively large space and time dimensions. The attendant logistic, economic and environmental considerations condition pest management decisions and are most often cited as severe limiting factors in the development and use of strategies and tactics. On the other hand, certain advantages for pest management accrue to forest crops. For example, (1) intermediate harvests allow sanitation, salvage and adjustments of stocking, (2)

Table 1. Some important disease and insect pest management problems of slash pine in Florida

| Management Unit[1] | Pest or Problems Name | | Damage | Current Pest Management |
	Common	Scientific		
Seed Orchard	Flower thrip	*Gnophothrips fuscus*	Female flowers killed, Cones destroyed and subsequently attacked by coneworms	Insecticide spray Establish orchards outside range of evergreen oak; protective fungicide on female flowers
	Cone rust	*Cronartium strobilinum*		
	Coneworms	*Dioryctria amatella*	Cones destroyed	Contact or systemic insecticides
	Seedbugs	*Leptoglossus corculus*, and *Tetyra bipunctata*	Seeds destroyed	Contact or systemic insecticides
Nursery	Damping-off and root rot fungi	*Macrophomina, Fusarium, Pythium*	Mortality and reduced survival and growth of planted seedlings	Chemical fumigation of nursery soil and chemical seed treatment
	Fusiform rust	*Cronartium quercuum* f. sp. *fusiforme*	Planted seedlings die, become deformed or succumb to stem breakage	Protective foliar fungicide
	Spider mite	*Oligonychus milleri*	Reduced survival and growth of planted seedlings	Miticide
	Nematodes	*Meloidogyne, Belonolaimus* and others	Reduced survival and growth of planted seedlings and subsequent attack by root rot fungi	Chemical fumigation of nursery soil
Plantation Regeneration	Reproduction weevils	*Pachylobius picivorus* and *Hylobius pales*	Seedlings girdled and killed	Delay of planting 1 year if harvested after June
1–10 yr.	Sawflies	*Neodiprion* spp.	Growth loss and mortality of extensively defoliated dormant trees	Persistent populations treated with insecticides or virus
	Fusiform rust	*Cronartium quercuum* f. sp. *fusiforme*	Mortality, unmerchantability and stem-breaking through the rotation	Plant rust resistant seedlings
10–20 yr.	Pitch canker	*Fusarium moniliforme* var. *subglutinans*	Dieback, growth loss and mortality	Recommendation not available
Thinning	Annosus root rot	*Heterobasidion annosum*	Growth loss, mortality and windthrow	Thin in proper season and use chemical or biological stumptop treatment
15 yr.	Black turpentine	*Dendroctonus terebrans*	Beetles or accompanying stain fungi kill trees, especially those under stress	Maintain tree vigor and timely salvage of infested trees
	Engraver	*Ips* spp.		
	Sawyer	*Monochamus titillator*		
	Ambrosia beetles	*Platypus flavicornis*		
>20 yr.	Brown cubical root rot	*Phaeolus schweinitzii*	Growth loss, mortality and windthrow	Timely salvage
>40 yr.	Red ring heart rot	*Phellinus pini*	Unmerchantable timber and stem breakage	Timely salvage and avoid wounds
Post-harvest	Borers	*Monochamus titillator*, etc.	Lumber degraded or culled	Rapid utilization of logs
	Blue stain of logs	*Ceratocystis ips* and *Ips* spp.	Stained lumber and poles degraded	Timely harvest, rapid utilization and proper storage
	Subterranean termites	*Reticulitermes* spp.	Wood destroyed and subsequently decayed by fungi	Poison soil, use wood preservative and approved building practices
	Decay of wood in use	*Poria incrassata* and other wood decay fungi	Wood destroyed	Drying, wood preservatives, moisture barriers, paint and proper building practices.

[1] Average rotation ages for pulpwood and saw logs are 25 and 40–60 years, respectively. New milling techniques make roundwood production feasible at 30–40 years.

the time of final harvest is flexible, (3) forest can contain significant amounts of diversity which mitigates pest outbreaks, (4) trees are often quite tolerant of large pest populations, this is especially true of diseases and (5) forest product-acceptability does not often depend on zero or little damage caused by pests.

In intensively managed forests additional tactics for pest management include: (1) choice of site and its preparation, (2) selection of species or genotype for desired growth characteristics and pest resistance, and (3) choice of planting time and density. These and other tactics are listed in Figure 1 according to managemant units appropriate for slash pine culture. These silvicultural management units are also pest management units. Although a few pests are common to several units, each unit has unique pests and, therefore, unique opportunities for pest management. Few, if any, of these pest management tactics are new, but collectively they provide the basis for developing integrated pest management strategies. Such strategies might utilize several tactics through time to manage one pest or seek to provide an overall plan to manage several important pests.

Unfortunately, the statement of the problem presented here gives only a few ingredients in place of a much needed recipe for implementation. A comprehensive plan for integrated forest pest management requires information on pest incidence and distribution, pest population dynamics, crop-loss models and effectiveness of pilot-tested management strategies and tactics [17]. There is a paucity of such information for the slash pine ecosystem; appropriate plantation inventories, life tables, loss models and demonstrations of management strategies are lacking. As such, *a priori* management, (e.g., prevention, a most important strategy for forest pest management), is often precluded and the recourse is to adopt remedial strategies.

1. Orchards

In slash pine seed orchards pesticides are an important component of pest management, and the development of systemics has enhanced pesticide utility. Orchard location is determined by many factors, but when possible pests should be considered, e.g., locating orchards away from live oak—the important alternate host—aids control of cone rust. With respect to coneworms, the most important pest of second year slash pine cones, it is suggested that control measures be concentrated on susceptible clones. Roguing or isolation of such clones and destruction of infected cones are possible. Similar treatment of cone rust-susceptible clones and infected cones might be effective since coneworms

are attracted to rust-infected cones. Collection of coneworm infested-cones in cages which allow egress or parasites—but not coneworm moths—is suggested here as a means of conserving parasites for natural biological control. Southern pine seed orchards are the only management unit for which appropriate life tables [6] and inventory analysis [3] are available. Even so, these life tables require updating to include newly recognized pests, especially seed and cone fungi [9].

2. Nurseries

In slash pine nurseries seeds are sown in early spring and seedlings are lifted in late fall and winter just prior to planting. Damping-off fungi and other root rot fungi are most damaging pests and are controlled by soil fumigation which also controls nematodes and weeds. Manipulation of other cultural practices including amending the soil, fertilizing, favoring mycorrhizae either naturally or artificially induced and locating nurseries in well-drained soils aid in damping-off control. Foliar pesticides are useful for control of fusiform rust and spider mites. In either case the development and use of systemics with residual effects subsequent to "out-planting" would be advantageous. Additional management for control of fusiform rust might include: locating nurseries away from the alternate oak hosts, sanitation of inoculum on oak in the vicinity of the nursery and the use of an irrigation system which does not wet the foliage, thereby reducing the critical period of leaf moisture. Finally, rust-infected seedlings should be rogued at the nursery during lifting.

3. Regeneration

The opportunities for and benefits of pest management are in many respects greatest during regeneration. Important decisions which affect pest incidence include: site selection and preparation, planting time and density and pine genotype. The appropriate time to assess pest hazard or risk is prior to regeneration, and the strategy should be to use "an ounce of prevention". Decisions, or the lack thereof, may affect pest damage and management throughout the rotation. For example, ignoring fusiform rust hazard can result in plantation destruction within five years and the need to replant with rust-resistant genotypes.

Reproduction weevils are perhaps the most important pests attacking newly planted seedlings. Delaying planting for one growing season after harvest allows deterioration of fresh stumps and other logging debris that serve as breeding habitats for weevils. The cost associated with this delay might be offset by planting fallow land to an annual crop.

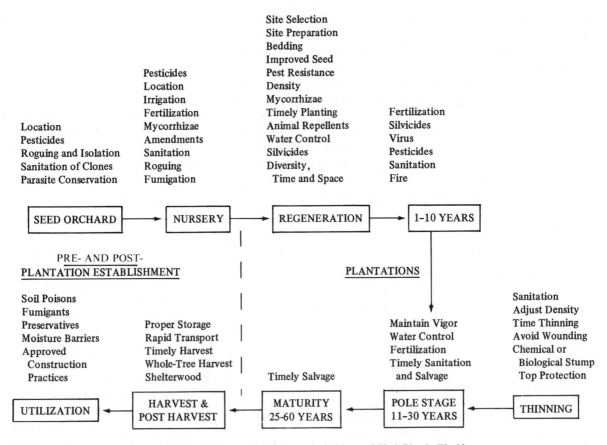

Figure 1. Some Prospects for Integrated Pest Management in Intensively Managed Slash Pine in Florida

Also stump and root harvest would seem a practical means of removing weevil breeding habitats. Other possibilities are discussed by Thomas [14]. Also, inoculum of root-rot pathogens, e.g., *Phaeolus schwienitzii, Heterobasidion annosum,* and *Clitocybe tabescens* might be reduced by whole-tree harvest. With regard to management of fusiform rust, oak suppression during site preparation and the maximizing of age and genotypic diversity during regeneration represent epidemiologically sound strategies [5]. Proper gene management, that is, purposeful, deployment of rust-resistant genotypes, is very important. When appropriate, fertilization of young pine seedlings should be delayed so as not to enhance plant succulence and, thereby, risk increased rust incidence.

On poorly drained sites some root problems can be avoided during regeneration by bedding or draining. Excess drainage can lead to moisture stress and susceptibility to bark beetles, weevils and diseases such as pitch canker later in the rotation when pole-size trees require more water. Following plantation establishment water levels can be restored by blocking drainage ditches.

4. Plantations, Thinnings, and Harvest

Some prospects for pest management in young, pole-stage and mature plantations are given in Figure 1. Likewise, thinning, harvest, post-harvest and utilization prospects are considered. No attempt is made to further detail each of these, but a few comments are appropriate. In general opportunities for pest management diminish as trees mature. Still there are some prospects even for direct control, but often the forest manager is faced with "a pound of cure".

Pine sawflies periodically cause severe defoliation of dormant, open-grown slash pine seedlings. Prospects for prevention include careful selection of site and planting density to assure survival and rapid crown closure. Cypress ponds provide reservoirs for sawfly natural enemies and should be retained where possible. Persistent outbreaks require spot treatment with virus or insecticides [18]. In older stands there may be opportunities for control of sawfly by ants.

Thinning, fertilization and water control, as previously mentioned, provide opportunities to maintain vigor in pole-stage plantations. This is critical where dense plantations exceed the maximum carrying-capacity of the site or during extended droughts. Reducing moisture stress can alleviate bark beetle problems which otherwise result in the death of significant numbers of slash pine during drought. Raising water levels in pole-stage plantations might also alleviate weevil and pitch canker outbreaks [13]. In fact, planting densities or stocking adjustments should be prescribed with carrying capacity and periodic drought in mind. Prior to harvest, sanitation and salvage are useful tactics for pest management. Rapid utilization of merchantable lightning-struck trees during the rainy season may be beneficial since these trees appear to serve as a refugium for bark beetles. While thinning can be useful in pest management, it is especially timely (anticipating a change to longer rotations and intermediate thinnings) to caution managers that significant loss can accrue to thinned stands attacked by bark beetles and *Heterobasidion annosum*.

There are significant root rot problems in older slash pine stands and recent information [2] suggests that younger plantations may be damaged as inoculum spreads from reservoirs in stumps residual to harvest. Also there is evidence [1] to suggest that root rots can predispose pines to bark beetle attack. Removing such stumps during harvest might prevent root rot and alleviate subsequent losses due to bark beetles and weevils. However, longterm effects of "whole tree" harvesting on site productivity may preclude this practice.

A final example applies to fusiform rust and maturing plantations. In high rust-hazard areas harvest of healthy plantations should be delayed, thereby avoiding losses associated with succeeding susceptible young plantations. A shelterwood system of regeneration would reduce rust in understory seedlings [11]. Such would not be acceptable routinely for short-rotation pulpwood production since prompt regeneration is desirable, but for small or isolated stands shelterwood management has the added advantage of reducing site preparation cost.

Literature Cited

1. Alexander, S. A. and J. M. Skelly. 1979. Pine Beetle Prog. Rept. Synop. (July 1-Dec. 31, 1978) U.S.D.A. For. Serv. Pineville, LA.
2. Blakeslee, G. M. and S. W. Oak. Plant Dis. Reptr. (in press).
3. Bramlett, D. L. et al. 1977. U.S.D.A. For. Serv. Gen. Tech. Rept. SE-13.
4. Dinus, R. J. 1974. *Proc. Am. Phytopathol. Soc.* 1:184–190.
5. Dinus, R. J. and R. A. Schmidt, (Eds.) 1977. *Management of fusiform rust in southern pines. Symp. Proc.* Univ. Fla., Gainesville.
6. DeBarr, G. L. and L. R. Barber. 1975. U.S.D.A. For. Serv. Res. Pap. SE-131.
7. Dohrenwend, R. E. 1978. Univ. Fla. Agric. Expt. Sta. Bull. 796.
8. Griggs, M. M. and R. A. Schmidt. 1977. p. 32–38. In *Management of fusiform rust in southern pines. Symp. Proc.* Univ. Fla., Gainesville.
9. Miller, T. and D. L. Bramlett. 1978. *Phytopathol. News* 12(9):207.
10. Phelps, W. R. and C. W. Chellman. 1976. U.S.D.A. For. Serv., State and Priv. For., Soeast. Area.
11. Rowan, S. J. et al. 1975. U.S.D.A. For. Serv. Res. Note SE-212.
12. Schmidt, R. A. 1978. p. 287–315. In *Plant Disease: An advanced treatise.* Vol. II. Horsfall, J. G. and E. B. Cowling (Eds.) Acad. Press, Inc., NY.
13. Schmidt, R. A. et al. 1976. Univ. Fla., Inst. Food Agric. Sci. Prog. Rept. 76-2.
14. Thomas, H. A. 1971. J. For. 62:806–808.
15. U.S. Dep. Agric. 1973. U.S.D.A. For. Serv. Agric. Handb. 445.
16. U.S. Dep. Agric. 1978. Forest statistics of the U.S., 1977. U.S.D.A. For. Serv. (Review draft)
17. Waters, W. E. and E. B. Cowling. 1976. p. 149–177. In *Integrated pest management* J. L. Apple and R. F. Smith (Eds.). Plenum Publ. Corp. N.Y.
18. Wilkinson, R. C. 1969. Univ. Fla., Agric. Expt. Sta. Sunshine State Agric. Res. Rept. 14(6):13–15.

37.3 Integrated Protection of Poplar Stands in Europe and Asia

G. P. Cellerino and G. Lapietra

Poplar Research Institute, P.O. Box 116, Casale Monferrato (AL) Italy

In manmade poplar forests, defense against pathogenic agents must be aimed not only at preserving the amount of product, but also its quality standards which are required by industrial utilization. In this particular type of forest, especially when the wood will be destined for the plywood industry, it is necessary to keep attacks by adverse causes, especially insects, within limits strictly lower than in natural forests. For instance, it is well known that a single larval gallery by *Saperda carcharias* L. may strongly lower the value of a poplar trunk, making it unfit for the most profitable utilization.

In its best-developed form, poplar cultivation goes through two phases: nursery and plantation.

In the nursery, cuttings from stool beds or young plants ("barbatelle") which have been grown for this purpose are planted. These young plants come from special nurseries ("barbatellai") where cuttings are placed very close together (even more than 8 plants/m²).

Poplar plantations are made with young plants aged 1 to 3 years (in temperate climate regions) or with long cuttings from the nursery (in hot, arid regions). Usually they are monoclonal plantations of equally aged plants. As such they are particularly exposed to attacks by parasites, which not infrequently may become dangerous during a specific stage of development of these plants which are genetically all equally susceptible.

The cultivation periods vary from 5 to 20 years according to the clone, pedoclimatic conditions, and spacing. The shortness of these rotations provides the poplar stands with typical characteristics of instability which are not conducive to dispersal of entomophagous insects and of extreme simplicity of the ecosystem.

In the regions of interest, the poplar is affected by several diseases and pests: viruses, bacteria, fungi, insects, weeds, and others. Losses caused by nematodes or by mites are uncommon in the regions studied and for the time being do not have any significant economic effects. The control of diseases and pest insects is based on (1) search and utilization of resistant clones, (2) utilization of sound propagation material, (3) phytosanitary selection during the growing period, (4) suitable cultivation practices, (5) biological protection, and (6) chemical treatments.

Search and utilization of materials which are resistant to adverse causes is of fundamental importance. Through genetical improvement, it has already been possible to obtain clones that are resistant to several dangerous pathogens. For example, plant breeders have utilized resistance characters of some *Populus deltoides* Bartr. to *Marssoninae, Venturiae, Melampsora alliipopulina, M. larici-populina, Taphrina,* and *Phloeomyzus passerinii;* of some balsam poplars to poplar mosaic virus (PMV) and *M. medusae;* and of some *Populus nigra* L. to PMV and *Xanthomonas populi.* Moreover, where poplars of the Leuce section are largely cultivated, they have exploited resistance factors of some *P. alba* L. to *Venturiae* and *Hypolyxon mammatum,* and those of *P. tremula* L. and *P. tremuloides* Michx. to *M. castagnei.*

Nevertheless, it has not yet been possible to combine in a single clone an acceptable degree of resistance to the most important adverse causes with good production qualities in different climatic conditions.

Nurseries

In the nurseries, the use of healthy material is of fundamental importance to obtain plants free from PMV. Moreover, it reduces passive spreading of several parasites which may dwell in the cortical tissues or in buds (for instance, *Marssoninae, Venturiae, X. populi*). Frequent inspections, particularly of "barbatellai," get rid of diseased plants without relevant negative economic effects.

Various cultural procedures have beneficial effects. For example, pruning, with disposal of the residues, reduce the inoculum of *Marssoninae, Venturiae, Dothichiza populea, Phomopsis* sp., *Cytospora chrysosperma,* and infestations of *Gypsonoma aceriana, Saperda populnea, Apriona cinerea,* and species of similar habits.

Soil cultivations may (a) lower the population of some insects which through a terricolous phase (for example, *G. aceriana*); (b) improve soil aeration increasing the reaction of the root system against rotting agents (*Rosellinia* sp., *Armillaria mellea*); (c) enhance resistance of plants to trunk scab, an alteration ascribed to hydronutritional unbalances, and to attacks by several weakness parasites (*D.*

populea, C. crysosperma, Phomopsis sp.); (d) reduce the inoculum of leaf diseases by burying infected leaves (*Marssoninae, Venturiae,* rusts, *Taphrina aurea*) and eliminating possible hosts of heteroic fungi (*M. allii-populina*); (e) restrain weed spreading and, consequently, their competition with poplar plants.

Rational irrigations protect poplar plants from attacks by several xylophagous insects, such as *Capnodis miliaris* and *Melanophila picta,* which are particularly dangerous in hot, arid regions. Moreover, irrigation combined with fertilization makes plants more able to endure other insects such as *G. aceriana* and *S. populnea.*

Sufficiently large spacings, by placing a larger soil volume at the disposal of the plants, reduce possible nutritional deficiencies and lower the risks connected with trunk scab and cortical necrosis. Moreover, by modifying the microclimate, they reduce the probability of attacks by fungus parasites (in particular, rusts and *Marssoninae* (and by leaf-sucking insects (*Monosteira unicostata* and *M. buccata* Horv.).

The natural biological protection from diseases and pests, unfortunately, is scarcely relevant and gives aleatory results. Nevertheless, in Turkey, attempts have been made to control poplar rusts by spraying poplar leaves with conidial suspensions of *Darluca filum* (Biv. ex Fr.) Cast. and *Ramularia uredinis* (Voss.) Sacc. In Italy, it is being tried with somewhat effective results to increase the action of some Ichneumonidae against *Paranthrene tabaniformis.*

The above-mentioned agronomic and biologic procedures reduce but do not eliminate the effects of adverse causes which may later lead to negative consequences in plantation stands. For the purpose of keeping the attacks below dangerous limits, which must be extremely low, these procedures could be integrated with chemical control. Use of chemical control will vary according to local conditions and to the resistance or susceptibility of single clones. Chemical products should be chosen from those which have low toxicity and persistence; they should not interfere with the local useful entomofauna. Broad-spectrum insecticides (derived from cyclo-esane, cyclopentadienes and some organophosphates) should be replaced, for instance, with diflu-benzuron against lepidopteran defoliators, and with trichlorphon against *Cryptorhynchus lapati* during its subcortical larval phase. The dates of treatments in different places should be determined according to the biology of the parasites and of the host plant. In some cases, by choosing suitably the active principle

and the time of treatment, it is possible to act simultaneously against several parasites. For instance, during summer in the Mediterranean countries, it is possible to combine the treatments against diseases such as *M. brunnea, D. populea, C. crysosperma,* and rusts with those against insect pests such as *G. aceriana, M. unicostata,* and *P. tabaniformis.*

Plantations

All the cultural practices must be employed which contribute directly or indirectly to reducing transplanting crises which are likely to develop when water losses due to evaporation and transpiration are not balanced by root absorption. These stresses, consequently, make the plants susceptible to the attacks of various cortical fungi, as well as several dangerous Buprestidae, such as *M. picta* and *Agrilus suvorovi populneus.* The propagation material must be well lignified and it should be planted when still dormant. If too dry, the material must be dipped previously in water and then planted in well prepared soil deeply enough to assure a sufficient water supply during the initial phase of the growth. For instance, long cuttings are used in Iraqi plantations while young poplar plants in Italy are sometimes placed 2-2.5 m in depth in dry soil where the water table is deep.

To contain the spreading of some dangerous xylophagous insects it is of vital importance to avoid using infected propagation material (for example, by *C. lapathi*). Whenever possible all plants which host *S. carcharias* and *Cossus cossus* L. should be eliminated from the neighborhoods of new plantations.

An interval of some years between a plantation and the succeeding one leads to a reduction of the inoculum mass of the agents of root rots and bacterial diseases in the poplar stands.

Soil cultivation, irrigation, and fertilization act in plantations similarly as in nurseries. In the Mediterranean regions it must be noticed that pruning performed during the summer predisposes the plants to attacks by *P. tabaniformis* and *C. cossus* by favoring the laying of eggs of these insects near pruning wounds.

Biological defense against fungous diseases by means of other organisms is very difficult; on the other hand, it can turn out to be effective in containing insects which become dangerous at high-level density populations. For instance, the action of many Diptera (Tachinidae) and Hymenoptera (Ichneumonidae and Braconidae) is often sufficient to keep some defoliators such as *Pygaera anastomosis, Stilpnotia salicis* and *Lymantria dispar* below the economic threshold of chemical intervention.

In plantations, also, chemical control is often necessary against serious fungus parasites, such as *M. brunnea* (in Europe, Japan, Korea); against cortical sapsuckers, such as *Ph. passerinii* (in Mediterranean countries); and against the most dangerous xylophagous insects, such as *C. lapathi* (in young poplar stands) and *S. carcharias* and *C. miliaris,* which cause serious losses even at very low populations.

It will be almost impossible to avoid chemical treatments against xylophagous insects, particularly in poplar stands for high quality wood production. More than in nurseries, the choice of chemicals must be directed toward active ingredients characterized by low toxicity, limited persistence, and, possibly, good selectivity. In addition to chemical insecticides (diflubenzuron, gardona, trichlorphon, malathion) several bacterial preparations of *Bacillus thuringiensis* Berl. found practical application against defoliators. In several cases treatments with different mixed products can act against several pathogens. For example, in Mediterranean countries, treatments with diflubenzuron and mancozeb, performed in June and July, may contain the attacks of defoliating and leaf mining Lepidoptera, *M. brunnea,* or eventually rusts.

By integrating the different control means in different combinations according to the various pedoclimatic, social, and economic characteristics of the single areas, it will be possible to keep the poplar stands in a satisfactory sanitary state by means of a limited number of chemical treatments.

More encouraging results from the integrated defense could be obtained by enlarging the present knowledge of resistance characters of poplars, of the biology and epidemiology of the parasites, and of the climatic conditions which affect them.

Acknowledgements

The information reported here originates from the authors' personal knowledge and from the contributions of specialists from various countries. The authors are particularly indebted to: M. Vural (Turkey); G. Gojkovic (Yugoslavia); M. Natercia, S. Santos, and C. D. Serrao Nogueira (Portugal); A. G. Mackay (United Kingdom); J. Pinon (France); D. S. Kailidis (Greece); E. Donaubauer (Austria); J. Luitjes (Holland); I. Haidenov (Bulgaria); K. C. Sahni (India); J. Halperin (Israel); M. Ismail Chandhry (Pakistan); S. K. Hyun and Yi Chang-keun (Korea), who have given information about the state of the art in their own and neighboring countries.

Selected Literature

Arru, G., 1975. Less ennemis des salicacees: Insectes et autres animaux. XV Sess. Int. Poplar Commn. FO:CIP/75/21(j) 28 pp.

Cellerino, G. P., 1971. La lotta contro le principali malattie delle piante forestali nel bacino mediterraneo. C.r. 3ᵉ *J. Phy. Phytoph.* Circum mediter. Sassari. 403–419.

37.4   Integrated Pest Management of Poplar Species[1]

T. H. Filer, Jr., J. D. Solomon, D. T. Cooper and M. Hubbes[2]

Silvicultural Practices

Proper site selection, good site preparation, correct planting practices, and 1st-year cultivation directly and indirectly affect survival rate of trees. Losses from canker fungi are minimized by cultural practices that increase tree vigor—poor tree vigor means more cankers per acre and greater mortality (Filer 1964, 1967).

Spacing

Tree spacing is dictated by the kind of wood—pulp or sawtimber—the plantation is producing. Regardless of initial spacing, fast-growing species such as

1. Discussion of pesticides here is not recommendation of their use. If pesticides are handled, applied, or disposed of improperly, they can harm humans, domestic animals, desirable plants, and pollinating insects, fish, or other wildlife, and may contaminate water supplies. Use pesticides only when needed and handle them with care. Follow directions and heed all precautions on the container label.

2. Filer, Solomon, and Cooper are stationed at the Southern Hardwoods Laboratory, maintained at Stoneville, Miss., by the Southern Forest Experiment Station, Forest Service—USDA, in cooperation with the Mississippi Agricultural and Forestry Experiment Station and the Southern Hardwood Forest Research Group. Hubbes is at Univ. of Toronto, 203 College St., Toronto, Ontario, Canada, M5A 1S1.

cottonwood must not be allowed to become crowded or stagnant. The canopy must not close, or growth will decline and insect and disease problems will increase.

Mechanical Weed Control

Clean cultivation not only increases yield but also reduces effects of insect and disease. The accepted early cultivation practice in cottonwood plantations is cross-discing, which has increased cottonwood height-growth 300% over that of trees in mowed plots; d.b.h., has increased 140%.[3] Higher nutrient concentrations were evident in leaf samples from disced plots. Not only does weed control promote tree vigor and growth, but it also simultaneously reduces the impact of cottonwood twig-borer (*Gypsonoma haimbachiana* Kft.). Discing must not be so deep that it damages lateral roots. Such damage induces top-dieback, thin crowns, and tree death (personal communication, Ray Gascon, Trans/Match).

Chemical Weed Control

Herbicides are commonly used in the northern USA and southern Ontario. Raitanen (1978) reported the use of several directed herbicidal sprays on hybrid poplar plantations in Ontario. Herbicides used for weed control in plantations older than 2 years appeared most economical and least damaging to roots. Raitanen reported a threefold increase in growth rate over control plots and a reduction in cultural costs from $150/ha to $60/ha.

Research at Stoneville, Miss., shows that weed control reduces pupation sites for many insects such as twig borers, leaf beetles, and poplar tentmaker (*Ichthyura inclusa* Hbn). Weed control also enables predators such as birds, especially woodpeckers, to locate pest insects more easily.

Observations at Stoneville indicate correlation between vine growth on *Populus* trunks and degree of poplar borer (*Saperda calcarata* Say) infestation; when weed control has eliminated vine growth, few borers are present. But vine growth on tree trunks provides shading that helps the self-pruning process and reduces epicormic branching. So, the benefits of controlling vines have to be weighed against the benefits of retaining them.

Sanitation

Cultural and sanitation practices can greatly reduce problems with insect and disease pests in *Populus* nurseries. All branch, terminal, and basal trimmings as well as cut stems from vegetative cutting operations should be burned (Cook and Solomon 1976; Filer 1976). This practice destroys many overwintering cottonwood twig-borers, clearwing borers (*Paranthrene* spp.), Oberea borers (*Oberea* spp.), and willow shoot-borers (*Janus abbreviatus* Say), and reduces the potential for infestation. Infested stumps and rootstocks serve as the principal reinfestation reservoir for the cottonwood borer (*Plectrodera scalator* F.) and clearwing borer. After three harvests, every stump in the infested nursery should be pulled and new borer-free cuttings planted. The stumps should be destroyed by burning before early April, so overwintering larvae of these borers are killed before they emerge as adults and reinfest the nursery (Cook and Solomon 1976; Solomon 1979). Clearing the entire nursery of all rootstocks before replanting eliminates the major source of reinfestation. The annual harvest of cuttings should be inspected, and infested cuttings should be culled and destroyed. Clean cultivation and destruction of fallen leaves eliminates hibernation sites for cottonwood leaf-beetles and removes much of the inoculum reservoir for leaf and canker diseases.

These practices will directly or indirectly affect many of the more important diseases and insects including leaf diseases and insect defoliators. In most cases, when trees are vigorous, pests present fewer problems. And vigorous trees can generally tolerate greater disease incidence and more insects.

Insect and Disease Control

Chemical Control

Chemical controls have been effective in control of cottonwood leaf-beetles and cottonwood twig-borers in nurseries and plantations (Morris 1960; Abrahamson et al. 1977). Carbofuran, a systemic insecticide that is applied to the soil and taken into the roots, may also reduce nematode infestation. Carbofuran improves growth response in the absence of insect manifestation.[4] Such increase could be due either to nematode control or to fertilizer effect. EPA-registered insecticides (carbaryl and chlorpyrifos), used in aerial or ground application, control cottonwood leaf-beetles and twig-borers. And control is not delayed as it is when systemic insecticides have been applied to the soil.

Pest control in stressed or weakened cottonwood stands can be important in reducing growth loss and

---

3. Kennedy, H. E., Jr. Soil and hardwoods influenced by cultural treatments. Prog. Rep., Southern Hardwoods Laboratory, Stoneville, Miss.

4. Francis, J. K. 1978. Irrigation of cottonwood to improve tree growth and increase effectiveness of a systemic insecticide. Final Rep. No. FS-SO-1111-2.16, Southern Hardwoods Laboratory, Stoneville, Miss.

mortality. Serious defoliation by *Septoria* leaf spot and poplar tent maker occurred in 1978 in 6- to 9-year-old plantations at sites in Arkansas and Mississippi. Some of these stands were already weakened from delayed or late thinning, dry weather, and heavy growth of vines and weeds. Defoliation at these sites caused serious crown thinning, dieback, and mortality exceeding 50% in some stands. Trees in plots sprayed with insecticide (carbaryl) and fungicide (copper oxide) for poplar tent maker and *Septoria* leaf spot control exhibited noticeably better crowns 1 year later than did unprotected plots. In 1 year, growth in sprayed plots averaged 0.45 inch d.b.h., compared to about half that in unprotected (defoliated) plots. Mortality ranged from 0-5% in sprayed plots, but was 3-38% in unprotected plots. Damage from defoliation is likely to persist even after defoliation has ceased; so we will continue to evaluate benefits of protection for 5 years.

Genetic Resistance

Selecting for pest resistance in *Populus* shows promise. *Melampsora* rust resistance, because of its high heritability, is perhaps the easiest resistance to obtain. A population of 1440 cottonwood clones collected from the lower Mississippi River and studied at Stoneville contained about 30 clones highly resistant to *Melampsora* rust (Cooper and Filer 1977), 10 clones resistant to *Septoria* leaf spot (Cooper and Filer 1976), and several clones with some apparent resistance to the cottonwood leaf-beetle (Oliveria and Cooper 1977). Abrahamson et al. (1977) and Caldbeck et al. (1978) also reported resistance to leaf beetle. Twig-borer resistance was found in a hybrid poplar by Woessner and Payne (1971). Genetic differences apparently exist among *P. deltoides* clones in their ability to recover from repeated defoliation by the poplar tent maker. Presumably, resistance to most other major pests can be found in *Populus*.

Jokela (1966) reported that some resistant clones of *P. deltoides* have been found that are not attacked by *Melampsora* or *Marssonina* in the central USA. He concludes that breeding for resistance to these diseases is both feasible and necessary.

Schreiner (1971) pointed out that, though *Populus* has many actual and potential disease and insect problems, its rich genetic diversity should provide the needed resistance. Finding individuals having necessary characteristics and resistance to the multitude of important pests is difficult. Repeated cycles of crossing and selection could be used to bring the desired characteristics together, but the time and expense required would be great. Today, we can select for the composite characteristic of sustained good growth under a range of conditions representing those likely to occur in commercial plantations. We should also retain several unrelated clones for genetic variety. Disease control must change the predisposition of hosts so they will be more resistant to a particular disease by (1) clonal selection for a particular site, and (2) implementation of appropriate silvicultural methods (Hubbes 1979).

Literature Cited

Abrahamson, L. P., R. C. Morris, and N. A. Overgaard. 1977. Control of certain insect pests in cottonwood nurseries with the systematic insecticide carbofuran. *J. Econ. Entomol.* 70:89-91.

Caldbeck, E. S., Howard S. McNabb, and Elwood R. Hart. 1978. Poplar clonal preferences of the cottonwood leaf beetle. *J. Econ. Entomol.* 71:518-520.

Cook, J. R., and J. D. Solomon. 1976. Damage, biology, and natural control of insect borers in cottonwood. Pages 272-279 in Bart A. Thielges and Samule B. Land, Jr., eds. *Proc. Symp. on Eastern Cottonwood and Related Species.* [Greenville, Miss., Sept. 28-Oct. 2, 1976.]

Cooper, D. T., and T. H. Filer. 1976. Resistance to Septoria leaf spot in eastern cottonwood. Plant Dis. Rep. 60:813-814.

Cooper, D. T., and T. H. Filer, Jr. 1977. Geographic variation in *Melampsora* rust resistance in eastern cottonwood in the lower Mississippi Valley. *Proc. Cent. States For. Tree Improv. Conf.* 10:146-151.

Filer, T. H., Jr. 1964. Outbreak of cankers on plantation-grown cottonwoods in Mississippi. Plant Dis. Rep. 48:588.

Filer, T. H., Jr. 1967. Pathogenicity of *Cytospora, Phomopsis,* and *Hypomyces* on *Populus deltoides. Phytopathology* 57:978-980.

Filer, T. H., Jr. 1976. Etiology, epidemiology, and control of cankers in cottonwood. Pages 226-233 in Bart A. Thielges and Samule B. Land, Jr., eds. *Proc. Symp. on Eastern Cottonwood and Related Species.* [Greenville, Miss. Sept. 28-Oct. 2, 1976.]

Hubbes, M. 1979. Some important diseases of poplars. Poplar Research Management and Utilization in Canada. For. Res. Inf. Pap. No. 102. Ontario Minist. Nat. Resour.

Jokela, J. J. 1966. Incidence and heritability of *Melampsora* in *Populus deltoides* Bartr. In H. D. Gerhold et al. (eds.) *Breeding Pest-resistant Trees,* p. 111-117. Pergamon Press, N.Y.

Morris, R. C. 1960. Control of cottonwood insects with a systemic insecticide. *J. For.* 58:718.

Oliveria, F. L., and D. T. Cooper. 1977. Tolerance of cottonwood to damage by cottonwood leaf beetle. Proc. South. For. Tree Improv. Conf. 14:213-217.

Raitanen, W. E. 1978. Energy, Fibre and Food: Agriforestry in Eastern Ontario. In For. for Food Agenda Item 7. 8th World For. Congr., Jakarta, Indonesia, Oct. 16-28.

Schreiner, E. J. 1971. Genetics of eastern cottonwood. U.S. Dep. Agric. Res. Pap. WO-11. 19 pp. Washington, D.C.

Solomon, J. D. 1979. Cottonwood borer (*Plectrodera scalator*)—a guide to its biology, damage, and control. U.S. Dep. Agric. For. Serv., Res. Pap. SO-157, South. For. Exp. Stn., New Orleans, La. [In press]

Woessner, R. A., and T. L. Payne. 1971. An assessment of cottonwood twig borer attacks. In *Proc. South. For. Tree Improv. Conf.* 11:98-107.

37.5 Integrated Protection of Oak Stands Against Pests in the USSR

F. S. Kuteev

*141200 All-Union Research Institute of Silviculture and Mechanization of Forestry (VNIILM)
Institutskaya 15, Pushkino Moscow Region, USSR*

Oak stands in the USSR are widely spread to the south from the main taiga range and grow largely in the forest and forest-steppe zones where mass outbreaks of pests are frequent.

Pest insects are often mentioned among the factors in the complex that causes oak dying. Most important are *Lymantria dispar* L., *Tortrix viridana* L., and *Euproctis chrysorrhaea* L. Stem pests form episodic infestation centers, chiefly in severly weakened stands.

In the system of integrated protection of oak stands against insect pests, the leading role belongs to silvicultural measures aimed at forming stands having a mixed composition and a complex structure and displaying a considerable biological resistance. That is why seed regeneration is preferred to sprout regeneration, resistant forms of oak (the form with late flush instead of the form with early flush, in particular) are chosen and seeded (planted), much attention is given to timely soil treatment in forest plantations, and young plantations and older stands are thinned at the proper time and intensity.

Mixed and high-stocked stands with dense understory are more rarely infested by pests. The deterioration of the condition of such stands is observed, as a rule, during depression resulting from severe changes in ecological conditions caused by natural or human factors.

The rehabilitation of oak stands is facilitated by biotechnical measures taken to attract and protect insectivorous birds. In recent years considerable work has been done in the USSR on the improvement of their habitat and conditions for reproduction. During selective sanitary cuttings hollow trees are left unfelled if they are not dangerously diseased. During the nesting season thinning in young stands is prohibited. Trees and shrubs that are used by birds for feeding and nesting are to be protected. Groups of special trees and shrubs are planted and nature refuges are established. Artificial feeding is practiced if necessary, and artificial watering places are provided in areas with no natural water reservoirs. For attraction of birds nesting in tree hollows, artificial nests are made that attract birds not only during the nesting season but serve them as shelter during nights and bad weather. Artificial nests allow the density of nesting birds to increase by 2-3 or even more times.

The number of manmade nests per unit area depends on the characteristics of stands, their geographical location, and the expected level of tree damage due to pests. Depending on these factors, bird species are defined that are desirable to be attracted. In oak stands in forest-steppe and steppe zones, with the stand age and degree of stocking taken into account, from 2 to 15 nests per ha are hung out in the areas where severe pest infestations may be expected.

Protection and immigration of forest red ants (*Formica rufa*) appeared to be useful. Artificial immigration has different aims: removal of ant hills from clearcut areas to prevent them from being destroyed, ant immigration to the areas of active and reserved infestations of defoliating insects, colonization of stands to increase their biological resistance and improve forest-growing conditions, and stimulating "donoring" in old ant hills for their growth activation.

Stand colonization by ants as well as bird attraction is a prophylactic measure. Being an important component of forest biocenoses, birds and ants regulate pest population densities only at low levels and are not able to eradicate a mass pest outbreak, especially when it acquires a pandemic character.

A much more important role in decreasing the population of oak defoliators is played by parasites, predators, and diseases. Active entomophages of the gypsy moth in the European part of the USSR are Braconidae (*Apanteles liparidis* Bouché, *A. melanoscelus* (Ratz.), Tachinidae (*Blepharinoda scutellata* R.-D., *Phorocera silvestris* R.-D.), Sarcophagidae (*Parasarcophaga harpax* Pand., *Pseudosarcophaga affinis* Fall., *P. monachae* Kram.), and Dermestidae. *Calosoma sycophauta* L. and *Xylodrepa quadripunctata* L. are dominating among predators.

Oak leaf roller, *Tortrix viridana* L., has a rather wide complex of entomophages, but the maximum degree of infection is observed at the pupal stage. During this period the pest appears to be most frequently invaded by *Phaeogenes invisor* Thunb., more rarely by *Apechtis rufata* Gmel. and *A. resinator* Thunb. The browntail moth, *Euproctis chrysorrhaea*, is invaded by parasites at all stages of development. *Telenomus phalaenarum* Mayr. parasitizes eggs; *Eupteromalus nidulans* Foerst.,

Pleurotropis complantusculus Ratzb., *Apanteles lacteicolor* Vier., first-instar larva; Tachinidae, in particular *Exorista libatrix* Panz. and *Ceromasia nigripes* Fall., late-instar larva. At the pupal stage the pest is invaded more frequently by *Pareudora praeceps* Meig. For the active suppression of the above pests attempts have been made to apply certain entomophages from the families Carabidae, Braconidae, Tachinidae, Sarcophagidae, and Dermestidae.

Fungus, bacterial, and virus diseases are the main natural regulating factors for pest population levels, especially during the collapse of mass outbreaks. High mortality due to natural infection is characteristic for the gypsy moth and some other defoliating insects. As a result of study of the biological activity of some species of fungi, bacteria (*Bacillus thuringiensis*), and nucleopolyhedrosis virus, some biological materials have been produced, and their portion in the system of forest protection increases each year.

Suppression measures against oak pests are applied only in the zone of constant mass outbreaks. They are usually undertaken in weakened stands, where the natural mechanisms regulating population dynamics are not able to confront an outbreak when it develops. Therefore, the stand conditions, and the presence of entomophages, entomopathogenic diseases, and other natural enemies checking the mass reproduction of pests are taken into consideration when choosing control measures.

Oak pest control is necessary in the case of severe damage and potential negative impact of defoliation on stand conditions. In the forest used for recreation as well as in the oak stands on floodlands, control is oriented at application of microbial materials (Dendrobacillin, Gomelin, Entobacterin, Insectin, Viron-ENSH).

In industrial forests, in addition to bacterial and virus products, chemical insecticides of the organic phosphor group (chlorphos, carbophos, pholazon, metation, etc.) are used, with dates and norms of application strictly followed. When the populations of parasites and predators are high, especially if the late form of the oak prevails in a stand and entomophages are not able to lower the numbers of pests to the level acceptable for management, treatments are carried out in the centers of infestation or in strips. If it is found that a considerable part of the pest population is diseased, control measures are either canceled or the dose of the insecticide used is reduced. Most frequently, stands are treated either with pure bacterial products or with the additions of small amounts of chemical insecticides. Of the list of formulations for ULV-spraying riciphon and carbophos are registered for application. The main method of application is aerial.

The efficiency of forest protection depends to a large extent on the level to which pest diagnostic survey is carried out. The results of observation on parasites, predators, and entomopathogenic microorganism infection allow qualitative definition of pest population conditions. Empirical application of suppression measures without considering the actual situation in the forest does not eradicate the mass outbreak of pests and often leads to relatively fast restoration of infestation centers. In this connection, integration of various means and methods with the action of the natural complex of factors regulating pest populations should become an indispensable measure in forest protection. Practice requires more profound research into pest population dynamics, development of optimal methods for diagnostic surveying, counting and predicting populations with the peculiarities of each species taken into account, wide application of computers for data processing and biological process simulation, and further improvement of micro-biological and chemical methods of control. The current achievements in science and technology permit fast solving of many aspects of the problem under consideration.

37.6 Integrated Protection of Oak Forests with Emphasis on Gypsy Moth and Oak Wilt

Daniel O. Etter, Jr., William E. Wallner, and David R. Houston

USDA Forest Service, Forest Insect and Disease Laboratory
151 Sanford Street, Hamden, CT 06514

Abstract

The oak forests of the USA are a very large natural resource. Their utilization varies substantially from place to place, both in kind and in intensity. In a purely economic sense, they are generally under-utilized. They receive only limited application of intensive management practices, including protection against pests. Many organisms are present that can, and sometimes do, cause extensive damage. In recent years, the most troublesome pest has been the gypsy moth. Practical management of this pest is extraordinarily complicated, largely because outbreaks extend over a mosaic of tracts belonging to different owners with disparate management objectives, lying in different political units with varying legal requirements and political concerns. Coordination is a major and central problem. For the gypsy moth, IPM concepts point away from the use of synthetic insecticides in a somewhat indefinite way but do not point to clear selections of specific management tactics. To illustrate some of the problems of systems integration, a currently hypothetical integration of gypsy moth and oak wilt management is revealing. Some interaction does occur, which implies a need for some integration.

Introduction

Oak Forest Resources and Management

Oak forests in the USA occupy some 71 million hectares, and the three main types—oak-hickory, oak-pine, and oak-gum-cypress—make up approximately half of the eastern forests (U.S. Forest Service 1977). Pest management fits within a hierarchy of components of the management of the oak forest resource; it is a part of forest protection which in turn is part of resource production. The benefits of pest management derive ultimately from the utilization of the oak forest resource. The degree and nature of oak forest utilization is far from uniform and reflects a wide range of forest ownership and land management objectives. Southern New England is an extreme case as shown by Kingsley (1974); most trees removed from growing stock were for land clearing, and over half of the forest was not utilized. Although timber utilization in this region is thought to be increasing, the forests are not yet heavily utilized for timber or wood products but for recreation and other forest-based amenities. Gansner et al. (1977) found only limited use of intensive forest management, including forest protection, in northern forests generally. The results of DeBell et al. (1977) indicate that this is part of a national pattern.

In the USA, the gypsy moth, *Lymantria dispar* (L.), is an exceptionally serious pest, mainly in oak forests. It is univoltine and overwinters in the egg stage. Natural wind-borne dispersal occurs generally before the newly hatched larvae become established and begin feeding on hosts. Flight of females has not been reported here; they attract males by means of a pheromone. The gypsy moth has not yet fully occupied its ecological range in the USA, and the region generally infested is steadily expanding. In a large-scale outbreak, several million hectares may be defoliated. The biological consequences of severe defoliation include a decline in tree vigor and internal changes which make trees much more susceptible to secondary attacks by the two-lined chestnut borer, *Agrilus bilineatus* (Web.), and the shoestring root rot fungus, *Armillaria mellea* (Vahl. ex. Fr.) Kummer. The secondary attacks lead to tree mortality following severe ($>$50 to 60%) defoliation, commonly 25% and sometimes 100% among oaks.

Systems Approach

Pest management is considered here from the viewpoint of systems engineering. A pest management system is defined as a complex of resources, including people, procedures, supplies, equipment, and facilities, organized to limit the economic and social effects of a pest. A complete characterization—the design—of a system consists of specifications for all aspects of the system, including basic descriptions (analogous to blueprints), goals and criteria, methods for controlling or managing the system itself, interfaces with other systems, and future objectives (Nadler, 1970). These concepts have been applied to gypsy moth pest management to outline a comprehensive system.

The basic findings are: (1) Both existing and proposed systems have a net-like structure with many connected parts, arranged in a hierarchy from local field activities to national activities; (2) The existing system resembles a loose confederation of separate, essentially independent systems with limited and generally informal coordination; (3) The existing system has no ongoing, formal evaluation of performance and capabilities and no regular planning of system improvements; and (4) No opportunities were found for major simplifications of system functions and operations.

System Design

Emphasis was placed upon comprehensiveness instead of fine details; such details might include spray nozzle design and spray aircraft speed and altitude. Opportunities will be more easily found later for adding details than for adding overlooked structural components.

System Functions

The functions of the comprehensive gypsy moth pest management system are organized into three hierarchical levels: level I, local functions including most field activities; Level II, functions generally statewide or regional in scope; Level III, functions national in scope. The more important functions are as follows:

Level I functions
Intervention

1. modify gypsy moth population densities and population trends—use of insecticides, biological control techniques, habitat modification, mechanical destruction of pests, etc.
2. reduce vulnerability of trees to effects of defoliation—fertilization, irrigation
3. alleviate losses—salvage dead trees
   Surveillance:
1. obtain and transmit data to facilitate intervention decisions—egg mass densities, eggs per egg mass, pupal sex ratios, stand compositon, tree size, distribution, tree condition, land use, etc.
2. obtain and preserve data to facilitate subsequent surveillance plans and decisions
   Public communications
1. inform and educate land owners, land managers, pest managers, and other concerned indivduals about the gypsy moth, its effects, and pest management options
2. determine public preferences concerning pest management options
   Logistics
1. provide trained personnel, supplies, and equipment when and where they are needed
2. train personnel as needed

Level II functions

1. produce most details of operational plans for Level I functions
2. manage, administer, and coordinate activities within states or regions
3. produce and implement state-wide or regional logistics plans
4. produce and implement plans for large-scale or long-term activities—analysis of satellite imagery, establishment of an exotic parasite, etc.
5. manage data and conduct system performance evaluation
6. carry out research, development, and system improvement

Level III functions

1. formulate national policies, strategies, goals and objectives
2. organize and conduct general public communications
3. manage data at the national level and conduct overall system performance evaluation
4. coordinate activities at the national level
5. manage and conduct research, development and overall system improvement

Given these functions for the pest management system, let us consider briefly ways of measuring how well they are carried out.

System Criteria

The most important measures for the gypsy moth pest management system derive from some characterizations of integrated pest management (IPM) (Clark et al., 1967; van den Bosch et al., 1974; DeBach, 1974; Bergland, 1977): (i) limit damage done by the pest to tolerable levels, (ii) limit the cost of the pest management system to tolerable levels, and (iii) minimize the risk of side-effects injurious to property, to humans, and to non-target organisms generally.

Technically, these can be restated as a single criterion: minimize a weighted sum of the damage costs, pest management system costs, and the risk of unwanted side-effects.

The IPM criterion enters into the system design in two different ways: (1) to assist in making a treatment decision in a particular case; and (2) to compare whole pest management systems, on the basis of their expected performance over some range of situations and, ideally, over some span of years. The two usages become essentially the same only if both tactical and strategic planning are restricted to a single gypsy moth season. In practice, suppression planning in the generally infested area does use a planning horizon of

624

one season; eradication projects involve longer planning horizons (Blacksten et al., 1978).

Under some conditions, use of the IPM criterion is relatively simple. For example, a reduction in the amount of pesticide used can be expected to reduce system costs and the risk of injurious side-effects, at the expense of a possible increase in damage done by the pest. If the increase in damage is tolerable, then the reduction in pesticide use will be an improvement in the pest management system. Application of the criterion is often not so easy in practice.

Formulating criteria is easier if four major goals are specified: (1) prevent establishment of gypsy moth populations in places remote from the current infested area; (2) prevent expansion of the generally infested area; (3) minimize overall impacts in areas of high gypsy moth population density; and (4) maximize the area within which gypsy moth populations are sparse and stable. Within existing and planned gypsy moth pest management systems, goals (1) and (2) above summarize the regulatory mission of the USDA Animal and Plant Health Inspection Service (APHIS) in gypsy moth pest management. Cooperative gypsy moth suppression projects, between states in the generally infested area and the USDA Forest Service (USFS), come under goal (3). Goal (4) is served by a variety of tactics and research and development efforts aimed at stabilization of gypsy moth populations at innocuous levels.

The four goals just pointed out suggest quantitative criteria for the overall functioning of the gypsy moth pest management system. For example, newly infested hectares per year is a measure of how far the system misses a goal of preventing expansion of the generally infested area. A long list of possible criteria can be produced with little effort. In practice, it is rarely possible to obtain clear, objective evaluations both of the criteria and of the associated weightings. Such evaluations most often reflect individual assessments and preferences; this is why public communications becomes an essential function of the gypsy moth pest management system.

Surveillance and Intervention

Specific objectives and procedures for surveillance and intervention depend upon which of the four major goals of the pest management system apply. For example, there is a national surveillance program, based on male moth trapping, intended to detect and delimit isolated infestations beyond the generally infested area, so that they can be eradicated. Within the generally infested area, the USFS goal is to prevent the occurrence of enough defoliation to cause serious losses. The major components of surveillance to support suppression are defoliation surveys, gypsy moth egg mass counts, and assessments of the vigor of the populations. In most cases, the surveillance results are used to formulate control plans for the next gypsy moth season after the observation is made. Ideally, the surveillance system would detect the beginnings of outbreaks and would intensify monitoring of populations in stands eligible for treatment on the basis of land usage. Stands that represent unusually favorable habitats for gypsy moths in New England have been characterized (Houston and Valentine, 1977; Valentine and Houston, 1979; Houston, 1979). Outbreaks appear to start in these stands, and data from these stands may provide the necessary warning to the surveillance system.

For the present and near future, the use of insecticides is a crucial pest-management tactic. Usually aerial application is preferred; the amount of pesticide used per unit area is usually less than for ground application, and, if the area to be sprayed is large enough, the total cost per hectare treated is less for aerial application. At this time, there are three chemical pesticides, a growth regulator, and two biological insecticides registered for aerial application to control gypsy moth. The chemicals also have some effect against other defoliators that may be present. Some other suppression techniques, such as release of sterile males, are under development but are not ready for use.

Basic annual cycles for surveillance and intervention are shown in Figure 1. It should be noted that not all activities are carried out at all places. Also, neither mating disruption nor parasite population manipulation is currently operational for suppression. Although existing and proposed systems depend on the use of insecticides, review of operations based on other techniques indicates that many details are similar and often identical.

Criteria for selecting an area for gypsy moth suppression or eradication with insecticides are not completely standardized. APHIS has some published criteria (USDA/APHIS 1976), but allows variations based on the circumstances. Individual states use guidelines to identify areas that are eligible for treatment when it is desired by the owners. Land use criteria are fundamental; they generally limit consideration to forested communities, high-use recreation areas, and high-value timber stands. These criteria are supplemented by more specific conditions of gypsy moth population density and vigor, and tree condition or recent site history. A typical goal is to hold defoliation to some level below 50%, particularly if the trees are already stressed, as from severe defoliation the previous year. The guidelines also generally exclude areas in which gypsy moth population vigor is expected to be low and which may have a relatively high density of

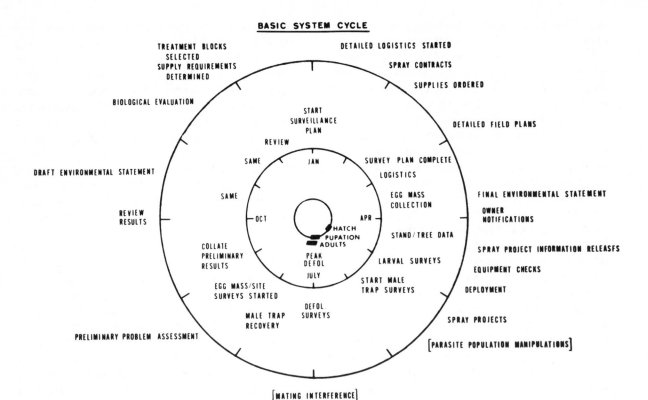

Figure 1. Annual Intervention and Surveillance Cycles

natural enemies. These assessments usually involve subjective judgments.

Possible Future Directions

Some consideration has been given to the possible characteristics of more advanced gypsy moth pest management systems—say 5 or more years into the future. It is anticipated that the more advanced systems will depend less upon chemical pesticides. Currently, two phases in the development are postulated. During the earlier phase, gypsy moth populations would be partially stabilized, and pesticides would be used to supplement, rather than supplant, natural mortality factors, as would other techniques such as sterile male release, use of the pheromone as a confusant, and manipulations of populations of natural enemies.

In the second phase, gypsy moth populations would be generally stabilized at innocuous levels. The stabilization would be least secure, if it were achieved at all, in stands whose natural stress and disturbances render them susceptible to outbreaks. In stands other than these susceptible foci, outbreaks would occur as a result of two kinds of events: major natural disturbances, such as ice storms, capable of disrupting the stabilized life system; and major disruptions caused by human activities. This suggests a need for emergency suppression and restabilization. Emergency suppression in such a system might

use a relatively nonspecific chemical or biological insecticide, since occasional suppression of other pests would also be desirable and the cost of maintaining a battery of completely specific agents might not be acceptable. The prospects for reliable natural regulation of gypsy moth at innocuous levels are uncertain. The biology and ecology of the insect suggest that it is an "*r*-strategist" and therefore it may not be possible to eliminate the risk of outbreaks completely (Southwood, 1977).

Broad-Scale Integration Problems and Opportunities

A weak administrative integration necessarily occurs at the higher levels of pest management system hierarchies. Opportunities and requirements for stronger forms depend upon the situation at lower levels, closer to the pest. The management system for any widely distributed pest will have an overall functional structure similar to that of the gypsy moth; specific details obviously depend upon the particular pest and its effects. Some integration of management of gypsy moth with that of concurrent lepidopterous defoliators is desirable and, in fact, unavoidable; defoliation mapping is an example. It is more revealing to examine problems and opportunities for systems integration in the case of a dissimilar pest. The oak wilt fungus, *Ceratocystis fagacearum* (Bretz) Hunt, was chosen as an example; the areas infested by the two pests are just beginning to overlap. A few years ago, oak wilt appeared to be a major threat to oak forests in the USA, especially to members of the red oak group. The disease has not spread as anticipated, and there is no substantial national oak wilt control program at this time. For present purposes, one is hypothesized. A recent review of the oak wilt situation has been given by Gibbs (1978).

Intervention

Salvage of dead trees was considered part of the intervention system for gypsy moth, because it could help alleviate losses caused by the pest. The concept applies equally to oak wilt. A plausible alternative would be to establish a generalized salvage program that interfaces with pest management systems, as an extreme form of integration. Silvicultural measures—species selection, in particular—should be integrated if they can be carried out practically. Members of the red oak group are much more susceptible to oak wilt than are members of the white oak group, whereas members of the white oak group are more susceptible to gypsy moth defoliation and its effects. The latter difference is slight compared to the former. In an oak stand threatened by both pests, and for which some species selection is feasible, the composition generally should be shifted toward the white oak group, or away from oaks entirely, taking into account the fact that a large white oak may serve as a focus of a gypsy moth infestation.

Surveillance

The primary interaction, where some integration is appropriate, occurs in the aerial surveys for defoliation and symptoms of oak wilt. The scales of interest are not similar; wilted crowns, perhaps bordering a patch of dead trees, are sought in an oak wilt survey, whereas a defoliated area may extend over thousands of hectares. The timing requirements are also different; peak defoliation may be observable only for about 3 weeks in July, because heavily defoliated deciduous trees usually refoliate; however oak wilt could be observed at any time when the foliage is green. Oak wilt would not be observable in the midst of severe defoliation. Defoliation would be observable on an oak wilt survey flight if the timing were appropriate. There is still another interaction. Oak mortality of 25% or more should be anticipated following heavy defoliation by gypsy moth. Such mortality would tend to mask the presence of oak wilt; the lack of detection could lead to additional, otherwise avoidable oak mortality. The interaction between defoliation and oak wilt

surveys implies a need for some integration of surveillance system designs.

Logistics

Subsystems for removal of dead trees can be at least partially integrated. Some differences in detail should exist because patterns of tree mortality can differ, and salvage of trees killed by oak wilt would be subject to some regulatory control to prevent international spread. Logistics also includes training, as it is defined here, and some integration of surveillance training would be advisable to improve distinction between defoliation and its aftermath and indications of oak wilt.

References

Bergland, Robert. 1977. Secretary's memorandum number 1929: USDA policy on management of pest problems. U.S. Dep. Agric. 12 Dec. 1977.

Blacksten, R., I. Herzer, and C. Kessler. 1978. A cost/benefit analysis for gypsy moth containment. Ketron Report KFR 161-78. Ketron, Inc., Arlington, VA.

Clark, L. R., P. W. Geier, R. D. Hughes, and R. F. Morris. 1967. *The ecology of insect populations in theory and practice*. Methuen and Co., Ltd., London.

Debach, P. 1974. *Biological control by natural enemies*. Cambridge University Press, Cambridge, England.

Debell, D. S., A. P. Brunette, and D. L. Schweitzer. 1977. Expectations from intensive culture on industrial forest lands. *J. For.* 75:10-13.

Gansner, D. A., O. W. Herrick, and D. W. Rose. 1977. Intensive culture on northern forest industry lands: trends, expectations and needs. U.S. Dep. Agric. For. Serv. Res. Pap. NE-371.

Gibbs, J. N. 1978. Oak wilt. *J. Arboric.* 3(5):351-356.

Houston, D. R. 1979. Classifying forest susceptibility to gypsy moth defoliation. U.S. Dep. Agric., Agric. Handb. No. 542.

Houston, D. R., and H. T. Valentine. 1977. Comparing and predicting forest stand susceptibility to gypsy moth. *Can. J. For. Res.* 7:447-461.

Kingsley, N. P. 1974. The timber resources of southern New England. U.S. Dep. Agric. For. Serv. Resour. Bull. NE-36.

Nadler, G. 1970. *Work design: A system concept.* Revised edition, Richard D. Irwin, Inc., Homewood, IL.

Southwood, T. R. E. 1977. The relevance of population dynamics theory to pest status. p. 35-54. In J. M. Cherrett and G. R. Sagar, eds. *Origins of pest, parasite, disease, and wood problems.* Halsted Press.

U.S. Department of Agriculture, Animal and Plant Health Inspection Service. 1976. Gypsy moth and browntail moth— Manual of administratively authorized procedures to be used under the gypsy moth and browntail moth quarantine. 805-904. U.S. Dep. Agric.

U.S. Forest Service. 1977. The Nation's renewable resources. An assessment, 1975. For. Resour. Rep. 21.

Valentine, H. T., and D. R. Houston. 1979. A discriminant function for identifying mixed-oak stand susceptibility to gypsy moth defoliation. *For. Sci.* 25(3):468-474.

Van Den Bosch, R., T. F. Leigh, L. A. Falcon, V. M. Stern, D. Gonzales, and K. S. Hagen. 1974. The developing program of integrated control of cotton pests in California. p. 337-394 In C. B. Huffaker, ed. *Biological control.* Plenum Press, New York.

Index of Authors